Biomolecular Computation for Bionanotechnology

For a listing of related Artech House titles, turn to the back of this book.

Biomolecular Computation for Bionanotechnology

Jian-Qin Liu

Katsunori Shimohara

ARTECH HOUSE
BOSTON | LONDON
artechhouse.com

Library of Congress Cataloging-in-Publication Data
A catalog record for this book is available from the U.S. Library of Congress.

British Library Cataloguing in Publication Data
A catalogue record for this book is available from the British Library.

ISBN-10: 1-59693-014-4
ISBN-13: 978-1-59693-014-8

Cover design by Igor Valdman

© 2007 ARTECH HOUSE, INC.
685 Canton Street
Norwood, MA 02062

All rights reserved. Printed and bound in the United States of America. No part of this book may be reproduced or utilized in any form or by any means, electronic or mechanical, including photocopying, recording, or by any information storage and retrieval system, without permission in writing from the publisher.
   All terms mentioned in this book that are known to be trademarks or service marks have been appropriately capitalized. Artech House cannot attest to the accuracy of this information. Use of a term in this book should not be regarded as affecting the validity of any trademark or service mark.

10 9 8 7 6 5 4 3 2 1

Contents

Preface ... xi

CHAPTER 1
Introduction: How to Go Beyond Traditional Computers ... 1
1.1 Scientific Motivation Versus the Needs of the IT Industry ... 3
1.2 Cutting-Edge Technologies for Building a Molecular Computer: From Nanobioscience and Nanotechnology to Nanobioinformatics ... 5
    1.2.1 Synthetic Biology ... 6
    1.2.2 Emerging Technologies for Protein Analysis: To Gain Information about Proteins, Protein Interaction, and Their Links to the Medicine ... 8
1.3 Preliminaries in Nanobioscience ... 9
    1.3.1 Gedanken Model ... 10
    1.3.2 Some Concepts in Biochemistry ... 11
    1.3.3 Systems Biology ... 12
    1.3.4 Perspectives on Innovative Technologies for Biomolecular Computing: Benefits from Breakthroughs of Molecular Biology in the New Millennium ... 12
1.4 Challenges from Real-World Applications ... 13
    1.4.1 Performances of Biomolecular Computing ... 13
    1.4.2 Technological Difficulties on Feasibility of Implementation of a Biomolecular Computer: Scalability, Reliability, and Controllability ... 13
1.5 Back to Molecular Informatics: How to Use Molecules to Represent Information ... 15
References ... 19

CHAPTER 2
The State-of-the-Art Molecular Biology and Nanotechnology ... 23
2.1 Genomics ... 23
2.2 Proteomics ... 26
2.3 Cellular Structure from the Viewpoint of Molecular Biology ... 29

2.4	Cell as a Nanobiomachine	31
	2.4.1 Moleware Mechanics for Cellular Nanobiomachine: Molecules Carrying Messages	33
	2.4.2 Molecular Informatics for Cellular Nanobiomachine	34
2.5	Signal Transduction and Signaling Pathways of Cells	35
	2.5.1 The Link Between the Signaling Pathway and Molecular Movement	37
	2.5.2 The Links Between Signal Pathways and Neuron Function	37
2.6	Measurement and Detection in Material Science: Towards Manipulation of Biological Molecules	38
2.7	Pharmaceutical Nanobioinformatics	41
	2.7.1 "Naïve" Thinking for Pharmaceutics	41
	2.7.2 Molecular Information Flow as a Possible Solution Towards Potential Application of Nanobioinformation Processing Systems	42
References		45

CHAPTER 3

Nanobiomachines for Information Processing and Communication: Exploring Fundamental Principles of NanobioICT   49

3.1	Mission of NanobioICT	50
3.2	Information Theory of NanobioICT: Shannon Meets Feynman	53
3.3	Embryonic Approaches to NanobioICT	56
3.4	A Glance at Informatics of Moleware Communication	67
3.5	An Informatics Form of a Molecular Viterbi Algorithm	76
3.6	Network Coding in Molecular Informatics	80
3.7	Quadruple Convergence	84
References		87

CHAPTER 4

Computing by Biomoleware: Diverse Methods from Diversified Materials   91

4.1	How to Build an Engineered Computational Nanobiosystem: Inspiration from Existing Nanobiomachines in Nature	92
	4.1.1 Nanobioworld Becomes Observable with the Help of Innovative Measurement Technology: Schrödinger's Cat Is at the Door	92
	4.1.2 Seeking a Movable Nanobiomachine: Postman in Moleware	94
	4.1.3 Methodology Learned from the Cell and Beyond	96
4.2	Information Processing in Artificial Nanobiosystems: An Odyssey Beyond the Blind Watchmaker	97
	4.2.1 Molecular Complex as Memory—Memorizing Instead of Braining	100
	4.2.2 Molecular Clock—The Heart of Synchronous Moleware	105
	4.2.3 Moleware Coding in Nanobiomachine—A Solution from the Cell	108

4.3	Computing by Nucleic Acids	114
	4.3.1 DNA Computing	115
	4.3.2 RNA Computing	121
	4.3.3 Surface-Based DNA Computing	123
	4.3.4 Nanobiotechnology for DNA Computing	125
4.4	Computing by Biochemical Reactions in Microbes	127
	4.4.1 Information Processing Mechanism of Microbes	127
	4.4.2 Computing by Gene Operations in Ciliates	129
	4.4.3 Moleware Microarray	132
References		136

CHAPTER 5
Theoretical Biomolecular Computing — 141

5.1	Basic Concepts in Computer Science for Molecular Computing	142
	5.1.1 Formal Language	143
	5.1.2 Automata	145
5.2	Formalized Molecular Computing	146
	5.2.1 H-System	147
	5.2.2 P-System	150
	5.2.3 Rediscovering the Informatics Structure of the Biomolecular Computing System: An Informatics View of the Formal Processes of the Biomolecular Computing H-System	153
5.3	How to Design Algorithms for a Molecular Computer	157
	5.3.1 Observing Complexity from Benchmarks	157
	5.3.2 Obtaining Efficiency from Pathway Designs: Algorithmic Design Through Examples	160
5.4	Touchstone for Nanobio-Oracle: Moleware Logic	171
	5.4.1 Consistency of Computing Operators and Feasible Experimental Supports: Verification of Logic Process	171
	5.4.2 Formalized Method for Moleware Logic	173
References		179

CHAPTER 6
Cellular Biomolecular Computing Based on Signaling Pathways: Kinase Computing — 181

6.1	Cellular Pathway: Another Ubiquitous Society in Another Universe	182
	6.1.1 Ubiquitous Cell Communication for Parallel Information Processing	182
6.2	The Molecular Switch as a Bridge Between Cell Communication and Molecular Computing	184
	6.2.1 Binary Information Representation by Molecular Switch	185
	6.2.2 Computing Formalized as an Automaton	188
	6.2.3 Example: Designing an Automaton for Kinase Switches Guided by GTPase	190

		6.2.4	Information Structure for Automaton-Based Computing	191
		6.2.5	A Computing Model Based on Pathway Units with Turing Computability	193
	6.3	From Automaton to Rewriting: Toward General Parallel Computing		199
		6.3.1	Formalization	199
		6.3.2	Transition from Hypergraphs to Bigraphs	203
		6.3.3	McNaughton Language, Confluent Rewriting, and Controlling with the Structural Characteristics of MSP-Automaton	205
		6.3.4	Designing a Rewriting Process by Pathway Units Based on MSP-Automata	209
		6.3.5	A Compiler: Translating Moleware Language into Programmer-Friendly Informatics Operators	210
		6.3.6	Systematically Understanding the Interaction Structure in Pathway Computing	212
		6.3.7	Generalized Form for Computing	212
	6.4	Blueprint of a Kinase Computer		214
		6.4.1	Quantitative Description for Biochemical Features	214
		6.4.2	Materials for Information Processing	217
		6.4.3	Controllability Under Protocols in Bioinformation	218
	References			221

CHAPTER 7
Comparison of Algorithms for Biomolecular Computing and Molecular Bioinformatics — 223

7.1	Formal Characteristics of Algorithms for Biomolecular Computing		224
	7.1.1	DNA Computing	225
	7.1.2	Surface-Based DNA Computing	225
	7.1.3	H-Systems	225
	7.1.4	P-Systems	226
	7.1.5	DNA Computing Method by Ciliates	226
7.2	Interactions in Molecular Bioinformatics Algorithms		227
	7.2.1	Example 1: Interaction of GTPases	229
	7.2.2	Example 2: Interaction of Kinases/Phosphatases	232
7.3	Common Points of Biomolecular Computing and Molecular Bioinformatics for Algorithms		239
	7.3.1	Example: Describing Cellular Pathways by Graph Rewriting	242
7.4	Exploring Logical Description for Molecular Bioinformatics Based on Formalization and Abstract Operations		245
References			250

CHAPTER 8
Emerging Nanobiotechnology in Multiple Disciplines — 253

8.1	The Tale of Two Media: Molecular Electricity and Biomolecular Signaling	253
8.2	How Small Can an Information Processing System Be Made?	254

8.3		Informatics of Porphyrin Systems	255
8.4		Transition from the Supporting Points to Integrations of Different Aspects of Molecular Information Processing	260
8.5		Cell Communication for Engineering Purpose	263
	8.5.1	From Bit Level of Information Representation to Observe Cellular Communication	265
	8.5.2	The Biophysical Effectors of the Molecular Information Flow	266
	8.5.3	Effects of Molecular Protocols by the Internal Components of Cells	266
	8.5.4	Control Nodes in Moleware Communication Networks	267
	8.5.5	Collision-Avoid: An Issue on Efficiency of Moleware Communication in Cells	268
References			272

About the Authors 275

Index 277

Preface

In the new millennium, the National Nanotechnology Initiative (NNI) and genomics/proteomics have revolutionized and broadened our vision by aggressively advancing our understanding towards the unknown bionanoworld in nature and will certainly reshape our everyday life in the future. We are amazed by the way the moleware mechanism and its corresponding molecular signaling processes, which are different from those in the silicon world, see the nanobioworld vividly with the advanced measurement equipment in bionanotechnology. In order to harness moleware structure for engineering purposes, the innovation of new unconventional paradigms of information processing, computation, and communication in moleware have become inevitable. A biomolecular information processing system outperforms its counterpart, in silico, in the aspect of parallel informatics operations carried out by biochemical reactions. In biomolecular computation, information is represented by the biomolecules such as DNA, RNA, proteins, and other molecules in cells. The biochemistry characteristics of biomolecules are thus the foundation of designing and testing a biomolecular computer. As Richard E. Smalley says in "Of Chemistry, Love and Nanobots," *Scientific American*, September 2001, Vol. 285, No. 3, pp. 68–69, "Chemistry is a waltz with its own step-slide-step in three-quarter time." With respect to the bioinformation flow within cells, moleware melody with parallelism has emerged from cellular communications indeed. Sarah A. Woodson said in "Assembly Line Inspection," *Nature*, Vol. 438, December 2005, pp. 566–567, "... But unlike man-made machines, which are built on assembly lines, these cellular machines assemble spontaneously from their protein and nucleic-acid components." Our irreplaceable mission is to compose the fascinating bionano-instrument, the cell, free of fantasy.

Motivated by the promises of biomolecular computation, we have carefully arranged this book into eight chapters. It begins with Chapter 1, which questions why we need biomolecular computation and how information can be processed in moleware. Chapter 2 presents the necessary and preliminarily knowledge in biological sciences. Chapter 3 summarizes the field of nanobioICT including new progresses in moleware communication. Chapter 4 discusses different paradigms of biomolecular computing by employing different moleware materials such as DNA, RNA, proteins, and other molecules, as well as their diverse features in biochemistry. Chapter 5 elaborates the theoretical aspects of biomolecular computation in terms of automata, formal languages, and algorithmic complexity. Hinting at the naturally existing bionanomachine, the cell, Chapter 6 presents kinase computing—biomolecular computing based on signaling pathways in cells. This unconventional information processing paradigm by signaling pathways is expected to bridge nanobiocomputing and nanobioinformatics. Chapter 7 clarifies the relationship between biomolecular information processing and molecular bioinformatics. Chapter 8 gives the perspective on multiple disciplines derived from information processing at the nano-size level, whose broad applications are expected to be found in bionanotechnology. The restless advances in bionanotechnology will definitely unveil the mysteries of nanobioworld. Roger Brent and Jehoshua Bruck said in "2020 Computing: Can Computers Help to Explain Biology?" *Nature*, Vol. 440, No. 23, March 2006, pp. 416–417: "Here, however, we imagine ways that formalisms from computer science might contribute to a deeper understanding of biological function. These approaches will not bear fruit without deliberate and difficult work." Biomolecular computation is one of the cornerstones for systematic, in-depth understanding of living cells in bionanotechnology.

We sincerely thank Kozo Kaibuchi, Shinya Kuroda, and Mutsuki Amano on signal transduction; Yasushi Hiraoka on cellular biology; Shinro Mashiko on nanotechnology; Kazuhiro Oiwa on molecular motors; Osamu Katai on information sciences; Masami Ito and Ferdinand Peper on computer science; Peter Davis on communication engineering; Grzegorz Rozenberg for his advice and comments on the computing method based on the signaling pathways of cells; Minoru Kanehisa and Tatsuya Akutsu on bioinformatics; anonymous referees for their remarks on improving the contents of the book; Min-Jie Wei for reading the manuscript of the book and improving it both in content and in writing; Wayne Yuhasz, Barbara Lovenvirth, Judi Stone, William M. Bazzy, Darrell Judd, Kevin Danahy, and Christine Daniele of Artech House for their pleasant cooperation on the planning and publication of the book.

CHAPTER 1
Introduction: How to Go Beyond Traditional Computers

The semiconductor-based computer is one of the greatest achievements in the twentieth century. Its power has affected almost every aspect of human society and it accelerated the growth of the information era in the 1980s. We are still expecting to reap huge benefits from traditional computers, at lower and lower cost and increased performance that matches the advances of semiconductor technology, according to Moore's law. As defined in Wikipedia [1], Moore's law predicts "the doubling of the number of transistors on integrated circuits (a rough measure) every 18 months." However, as the IC chips used to build computers continue to shrink, those radical changes in physical size bring up new problems, both in theory and in practice; such as how to minimize energy consumption or how to release the heat generated by transistors within an extremely small space. What we do not know is when Moore's law will expire [2, 3]. Biomolecular computers (e.g., DNA/RNA computers), molecular computers made by molecular electronics, and quantum computers are thus proposed as alternatives to semiconductor computers.

Among the many predicated types of computers, biomolecular computing by nanobiotechnology is one of the most promising. The material commonly used for building a biomolecular computer is DNA molecules. In 2000, Laura Landweber testified to the merits of the DNA computer in *Beyond Silicon Computing* [4].

> Ever since scientists discovered that conventional silicon-based computers have an upper limit in terms of speed, they have been searching for alternate media with which to solve computational problems. That search has led us, among other places, to DNA. The advantage of DNA is that it is tiny, cheap, and can react faster than silicon.

The development of DNA computers is tightly connected with molecular DNA technology. Anne Condon, professor of computer science at the University of British Columbia, emphasizing the significance of applying molecular nanotechnology to DNA computers and DNA machines, states [5]:

> Physical and computational scientists can design rigid DNA structures that serve as scaffolds for the organization of matter at the molecular scale, and can build

simple DNA-computing devices, diagnostic machines and DNA motors. The integration of biological and engineering advances offers great potential for therapeutic and diagnostic applications, and for nanoscale electronic engineering.

Thanks to nature we have proteins and cells in addition to DNA. Drawing on rich evidence and sound theory, National Academy of Sciences President Bruce Alberts envisions the cell as a collection of protein machines [6]. We can use biomolecules to enter the nanobioworld, keeping in mind Feynman's dogma, "There's plenty of room at the bottom." The emergent nanobiotechnology, is what Gary Stix calls "little big science" or "nanoscale science and technology" that is being pioneered by the National Nanotechnology Initiative (NNI) project [7].

Biomolecular computation is just such "little" stuff labeled as an unconventional computing paradigm. The term *unconventional computing* implies that the biomolecular computing process as well as the physical mechanism are different from those used in traditional semiconductor computing. In biomolecular computation, biomolecules represent information and biochemical reactions carry out the computing process. Implementation of biomolecular computers will be different from that of conventional computers. Exploration on how to integrate *nano* with *bio* to build a new generation of computers by new materials and new methods is the central idea in this book.

Before our formal discussion on biomolecular computation, we will briefly summarize the tendency of computing paradigms in general. As we know, computer science started with the simple job of calculation and evolved into the field of scientific computing. With the automation in industries, computers have been applied to instruments, control, and manufacturing. Computers are not limited to computing in the traditional, narrowly defined meaning of the word, but today handle information in general. Applications such as Internet browsing and communication technology have greatly changed our social environment. In social informatics, friendly human-machine interfaces are built to process nonnumerical information such as speech, images, and language, in which the computer acts as an information processing tool. Supercomputers are also developed for high-performance computing. As we have seen, the ranking list of supercomputers is refreshed continuously and looks like a race [8]. In June 2006 [9], TOP500 Supercomputer Sites (http://www.top500.org/) announced, "The No. 1 position was again claimed by the BlueGene/L System" [10]. Nowadays, with their great power, computers are used not only to simulate things in the macroscopic world such as the Earth and earthquakes, but also things in the microscopic world such as cells. Roger Brent and Jehoshua Bruck [11] even propose using computer logic to explain biological phenomena [12]. The future of computers may be an open-ended evolution, whose types are diverse and unpredictable. However, these applications do not revolutionarily change the computer itself. The computer is at a crossroads, and its destiny is now driven by the pace of nanoscale science and technology.

As we know, modern electronic computers are built based on von Neumann architecture and Turing machine theory. Designing a computer that can go beyond the limits of Turing machines, that is, a super-Turing machine, is our goal. For example, Claude and Pâun [13] have proposed a membrane computing model "theoretically capable of trespassing the Turing barrier," asking whether biologically computing agents are capable of accomplishing Turing-uncomputable functions. Their answer is that it is *theoretically possible*. The significance of this is not limited to the theory of computing—it implies that biomolecular computing could hold a seat at the bottom of nanobioworld with its own merits, even though we are not familiar with it now.

1.1 Scientific Motivation Versus the Needs of the IT Industry

Scientific motivation and the needs of the IT industry are at opposite ends of the computing technology continuum. On the one hand, a rich store of knowledge in modern computer science provides very strong support for innovation. On the other hand, the industry's diverse, practical needs demand new tools now.

Information processing technology has undergone continuous development to satisfy the needs of industry. The invention of the first electronic computer was followed by a boom in artificial intelligence (AI), aimed at moving computers to ever more human-like reasoning. AI in philosophy is a kind of extension of the human brain. Butler said [14], "the Web is already full of knowledge-intensive components."

In the history of computer and information processing, the representative streams after AI are artificial neural networks (ANN) or neural networks (NN), evolutionary computation (EC), artificial life (AL), and natural computation (NC). ANN is a bold effort to explore the parallel computing capability inspired by the neural information processing. EC is a kind of algorithm hinted by the real evolution in the biological systems. AL at the outset is oriented to exploring life-like phenomena and corresponding mechanisms by computers rather than biological experiments. The field of NC emphasizes the nature-inspired style of computation to design new problem-solving algorithms with the help of the principles in nature.

Thoughts behind these approaches are to learn from nature and to build artificial systems inspired by it, which includes biology. Being different from the ways of nature, the overwhelming IT society desperately summons new ideas that can be put into action. The IT society itself also incubates new fields of information processing, which include:

- Autonomic computing;
- Proactive computing;
- Ubiquitous computing;
- Pervasive computing;
- Universal communication.

Above paradigms are different in concept but somehow connected in content. For example, the relationship between autonomic computing by IBM and proactive computing by Intel overlaps. The intersection of autonomic computing and proactive computing is ubiquitous computing, technically supported by distributed information processing [15]. The new information processing paradigms are innovated with advances of IT technology because IT products have to be rapidly upgraded in performance to occupy the increasing market. In order to provide technical solutions for the IT industry regarding mobile computing, dependable computing, network computing, and other streams of computing focused on practical applications of computers, new types of computing and communication paradigms have to be proposed to consolidate the function of systematical optimization in information processing systems with more efficient response to the dynamical environment where computers are applied. The autonomic, proactive, ubiquitous, pervasive, universal are just those features oriented to the services at computer terminals.

Above paradigms also somehow have their unique characteristics. According to the definition of "autonomic computing" given in [16], an autonomic computing system (AS) consists of eight elements. The most fundamental element is [17]:

> An autonomic computing system needs to know itself—its components must also possess a system identity.

In software engineering, the components refer to the modules in software. The system identity is the index describing these modules. It is radical to discriminate the computer software and computer virus. In the real world, computer viruses and junk e-mails frequently bother us. One of the biggest difficulties for a computer to gain immunity is it cannot detect viruses at all times. If the computer cannot really recognize itself, it is impossible for us to quantitatively design the criterion for judging whether a computing system is autonomic or not, and thus cannot introduce the criteria into the computer systems as well as the autopoietic systems in theory. Parasitic computing [18] is an interesting new concept for computer security. Diagnosing itself seems to be an eternal puzzle for a computer, which is not only limited to the domain of IT technology, but also to the domain of philosophy when we remember Descartes' famous words, "*Cogito ergo sum*" ("I think, therefore I am") [19].

Compared with AS, which emphasizes the concept of "system," the idea of proactive computing is closer to the process of information flow in computers. Proactive computing includes connecting with the physical world, deep networking, macro-processing, dealing with uncertainty, anticipation, closing the control loop, and making systems personal [15]. From the intersection of the above two fields, the ubiquitous computing or ubiquitous communication has the features anywhere and anytime. *Anywhere* refers to the mobility of the advanced IT facilities that allows users to get convenient services of communication so that users can go beyond the geographical limitation. *Anytime* refers to users being offered with instant services. With the progresses of advanced technology of media and telecommunication networking, pervasive computing has emerged as a new interdisciplinary field supported by the cyberinfrastructure and the advanced information technology (IT) [20].

As Makoto Nagao points out, now we are in the age of "universal communication" (UC) [21]. UC is a product of the application of information and communications technology (ICT) [22] in social life and provides us with the advanced communication technologies and various types of medium services ranging from the basic signals of speech and images, to a high degree of virtual reality and different-language communication. ICT can be regarded as a new progress in the integration of IT and communication technologies. With the technological support of ubiquitous computing, universal communication systems are extending the boundaries of human ability. ICT is one of the best terminologies to unify the earlier-mentioned streams of new information processing paradigms. From the viewpoint of logical speculation, autonomic and proactive attributes are the outcomes of technological innovation of the IT industry at system level. Pervasive, ubiquitous, and universal attributes are the long-term goal of IT industry. As we know, semiconductors have been one of the most important research subjects in material science since the 1950s because of their application in electronic devices and computers. Material, no doubt, is one of the deciding factors for the success of a new generation of computing technology. The high performance of nanomaterials from nanotechnology is encouraging us to exploring the possibility of using moleware (i.e., molecular-ware) to build the new generation of information processing systems. Among the branches of ICT, nanobioICT is directly related to the future generation of information processing systems including biomolecular computers. In approximately 10 or 20 years, nanobioICT is supposed to be complementary to semiconductor technology, electronics-based communication, and cyberinfrastructure already well developed. For example, the smell-sensor in UC is one of the most successful applications of biomolecular information processing. By this time, we hope biomolecular computers motivated from material science will bring human beings more superb services and enjoyment that traditional IT can not.

1.2 Cutting-Edge Technologies for Building a Molecular Computer: From Nanobioscience and Nanotechnology to Nanobioinformatics

Nanobioscience, which is integrated by two branches, nanoscience and bioscience, is an emerging area with knowledge of the biochemical process of moleware on the nanometer level. Nanotechnology based on the theory of nanoscience mainly provides us tools to build nano-sized artificial systems in terms of various materials existing in nature and artificially synthesized by human being. Bioscience mainly tells us how the biological functions are generated in biological systems. *Molecular biotechnology*, which is a biochemistry technology from bioscience, and *nanotechnology* are two major supporting technologies expected to contribute a great deal to the development of biomolecular information processing systems by means of engineering.

The test tube is commonly used in laboratories to study biological systems. Thus, we can conclude that "wet" is the most obvious feature of biochemistry experiments. Corresponding to hardware that has become the ABC knowledge, the term of "wetware" is used to describe the molecular structure and signaling process of biological systems under the wet conditions. Wet feature is intrinsic because it is a necessary condition of cells, survival. If we want to use them to make any circuit or device, we have to realize a kind of biological systems that can be connected to a certain fixed structure. The most commonly used fixed structures are glass and gold surfaces, to which the DNA molecules/DNA molecular sequences are attached for the immobilization of DNA molecules. Connecting biomolecules with surfaces is a cutting-edge [23] technology to interface between wetware of biomolecular computing and the user terminal. The "wet" cells, survival biological systems, such as proteins and cells, seem to have the natural linkage with medicine. Fawwaz T. Ulaby said, "With the human genome deciphered, ... and nanotechnology able to design molecules for treating diseases and probing biological systems at a far greater level of detail ..." [24].

Motivated by surprising progresses in molecular biology as well as the potential applications in molecular medicine, the edge-cutting technologies are expected to be used for building biomolecular computing systems. They can be classified into the following different levels:

- Operating on individual molecules;
- Operating on set of molecules;
- Operating on moleware in a systematic way.

The current molecular nanotechnologies allow people to operate on a single DNA molecule as well as synthesize spatial structure of DNA molecules, which is the basis for building DNA computers by aggregating DNA molecules on the basis of the self-assembly principle. DNA sequences can also be operated in the forms of single strand as well as double strand. Different DNA sequences can be connected in test tubes. One of the most noticeable advances in DNA molecular technology is the ability to construct two-dimensional structures of DNA molecules [25]. DNA molecules are not only promising for the molecular computer, but also for the nanomachine since DNA is a kind of fuel and medium for electrical conducting in moleware [26, 27].

We have several different choices for building biomolecular computing systems using different biomolecules. DNA is relatively easy to be operated in test tubes. Other candidates such as proteins are also promising with regard to their characteristics in bioinformatics and biochemistry. However, we have to consider one problem, which is the interactions among them. The interactions of proteins are physical and biochemical constraints on computing by proteins.

At the system level of biomolecular information processing, genetic circuit, which behaves like a kind of circuit, is a form of artificial systems built by cells and controlled by gene regulation. Genetic circuit is realized by molecular signaling processes in cells under the activation of specific genes. The signals in genetic circuit are the molecules represented by symbols, molecular concentration, and so on. In genetic circuits, molecular signals are reliable. Genetic circuit is also a basis for systematical design, control, and testing of moleware systems in which gene operations are needed. Gene operation is very useful in many experiments of molecular biology.

1.2.1 Synthetic Biology

Synthetic biology [28] is a new field in molecular biology. It deserves to be studied for its possibility in molecular therapy. The methodology of synthetic biology is to use the molecular signaling of cells and control them in a systematic way. Here we mention it, hoping the new technology of synthetic biology is helpful for building a biomolecular information processing system. The engineered cell communication system is an example of synthetic biology. Synthetic biology refers to integrating the molecular components into a moleware system under the control of engineering in order to produce a specific function as designed in advance. Here, "synthetic" implies a molecular signaling system built in terms of synthesizing the moleware components extracted from naturally existing systems. Synthesis is in contrast to analysis in concept. Synthesizing a complete or partially artificial molecular system requires knowledge from the analysis of the biological system in nature, and in turn, the synthesized system is a tool for understanding the biological system in nature. This methodology is consistent with the essence of the natural biomolecular system, as well as the philosophy of systems biology on proteomics.

Chen and Weiss [28] have realized a synthetic biology system made by yeast in which proteins extracted from plant *Arabidopsis thaliana* are embedded. The system consists of a sender cell and a receiver cell. The messenger is IP (isopentenyludenine)—a cytokine sent out from the sender cell through cell membrane. When IP arrives at the surface of a receiver cell, it has a biochemical reaction with the receptor AtCRE1 (a cytokinin on the cell membrane). This biochemical reaction can activate signaling cascades within the cell. The sending operation is controllable as well as the receiving operation, and therefore, the communication process in the synthetic system is controllable as designed in advance and is an engineered molecular signaling process. The SLN1-YPD1-SSK1/SKN7 phosphorylation pathway in yeast is used to produce the fluorescence effect as the signal-output of the receiver cell. SLN1 is a kinase, YPD1 is a protein, and SSK1 and SKN7 are regulators. SKN7 is attached to the GFP (green fluorescent protein) which can be detected by fluorescence technology. Chen and Weiss have developed two types of synthetic biology systems. One is explained earlier, and the other is an engineered cell communication system in which the sender and the receiver are within the same cell when the signal feedback mechanism is used.

The experiment by Knight and Weiss [29] as well as the experiment by Chen and Weiss [28] both show the strong potential in employing cells as the material to innovate an artificial moleware system to realize engineered intercell communication through instruction (control of the molecular signaling processes in cells).

Bacteria, yeast, and plant *Arabidopsis thaliana* are well-studied sample species in molecular biology. The experimental tools for molecular operation on these species are relatively mature, so we can select them as the material to realize an engineered cell communication system in laboratories. The experimental method for controlling the signaling process of bacteria in plasmids is also relatively mature. Through operating the gene in plasmid, the specific response of bacteria can be produced (for example, the bacteria to emit light). As reported [28], it is easy to extract molecules from the sample plant *Arabidopsis thaliana* and to embed them into the bacteria with low cost. Culturing the bacteria from either several hours to a few days makes engineered cell communication more reasonable in practice.

The complexity of the genotype and phenotype of cells varies among different species. The number of genes and the genome size directly decide the availability of molecular operation on genes (e.g., promoter). Because DNAs, RNAs, and proteins have different structural and biochemical features, their efficiency of bimolecular computing systems by different design schemes of molecular operators is different. Either way, there is a high possibility of using the cell as the material to build molecular information processing and communication systems through controlling molecular signals at the molecular level. Consideration will therefore need to be given to the aspect of cellular signaling in order to synthesize different molecules into one biomolecular computing system. The synthetic biology is just the beginning of the ambitious exploration of biomolecular information processing systems.

1.2.2 Emerging Technologies for Protein Analysis: To Gain Information about Proteins, Protein Interaction, and Their Links to the Medicine

When we use proteins to build a biomolecular computing system, we have to know how to manipulate proteins in experiments. In proteomics, various technologies on protein manipulation and protein analysis are developed. With the help of these technologies, proteins can be detected, making it possible to use them to represent information in computing. Designing protein-based computing algorithms requires the information on protein-to-protein interaction.

1.2.2.1 See-and-Touch Proteins: An Approach to Access Proteins

A basic method of protein detection is mass spectrometry [30]. The principle of this method is different proteins with different masses produce a different mass spectrum. By comparison with its mass, the correspondence between the protein and its mass can be quantitatively formulated so that we can recognize it. The mass spectrometry technology has been improved with new functions to the applications (for example, MALDI-related technology). The technologies for us to see proteins include mass spectrometry, X-ray crystallography, nuclear magnetic resonance (NMR), and fluorescence resonance energy transfer (FRET).

The direct way to get the information about proteins is to obtain the images of protein structures. Two technologies—X-ray crystallography and nuclear magnetic resonance (NMR)—have been used to see the structures of small proteins [31]. The X-ray crystallography technology uses X-ray to image the protein structure. NMR technology, whose principle will be explained in Chapter 2, is to use the quantum effect to get the image of the protein structure. The two technologies are complementary [31].

> Some proteins will give nice spectra but did not provide crystals, and some proteins that gave poor spectra did yield crystal structures.

The physical structure of the protein is configured by the spatially connected atoms and can produce the spectra effect. The information on the site or domain of proteins tells us what kinds of biochemical reactions could happen among the proteins. The term "spectra" means the spectra signals detected by the related equipment. The direct application of these imaging technologies is understanding the protein structure and predicating protein function.

1.2.2.2 Figure Out Protein-to-Protein Interactions

If only we knew the atom-to-atom connection of protein structure, information about motif patterns of proteins (which means the combinatorial form of the building blocks of protein structure) would be obtained. Here, pattern means the macro-level description of the protein structure. The reason we call it macro is the motif contains more information than the site and is a bigger unit than the site. The motifs tell us the information of protein interaction and what kinds of proteins can cause biochemical reaction with specific function on cellular activity. The motif knowledge can be used to infer protein interaction [32], which belongs to the bioinformatics method (inferring protein interaction by knowledge of protein structure. In order to verify what is really happening in the interacting proteins, seeing the protein interaction is the best way. Fortunately, this technology has become reality. Recent reports tell us [31]:

> Intracellular protein-protein interactions form the basis of most biological processes. Structural aspects of these reactions can now be analyzed in living prokaryotic cells and in atomic detail by nuclear magnetic resonance spectroscopy.

1.2.2.3 The "Inherited" Linkage of Protein Informatics and Medicine

Owing to the tradition of biology, the protein study is tightly related to medicine in which the detection technologies serve various applications. Medical applications of cell biology require detection tools to observe the proteins in cells to understand their biochemical functions. The technology of measuring interacting proteins is significant in medicine. For example, "Phosphorylated derivatives of phosphatidylinositol (PtdIns), known as phosphoinositides (PIs)," have been reported with the remarks [33]:

> Strategies and techniques for quantitative and qualitative measurement of PIs, for characterization of specific PI-binding proteins and for determination of PI kinase and phosphatase activities in vitro and in vivo.

Holger Lorenz, Dale W. Hailey, and Jennifer Lippincott-Schwartz point out [34], "Understanding the cell biology of many proteins requires knowledge of their in vivo topological distribution. Here we describe a new fluorescence-based technique, fluorescence protease protection (FPP), for investigating the topology of proteins and for localizing protein subpopulations within the complex environment of the living cell." Considering the biological functions of proteins, biomolecular computing systems made by proteins naturally connect to medicine. The real-time information about the locations of proteins in cells is radical for us to get and use in cellular information processing. The location of proteins in cells is the information to check where and what types of molecular information processing occurs in cells, and is an important type of data for the analysis of molecular signaling mechanisms in cells in order to design a biomolecular computer.

1.3 Preliminaries in Nanobioscience

In this book, we are mainly discussing biomolecular computing, in which molecules are used to represent information and biochemical reactions to carry out information processing. Molecular manipulation is crucial to the success of the implementation of a biomolecular computer, which is studied within the framework of nanobioscience.

As we know, nanotechnology is oriented to creating very small devices using the knowledge as well as technology of material science. In addition, nanotechnology is needed to build a biomolecular computer, and biological approaches developed for understanding the mechanism and function of biological systems are also needed since biomolecules are used as the components of a biomolecular computer. The material science and biological science belonging to the new emerged field called nanobioscience, therefore, get together in the case of biomolecular computing.

Our whole story will start from "nano." In concept, the prefix "nano" means 10^{-9}. One nanometer is $1/10^9$ of 1 meter. The terminology of "nanotechnology" was first proposed by Taniguchi [35]. The big initial event that makes "nano" become extremely hot is NNI. National nanotechnology initiative (NNI) is a state project on a huge scale launched in the United States and it has a great impact on the world [36, 37].

> Nanotechnology is the understanding and control of matter at dimensions of roughly 1 to 100 nanometers, where unique phenomena enable novel applications.

NNI also promotes broad applications that include [37]:

> One area of nanotechnology R&D is medicine. Medical researchers work at the micro- and nano-scales to develop new drug delivery methods, therapeutics and pharmaceuticals.

The promise of nanoscale science, engineering, and technology (NSET) [36–38] as what we see today was foreseen by Feynman several decades ago [39] and popularized by Eric Drexler [40–42]. Nowadays, the nanotechnology has covered circuits, devices, and systems at the nano-size level, which is constantly being improved by the technology progresses in molecular electronics.

1.3.1 Gedanken Model

Experiments in material engineering and computer-aided simulation are conventional methods for nanotechnology. Modeling is a bridge between experiments of material science and implementation of biomolecular computing. The term "Gedanken model" means "the model proposed in terms of thought experiment" [43, 44], which is proposed or studied according to the scientific principle under the condition as it could be. Normally this kind of model can be tested by the calculation or simulation based on physics mechanisms if the experimental tools to completely verify the model are not available for the time being. Gedanken was originally coined for the thought experiment [45].

> A thought experiment (from the German term Gedankenexperiment, coined by Hans Christian Ørsted) in the broadest sense is the use of an imagined scenario to

help us understand the way things really are. The understanding comes through reflection on the situation. Thought experiment methodology is a priori, rather than empirical, in that it does not proceed by observation or physical experiment.

In moleware, covering both physics and chemistry features, molecules are a linkage between the "nano"-side and "bio"-side of nanobioscience.

1.3.2 Some Concepts in Biochemistry

Molecule is the most fundamental concept in biology. As Wikipedia [46] defines,

> In chemistry, a molecule is generally an aggregate of at least two atoms in a definite arrangement held together by special forces. ... Generally, a molecule is considered the smallest particle of a pure chemical substance that still retains its composition and chemical properties.

DNA refers to deoxyribonucleic acid [47] and is genetic matter in a cell. Normally, the DNA molecule is with a helix structure. RNA refers to ribonucleic acid [48–50].

> A protein (from the Greek protas meaning "of primary importance") is a complex, high-molecular-mass, organic compound that consists of amino acids joined by peptide bonds. Proteins are essential to the structure and function of all living cells and viruses.

> The cell is the structural and functional unit of all living organisms, and is sometimes called the "building block of life." Some organisms, such as bacteria, are unicellular, consisting of a single cell. Other organisms, such as humans, are multicellular (humans have an estimated 100 trillion or 10^{14} cells; a typical cell size is 10 μm, a typical cell mass 1 nanogram). Cell is regarded as "the minimum unit of life."

As a science, biology was originated to understand the biological phenomena. The test tube, a glass container, is often used for biological experiments to operate the biological materials and to observe their characteristics. The term *in vitro* means "in test tube." The material in test tubes is called assay. The assay is used to test the biochemical reaction. The biological research requires various types of instruments. One of the most commonly used instruments in molecular biology is the microarray, which is a kind of extension of test tubes. Actually, the microarray is an array that consists of multiple assay containers and can be observed by microscopes. There are several different types of microarrays including DNA microarrays, protein microarrays, tissue microarrays, and others [51].

By applying microarrays in the biochemical experiment with automation, individual molecules and their relations are studied. Nowadays, the automation technology in biology can provide us rich amount of data to speculate the problems. How do molecules work in the cell and what is the relationship between the molecular signaling mechanism and the corresponding biological phenomena at the macro level. The size of DNA is about 10 nm. Owing to the natural link existing among DNAs, RNAs, and proteins, systematical methodology is recommended to help us to understand the mechanism of cellular signaling and to use it for designing a biomolecular computer.

1.3.3 Systems Biology

In addition to the experimental tools and instruments in biology, *systems biology* is a new methodology both in theory and in practice. The methodology of systems biology is to study the biology in terms of system science. As a complex system, the biological system such as cell involves multiple relations concerning many objects including DNAs, RNAs, proteins, ions, and others.

E.T. Liu [52] says:

> Systems biology seeks to explain biological phenomenon not on a gene-by-gene basis, but through the interaction of all the cellular and biochemical components in a cell or an organism.

The *objects* studied in systems biology are biomolecules such as DNAs, RNAs, proteins, and biological systems such as cells and organelles. The focus is put on the interaction among these molecules as well as the function caused by the integration mechanism of these molecules.

The major *method* of systems biology is to model the biological objects as a system; to identify the relation among the components of the system; to simulate the collective behavior of the biological system; and to verify the hypothesis by experiments.

Tools in systems biology can be classified into two categories in general: the computational tool and benchwork tool, which are complementary. Most of the computing work is on the simulation of the complex relations among molecules in which the involved parameters are not easily obtained by the wet experiment. We all know using a computer is a faster and cheaper way to get simulated information about the biological objects, but the wet experiment is the final judgment for simulation. The technical work of systems biology includes: (1) system analysis (nonlinear mechanism of interactions among the systems) for system synthesis; (2) system identification and parameter estimation; (3) control schemes for the nonlinear objects; (4) signal analysis and signal processing (filtering, noise reduction); and (5) random signal processing and statistics analysis for biological systems.

Applications of systems biology are broad enough to encompass medical science owing to the historical relation between biology and medicine. With the biochemistry being integrated with nanotechnology, the nanobioscience is hoped to be applied to new generations of information processing systems where the interactions among molecules are studied in terms of systems biology.

1.3.4 Perspectives on Innovative Technologies for Biomolecular Computing: Benefits from Breakthroughs of Molecular Biology in the New Millennium

Reviewing the technological progresses on frontier fields in recent years, one of the breakthroughs across the century is the success of the human genome project (HGP). Following the HGP, the emergence of new proteomics and "omics" technologies further renewed the record of life science. The significance of proteome information on organelles is it gives us more deep knowledge on the relations among the proteins so that the biological functions of cells can be traced. The technology of stem cell research has had a breakthrough by which the idea of the ES cell (also called universal cell) [53] arouses the passion of regenerative medicine.

Aiming at biomolecular computing, two innovative technologies emerged and are evolving called single molecule technology and suprachemistry. Single molecule technology makes it possible to manipulate an individual molecule. Using this technology, we can recognize and observe a single molecule, as well as control its biochemical state. In biomolecular computing, one molecule (protein) can be used to represent one-bit of information. Based on the information representation, word, which is commonly used in electronic computers, is also designed for biomolecular computers. A word consists of several bits. In biomolecular computers, word is represented by a molecular complex (compound)—an aggregate of multiple molecules. In living cells, there exist molecular complexes (compounds) of proteins. Through supramolecular chemistry, molecular compounds are artificially synthesized as designed in advance. Of course, the engineering method for synthesizing molecular complexes is still challenging. With the word made by molecules that is the basis of programming for biomolecular computing, it will be possible to design a prototype of a biomolecular computer.

1.4 Challenges from Real-World Applications

With the progresses of semiconductor technologies, the speed of this kind of computer is becoming faster and faster, while the size is becoming smaller and smaller. Developers of computers are competing with each other on the speed, size, and information processing ability because they are the most important indexes for us to judge.

1.4.1 Performances of Biomolecular Computing

In the case of DNA computing, which is still in its beginning stages, Leonard Adleman has a calculation result [54] that shows the advantages of DNA computing on size, energy consumed, and information processing ability compared with the semiconductor-based computing technology. Laura Landweber also has a remark on the merits of DNA computing [4]. Speed is dependent on where we would like to apply the biomolecular computer and what kind of biomolecular computer it is. Nano-size is unquestionably the most prestigious point of biomolecular computing. The information processing ability—the capability of massive information processing—is provided by the parallelism of biomolecular information processing.

1.4.2 Technological Difficulties on Feasibility of Implementation of a Biomolecular Computer: Scalability, Reliability, and Controllability

Since different biomolecules are used to build biomolecular computers, we have to discuss these biomolecules in information representation respectively. If information is represented by *DNA* sequences, we have to consider how to select DNA sequences to represent information efficiently and how to reduce the errors in the DNA computing process. The former should be studied in terms of scalability because we expect biomolecular computers to solve large scale problems by releasing the power of its massive parallelism. The latter is on reliability, which is fundamental for computers.

The massive parallelism in DNA computing is realized by using a pool of DNA sequences. As we know, the number of DNA sequences needed for DNA computing is huge. Juris Hartmanis pointed out in 1985 that the number of DNA molecules used is extremely huge when a standard DNA computing algorithm is applied for the large scale NP problem solving [55]. How to cut the number of DNA molecules used in DNA computing is an open problem. In order to solve this problem, we have to innovate new "efficient algorithms" that only use a limited number of DNA molecules and develop practical technology to realizing it.

The most challenging obstacle in DNA computing is getting the reliability that the semiconductor computer has already reached. During DNA computing processes, unwanted biochemical reactions often occur, which may result in part of resultant DNA sequences not being the required ones. This causes the error in DNA computing. If the above problems cannot be solved, a system of DNA computing is not yet complete.

The protein complex (i.e., protein compound) consists of multiple proteins. In the case of applying proteins to biomolecular information processing, each protein within the protein complex is defined as the minimum unit of information to represent one symbol or variable. As we know, the protein can have two states in biochemistry. The protein may be attached by a label molecule or not. These two states of the protein are used to represent the values of the variable in which the "label molecule" refers to the molecule attached to the protein. A simple way is to define the status of the protein attached with a "label" molecule as 1 in binary value or true in logical value and the status of the protein without any attached "label" molecule as 0 in binary value or false in logical value.

Synthesizing proteins into protein sequences is not as easy as it is in DNA molecules. However, the naturally existing protein complexes are good candidates for information representation. The protein complexes and their corresponding states (attached with molecules or not) are available in the living cell with a great number. A limited number of enzymes can efficiently control the cellular pathways. This gives us the hope of conducting information processing by proteins with the control of cellular pathways. The biochemical reactions among proteins under the activation of enzymes obey the rule of lock-and-hole. The enzymes act as the key and the corresponding proteins act as the hole. In order to achieve a specific biochemical reaction, we need to control the enzyme activation so that the aggregation of proteins will be formed as desired, which looks like the docking process of the space station.

In order to realize the information representation by the protein complexes mentioned above, control of the protein signaling process is needed. In the case of information processing by protein complexes, the problems of scalability and reliability are thus transformed into the problem of controllability on protein signaling processes because the scalability and reliability of biomolecular information processing by protein complexes are based on the controllability of the corresponding biomolecular signaling processes.

Technically, the degree of difficulty of controlling proteins for information processing is dependent on the experiment mode of in vitro and in vivo. The in vitro mode for controlling protein complexes and their biochemical reactions is comparatively easy, considering the current technology in laboratory experiment. The specific proteins are purified and cultivated before being put into test tubes. The biochemical reactions will then be controlled according to the plan designed in advance in order to get the final product of protein compounds, whose information is the result of information processing. Using the in vitro mode, the information flow (mainly data flow) can be controlled in the test tubes. The in vivo condition, in terms of living cells, controlling the molecular signaling processes by cellular pathways is complex and difficult. Because the signals in cells are coupled, it is necessary to monitor the biomolecular information flow as we wanted in advance which is the designed information flow. Using a single cell as the information unit, the entire computing task will be carried out in multiple cells. The signaling mechanism of intracell communication is complex. It is difficult to control individual biochemical reactions within living cells, but it is possible to observe and control a limited number of biomolecular signaling processes around the cellular membrane for biomolecular computing by single molecule technology. The controllability of signaling processes is very important for the success of the efforts on the biomolecular computing systems.

1.5 Back to Molecular Informatics: How to Use Molecules to Represent Information

Biomolecular computers are expected to be built complementary to electronic computers mainly because the application of biomolecular computers in life science is with great significance. For example, as Shapiro and Beneson presented, a DNA computer is developed for the purpose of cancer diagnosis [56].

Biomolecular signaling is the linkage between biomolecular computing and biological function in cells. The prerequisite for building a biomolecular computer is how information processing is carried out by biomolecules such as DNA, RNA, and others, in which their own special biochemistry features exist. In biomolecular computing, the programming process requires controls on molecular operations for information representation. The form of symbol or variable is fundamental for programming processes in molecular computers.

Generally speaking, the molecules and molecular complexes (such as DNA sequences and proteins) can be used to represent the specific symbols or variables and their states can be used to represent data: the value of the variable if we define the correspondence between the molecules and the variables as well as the correspondence between the states of molecules and the values of the variables. Based on the symbolic representation by molecules, the correspondence between molecules and information/data can be established. The aim of this arrangement is to describe the relationship between the form of information processing and the corresponding biochemical reactions. The molecular pattern made by nanotechnology is one way to represent information. The molecular concentration is another way to represent information. The patterns that consist of porphyrin molecules, a kind of organic molecule (the atomic structure of porphyrin and its physical features are discussed in Chapter 8) show the power of nanotechnology for information processing. Porphyrin molecules can be connected to form a line, a triangle, and a rectangle-shaped structure. The derived structural patterns are monomers, dimers, trimers, and tetramers. The monomers can be connected to construct a primitive nanowire. The structure made by one porphyrin molecule is a monomer. The linkage between N and H atoms within porphyrin molecules is used to connect different porphyrin molecules. There are four corners in one porphyrin molecule, which are the places for the linkages between N and H atoms. Two porphyrin molecules are connected to form the dimer structure. Three porphyrin molecules are connected to form the trimer structure and four porphyrin molecules to form the tetramer structure. The center of the porphyrin can be embedded by a metal element. As shown by the example (two nodes in a line) in Figure 1.1(a), the center locations of the connected porphyrin molecules give us the pattern information after the metal elements are embedded into the centers of porphyrin molecules. Different metal elements are used to define different symbols for the purpose of informatics. The spatial relation of monomers, dimmers, trimmers, and tetramers generates corresponding patterns [see Figure 1.1(b)]. These patterns are the building blocks to describe spatially constrained information such as a two-dimensional formal language in computer science. Formal language that we will discuss in Chapter 5 is an abstract form of theoretical models of computation that can be used to describe information processing systems. The commonly used data structure for a formal language is a string in one dimension or a graph in two dimension. A string consists of symbols. A graph consists of nodes called vertexes and links called edges. The vertexes and edges are represented by symbols. The triangle and tetramer forms in a two-dimensional formal language mainly describe the topological relation among the symbols. After we define the metal elements centered in the porphyrin molecules as symbols, we can get a spatially constrained formal language such as the one given in Figure 1.1(c).

In terms of molecular concentration, the dynamical process of biochemical reactions can be designed and controlled to carry out certain kinds of information processing. Assume that molecule A is used to represent variable X and molecule B is used to represent variable Y, the "state of X and Y" is defined as follows:

The state of X/Y = 1, if molecule A/B is attached with molecule L

The state of X/Y = 0, if molecule A/B is not attached with molecule L

These two states can be detected by biochemical technology according to the criterion whether or not the molecules A and B are attached with molecule L, respectively.

Let the product of the biochemical reaction of A and B be molecule C, which is represented by variable Z.

$$C = \text{biochemical reaction (A, B)}$$

1.5 Back to Molecular Informatics: How to Use Molecules to Represent Information 17

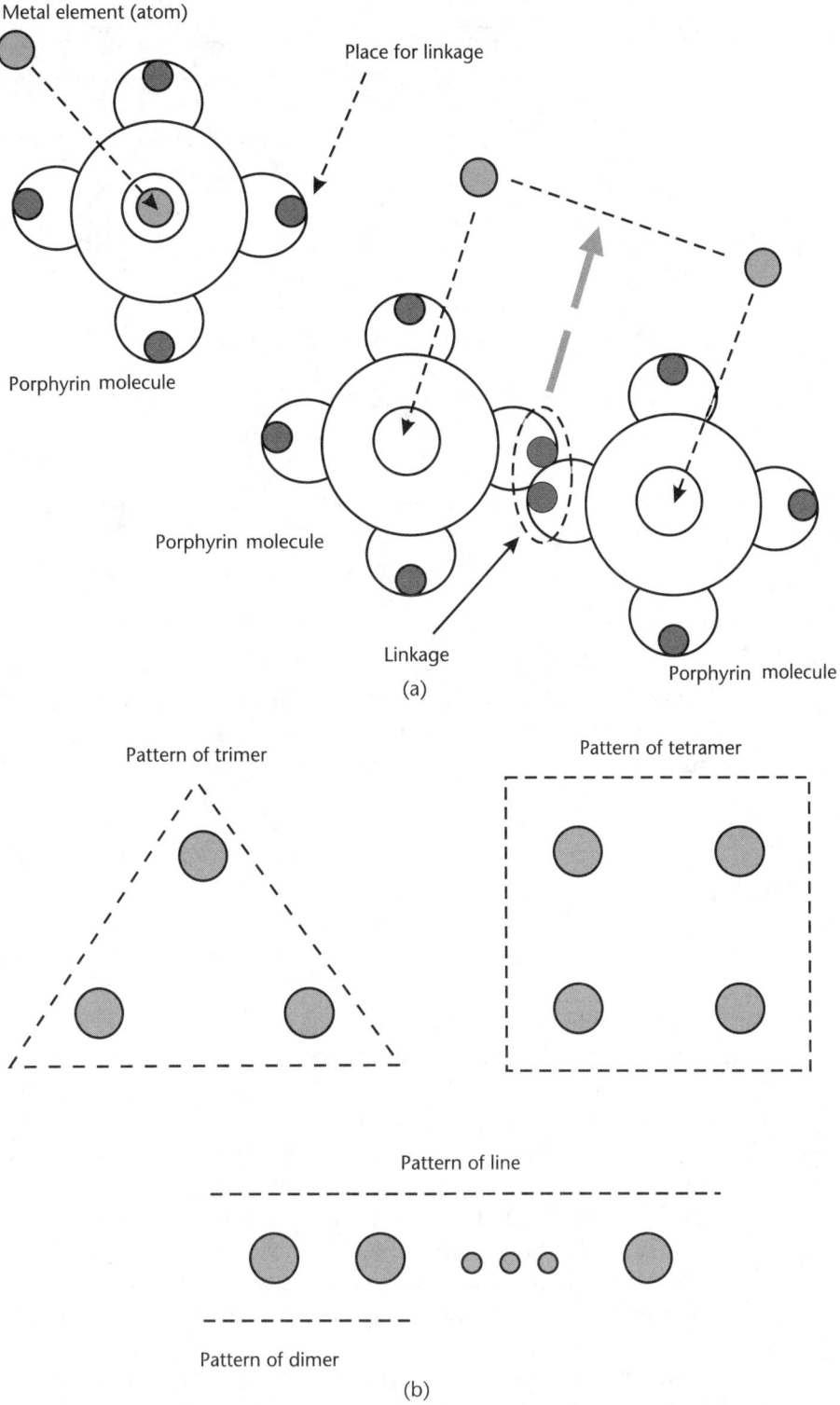

Figure 1.1 Informatics attributes of connected porphyrin molecules: (a) examples of connecting porphyrin molecules; (b) patterns of connected porphyrin molecules; and (c) formal language representations for patterns of connected porphyrin molecules.

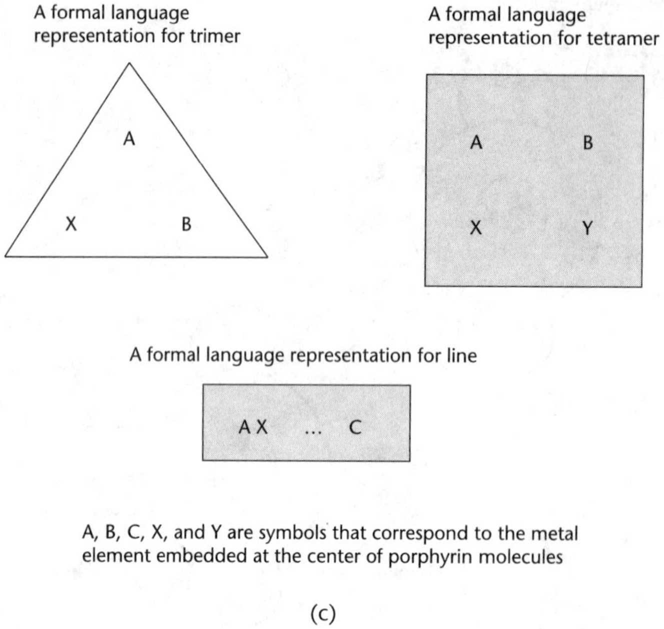

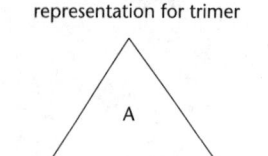
A formal language representation for trimer

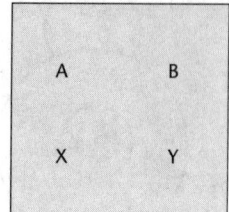
A formal language representation for tetramer

A formal language representation for line

A, B, C, X, and Y are symbols that correspond to the metal element embedded at the center of porphyrin molecules

(c)

Figure 1.1   (continued.)

If the molecule C is attached with molecule L, $Z = 1$. If the molecule C is not attached with molecule L, $Z = 0$.

Assume that $X = 1$, $Y = 0$, then the biomolecular signaling process implies that

$$X + Y \to Z$$

that is,

$$1 + 0 \to 1$$

where the molecule C is attached with molecule L.

To realize an individual computing process as mentioned earlier, the corresponding biochemical reactions should be carried out as a procession designed in advance. If we define multiple variables $X1, X2, \ldots, Xn$ ($n = 1, \ldots, n$, n is a natural number) by multiple molecules (under the condition that n proteins are available), the earlier-mentioned process can be extended from two variables to n variables. In a specific computing process, the input and output are deterministic. The ad hoc chemical signaling process in biological systems will be regulated to satisfy the deterministic condition, which is the basis of a biomolecular computer. The computing process of a biomolecular computer is thus different from that of a semiconductor computer.

Considering the strong potential in parallelism of biomolecular signaling, biomolecular computing is expected to outperform conventional computing in theory and in practices. In order to realize these potentials, we have to overcome obstacles in computation: feasibility, reliability, and controllability of biomolecular computing based on the biochemistry foundation of molecular signaling. We have enough reasons to be enthusiastic about the transition from the idea of computing by biomolecules to any reliable form of biomolecular computing system. The hope is brought by the promising nanobiotechnolgy. The *Molecular Science Manifesto* says [57]:

> And even though we are very clumsy in our use of the tools right now, and even though molecular biology has made only a small portion of them available to us so far, we can already use them to build a computer. And if you can build a computer, then presumably many other exciting things can be built. So, this is the challenge of molecular science: take the tools and build something great.

Benefited from nanobioscience and nanobiotechnology, biomolecular computing with the biochemical features inherited from the biological systems bear the mission of programming nanobiosystems for molecular medicine in order to contribute to human health at one of the last frontiers on life science: nanomedicine [13]. One of the most important points on the roadmap for biomolecular computers is to improve the technological approaches that are empirical for the time being. Nevertheless, by means of nanobiotechnology it may not take as long a time as the conventional computer once experienced for a complete implementation in the foreseeable future.

References

[1] http://en.wikipedia.org/wiki/Moore's_law#Formulations_of_Moore.27s_law.

[2] Lloyd, S., "Ultimate Physical Limits to Computation," *Nature*, Vol. 406, August 2000, pp. 1047–1054.

[3] Warren, P., "The Future of Computing—New Architectures and New Technologies," *Nanobiotechnology, IEE Proceedings*, Vol. 151, No. 1, 2004, pp. 1–9.

[4] Landweber, L., "Beyond Silicon-Based Computing: Quantum and Molecular Computing," September 12, 2000, http://www.house.gov/science/landweber_091200.htm.

[5] Condon, A., "Designed DNA Molecules: Principles and Applications of Molecular Nanotechnology," *Nature Reviews Genetics*, July 2006, pp. 7, 565–575.

[6] Alberts, B., "The Cell as a Collection of Protein Machines: Preparing the Next Generation of Molecular Biologists," *Cell*, Vol. 92, No. 3, February 1998, pp. 291–294.

[7] "Small Wonders, Endless Frontiers: Review of the National Nanotechnology Initiative," http://www.nano.gov/html/res/smallwonder.html.

[8] TOP500 Supercomputer Sites, http://www.top500.org/.

[9] http://www.top500.org/lists/2006/06.

[10] Stafford, N., "Top Computer Hangs on to Its Title," June 2006, http://www.nature.com/news/2006/060626/full/060626-12.html.

[11] Brent, R., and J. Bruck, "2020 Computing: Can Computers Help to Explain Biology?" *Nature*, Vol. 440, March 2006, pp. 416–417.

[12] Nanomedicine: NIH Roadmap for medical research, http://nihroadmap.nih.gov/nanomedicine/.

[13] Calude, C. S., and G. Pun, "Bio-Steps Beyond Turing," *Biosystems*, Vol. 77, No. 1–3, November 2004, pp. 175–194.

[14] Butler, D., "2020 Computing: Everything, Everywhere," *Nature*, Vol. 440, March 2006, pp. 402–405.

[15] Want, R., T. Pering, and D. Tennenhouse, "Comparing Autonomic and Proactive Computing," *IBM System Journal*, Vol. 42, No. 1, 2003, pp. 129–135.

[16] http://www.research.ibm.com/autonomic/.
[17] http://www.research.ibm.com/autonomic/overview/elements.html.
[18] Barabási, A. L., et al., "Parasitic Computing," *Nature*, Vol. 412, August 2001, pp. 894–897.
[19] http://en.wikipedia.org/wiki/Cogito_ergo_sum.
[20] Freeman, P. A., et al., "Cyberinfrastructure for Science and Engineering: Promises and Challenges," *Proceedings of the IEEE*, Vol. 93, No. 3, March 2005, pp. 682–691.
[21] http://www.wsis-japan.jp/doc_pdf/D-18DrNagao.pdf, http://www.nict.go.jp/publication/NICT-News/Eng/2006/NICT_2006JAN_E.pdf, http://www.nict.go.jp/about/message_e.html.
[22] http://www.nict.go.jp/about/message_e.html.
[23] Barth, J. V., G. Costantini, and K. Kern, "Engineering Atomic and Molecular Nanostructures at Surfaces," *Nature*, Vol. 437, September 2005, pp. 671–679.
[24] Ulaby, F. T., "Engineering in the Age of Biology," *Proceedings of the IEEE*, Vol. 94, No. 5, May 2006, pp. 863–864.
[25] http://seemanlab4.chem.nyu.edu/.
[26] Yurke, B., et al., "A DNA-Fuelled Molecular Machine Made of DNA," *Nature*, Vol. 406, 2000, pp. 605–609.
[27] Lewis, F. D., et al., "Direct Measurement of Hole Transport Dynamics in DNA," *Nature*, Vol. 406, July 2006, pp. 51–53.
[28] Chen, M. T., and R. Weiss, "Artificial Cell-Cell Communication in Yeats Saccharomyces Cerevisiae Using Signaling Elements from Arabidopsis Thaliana," *Nature Biotechnology*, Vol. 23, No. 12, December 2005, pp. 1551–1555.
[29] Weiss, R., and T. F. Knight Jr., "Engineered Communications for Microbial Robotics," *DNA6, Lecture Notes in Computer Science*, Vol. 2054, A. Condon, and G. Rozenberg, (eds.), Berlin, Heidelberg: Springer-Verlag, 2001, pp. 1–16.
[30] Doerr, A., "A New Avenue for Mass Spectrometry," *Nature Methods*, Vol. 3, 2006, pp. 72–73.
[31] Selenko, P., and G. Wagner, "NMR Mapping of Protein Interactions in Living Cells," *Nature Methods*, Vol. 3, 2006, pp. 80–81.
[32] Balla, S., et al., "A Tool for Investigating Protein Function," *Nature Methods*, Vol. 3, 2006, pp. 175–177.
[33] Nikolaev, N. O., S. Gambaryan, and M. J. Lohse, "Fluorescent Sensors for Rapid Monitoring of Intracellular cGMP," *Nature Methods*, Vol. 3, 2006, pp. 23–25.
[34] Lorenz, H., D. W. Hailey, and J. Lippincott-Schwartz, "Fluorescence Protease Protection of GFP Chimeras to Reveal Protein Topology and Subcellular Localization," *Nature Methods*, Vol. 3, 2006, pp. 205–210.
[35] Taniguchi, N., "On the Basic Concept of 'NanoTechnology,'" *1974 Proc. ICPE*, http://www.nanoword.net/library/nwn/1.htm.
[36] http://www.nano.gov/html/about/home_about.html.
[37] http://www.nano.gov/html/facts/whatIsNano.html.
[38] Roco, M. C., "Worldwide Trends in Nanotechnology," *Semiconductor Conference*, 2001, CAS 2001 Proceedings International, Vol. 1, October 2001, pp. 3–12.
[39] Feynman, R. P., "There's Plenty of Room at the Bottom: An Invitation to Enter a New Field of Physics," *Engineering and Science*, February 1960, http://www.zyvex.com/nanotech/feynman.html.
[40] Drexler, K. E., *Nanosystems: Molecular Machinery, Manufacturing, and Computation*, New York: John Wiley & Sons, 1992.
[41] Drexler, K. E., C. Peterson, and G. Pergamit, *Unbounding the Future: The Nanotechnology Revolution*, New York: William Morrow, 1991.
[42] Drexler, K. E., *Engines of Creation: The Coming Era of Nanotechnology*, New York: Anchor Press/Doubleday, 1986.

[43] Zhirnov, V. V., et al., "Limits to Binary Logic Switch Scaling—A Gedanken Model," *Proceedings of the IEEE*, Vol. 91, No. 11, November 2003, pp. 1934–1939.

[44] Garzon, M. H., R. J. Deaton, "Biomolecular Computing and Programming," *IEEE Trans. on Evolutionary Computation*, Vol. 3, No. 3, September 1999, pp. 236–250.

[45] http://en.wikipedia.org/wiki/Thought_experiments.

[46] http://en.wikipedia.org/wiki/Molecule.

[47] http://en.wikipedia.org/wiki/DNA.

[48] http://en.wikipedia.org/wiki/RNA.

[49] http://en.wikipedia.org/wiki/Protein.

[50] http://en.wikipedia.org/wiki/Cell_%28biology%29.

[51] http://en.wikipedia.org/wiki/Microarray.

[52] Liu, E. T., "Integrative Biology and Systems Biology," *Nature*, March 29, 2005.

[53] Insight: Stem-cell Biology, *Nature*, Vol. 441, 1059 (June 2006), A special issue on this topic: 10 papers.

[54] Adleman, L., "Molecular Computation of Solutions to Combinatorial Problem," *Science*, November 1994, pp. 1021–1024.

[55] Hartmanis, J., "On the Weight of Computations," *Bulletin of the EATCS*, Vol. 55, 1995, pp. 136–138.

[56] Shapiro, E., and Y. Benenson, "Bringing DNA Computers to Life," *Scientific American*, Vol. 294, No. 5, May 2006, pp. 45–51.

[57] http://www.usc.edu/dept/molecular-science/index.html.

CHAPTER 2
The State-of-the-Art Molecular Biology and Nanotechnology

This chapter, aimed at summarizing the latest knowledge and techniques about molecular biology and nanotechnology, will serve as an introduction to understanding the molecular information processing systems built by nanobiotechnology.

2.1 Genomics

Life is one of the most mysterious phenomena in nature. It has caused lots of attention in human history and biology is the science dedicated to studying it. Nowadays, the term life science is frequently used to cover biology and its related fields concerning biological systems.

One of the most astonishing achievements in life science at the beginning of the new millennium was the human genome sequencing. Genome is defined as a complete gene (the genetic material that consists of DNA molecules) sequence from which the biological function is interpreted. In additional to the human genomes being sequenced, the genomes of plants, such as rice and bacteria like *Helicobacter pylori*, are also sequenced. The technological tools for genomics have been greatly enhanced by automation devices such as robots.

Genome information is rich and requires handling by high performance computing technology. The complexity of genomic information stems from complex relationships among multiple genes, which can be described in terms of a gene regulation network and be simulated in computational biology. Through pathway networks, we can understand the gene function more systematically. The central dogma, which is the rule that defines the biological activities of living beings, is the basic knowledge for us to understand the informatics in biology.

Biomolecular flow starts from DNA molecules. Biochemical information stored in DNA molecules is transcribed into the RNA molecules. There are four types of DNA molecules denoted as A, T, C, and G, and four types of RNA molecules denoted as A, U, C, and G. The mapping process from DNA sequences to RNA sequences is called transcription. The mapping process from RNA to protein is called translation. The major information in cells influenced by evolution processes is mainly stored in three kinds of molecular media: DNA, RNA, and proteins as shown in Figure 2.1.

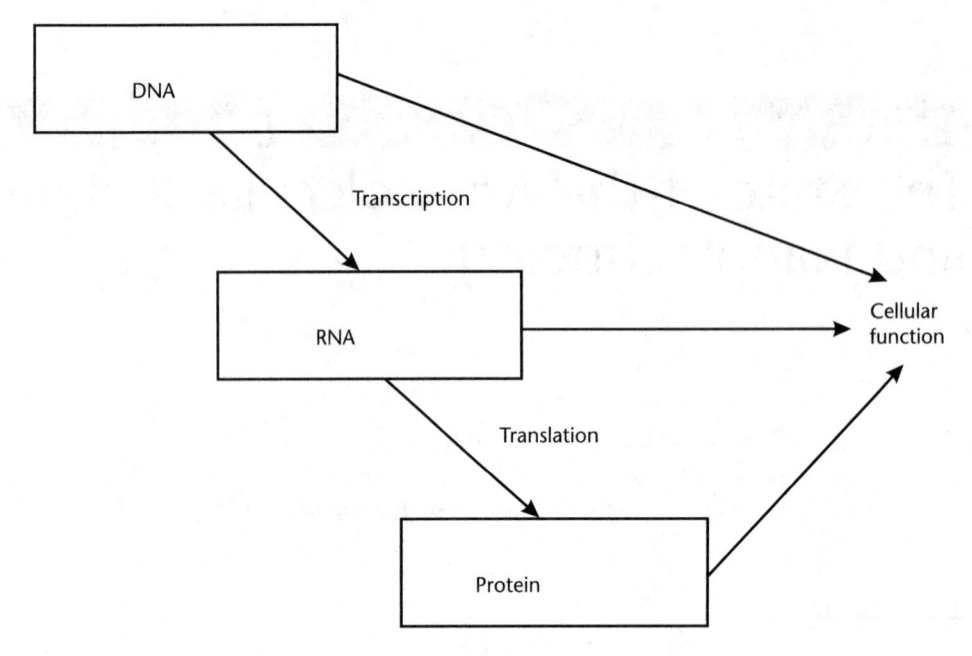

```
      ┌──────────┐
      │   DNA    │
      └──────────┘
         │    ╲
  Transcription ╲
         ↓        ╲
      ┌──────────┐   ┌──────────┐
      │   RNA    │──→│ Cellular │
      └──────────┘   │ function │
         │           └──────────┘
     Translation         ↑
         ↓               │
      ┌──────────┐       │
      │ Protein  │───────┘
      └──────────┘
```

Figure 2.1 Central dogma of molecular biology.

The protein consists of amino acids that include Ala, Arg, Asn, Asp, Cys, Gln, Glu, Gly, His, Ile, Leu, Lys Met, Phe, Pro, Ser, Thr, Trp, Tyr, and Val which is denoted as A, R, N, D, C, Q, E, G, H, I, L, K, M, F, P, S, T, W, Y, and V.

Before the information of the complete genome sequence was available, the individual level of a gene was a major study form in laboratories for a long time. The day of the realization of the complete genome sequence [1] came earlier than many people expected. Data on genome sequences is increasing everyday.

The genome information is described in a great book written in a language consisting of A, T, C, and G. The genome book is a code book for life. Reading this book is not easy because the genome language is unfamiliar for us as human beings, thus is a grand challenge. In order to understand the meaning of the huge amount of genome information, it is imperative that we should figure out the grammar and vocabulary in it.

The central task of genomics is gradually going to the genome bioinformatics in terms of systematic methodology. To find the biological function of the gene from genome data is an urgent task, especially now that the "wet" data of human genome is almost complete. One of the most important missions of genome informatics is to discover the biological function of the gene that has the direct application in medical science and drug design in terms of medical genomics. The scope of genome informatics covers the work to acquire data from experiment, handle the data to translate it into a readable form, analyze the informatics data to extract the useful information from junk-embedded databases, formulate a model for the biological description of genome, and to discover the potential relation of genome and medical applications. Basically, how to realize the "mining" process to obtain systematic genomic information from sea-like data of wet genome is the kernel of genome informatics.

By using electronic computers, the database can be stored and accessed. The artificial intelligence technology has been applied for handling the data form of knowledge. The problem is that the gap between the knowledge represented in natural languages and the corresponding data represented in computer programming languages makes it very hard to efficiently transform the human knowledge into computer data. The difficulty of obtaining complete knowledge for practical medical treatment in genome informatics is similar to the one in artificial intelligence. Almost all kinds of information technology have been introduced into bioinformatics including pattern recognition, signal processing, data mining (knowledge discovery in a database), formal language, unconventional computing algorithms, and conventional computing tools.

The task of the genome informatics is to explore the linkage between known data and unknown data. The relationship among genes is reflected in the information codes, molecular structure, and biochemical features, as well as cellular function. One of the most fundamental methods towards the discovery of a systematic relationship among genes is clustering. Clustering is a process to gather the data samples into certain classes with expected attributes (see Figure 2.2). Let X = {X1, X2, ..., Xn} be the sample set, the clustering process is to generate the classes Y1, Y2, ...,Ym which consist of the samples in X (i.e., Y1∪Y2∪...Ym = X). The criteria of clustering are that the elements that belong to the same set should satisfy the measure of homogeneity or uniformity under a certain threshold and the difference of the elements in different sets should satisfy the measurement of separation under a certain threshold. The essence of the clustering criteria is the natural characteristic of the data distribution. Of course, the criterion is *ad hoc* and varies greatly in concrete applications. Regarding checking the clustering result by informatics means, the DNA microchip is used to check the effect of gene expression. It in no strict sense looks like a blood test. Now, the automated technology is very helpful to speed up this work.

The purpose of applying clustering to genome informatics is to find the similarity of genes that is expected to offer some hints on possible biological function. Discovery of the individual gene with a specific biological function will be helpful to the drug design leading to gene therapy. The molecular operation on a limited number of genes is technically practical. The gene regulation network is a more complex network with a huge amount of information of bioinformatics, which will be discussed in Chapter 7.

The direct outcome of genome informatics is the quantitative explanation of gene attributes. The quantitative analysis of genome informatics is very powerful for gene function predication [2], for example, the genome-based informatics studies have [3] quantitatively described the reduction phenomena of genome where some genes overlap, which is important for us to understand the relations among genes.

The genome sequence is a kind of result of biological evolution processes. The related evolution knowledge about genes, such as how some genes are related in their function, is obtained from the evolution history of the species. These relations provide us with the functional links between genes in the form of a tree or graph structure. This knowledge about the relations could be used to reduce the searching space to cut the cost of the related computing process. Knowledge and data are crucial for the success of genome informatics.

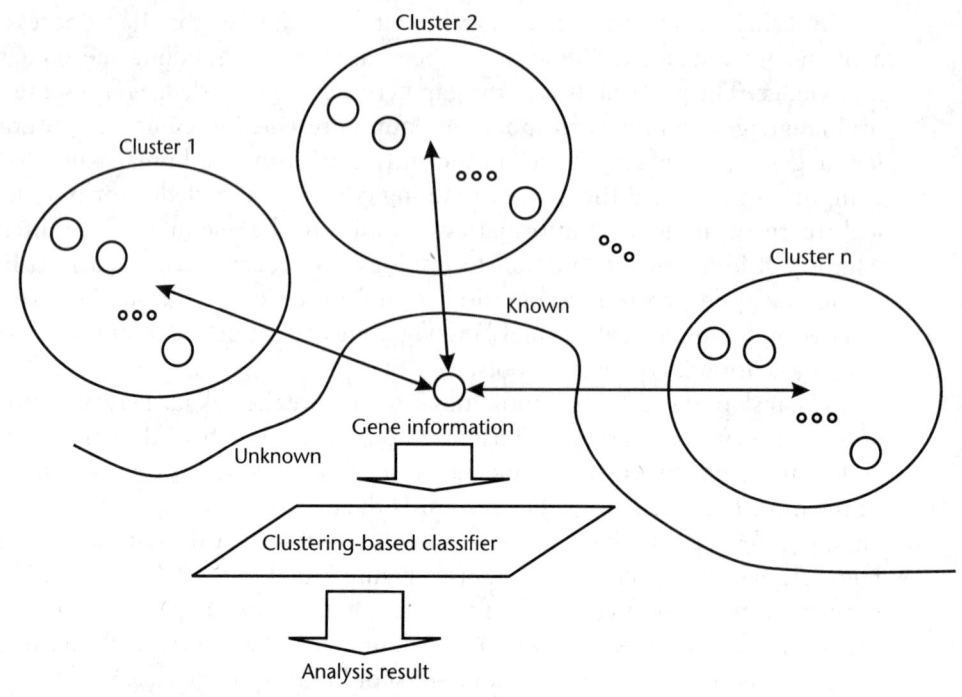

Figure 2.2 A brief of clustering for genome informatics.

In addition to powerful IT tools, we urgently need information resources of the data ranging from the gene to pathways of cells. For example, KEGG is a bioinformatics resource discovered by the Kanehisa Laboratories at the Bioinformatics Center of Kyoto University and at the Human Genome Center of the University of Tokyo [4]. It provides a rich amount of data ranging from gene to pathways.

Nowadays, as we know, most of the work in molecular biology has been shifted to the proteomes as the human proteomics project is being rapidly developed [5]. Of course, there still remain a lot of concrete themes in genome informatics that need to be investigated (for example, the transcription process [6]). Genomics and proteomics are connected tightly. Metaphysically speaking, their relationship looks like a helix.

2.2 Proteomics

Proteomics is a new emerging field focusing of the research of the molecular structure and biological function of proteome, which is all the proteins that correspond to the complete genome sequences. Two representative hot topics in proteomics are the function predication and the protein interaction network [7–13]. The protein function predication is not a completely new theme. However, its practical technology has been greatly strengthened recently. The similarity in spatial structure of proteins is a major clue to compare a protein whose function is known with a synthesized protein whose function is unknown. Derived from this basic idea, various technologies have been developed. The stochastics models and random signal processing methods have been applied in the function predication. The limitation of the searching method for protein function predication is expected to be reduced by guidance of the knowledge.

The philosophy and methodology of the proteomics could be helpful for molecular bioinformatics analysis that is expected to give us hints to discover protein functions. To understand the protein interaction networking, which is a milestone in proteomics and is contributing to function genomics in the post-genomics era, is one of the central themes in proteomics. Up until now, the reported protein interaction networks [7–13] are mainly obtained by experiments. Starting from the skill and knowledge, the bottom-up methods of wet experiments are adopted in benchworks. As wet data increased, computer calculation emerged as a necessary tool. With the integration of the wet experiments and dry data processing, a big stride forward has been made in protein interaction. The species whose protein interaction has been reported include yeast *Saccharomyces cerevisiae* [7], malaria parasite [8], and so forth. One of the direct uses of the knowledge about protein interactions in proteomics is to systematically study the protein's signaling function. For example, the epidermal growth factor receptor (EGFR) [9], an important protein in cellular signal transduction, is investigated in terms of protein interaction network.

Success of the proteomics experiment depends on advanced detection technology and corresponding equipment for proteomics. The biochemical detection tools for proteins vary in the aspect of biochemical (assay, microarray, and so forth), optical (e.g., FRET), electrical, and physical processes (e.g., MRI). Nowadays, "Probing real-time protein interactions" is the mainstream in molecular biology [11] to recognize the relation among proteins. One of the successful examples to use the structural feature to infer the protein interaction network is the work reported in [12], that is, three dimensional structure information of proteins is used to infer which kinds of proteins can be interacted to construct a protein interaction network.

At the proteome level, the kernel of proteomic networking is to find functions of proteins that correspond to genome-wide information. It is just beginning towards the goal of a complete proteomic network. In the aspect of application of proteomics, prediction of protein function is a task ranked with high priority. The atomic structure and atom-to-atom bond are the bases of the prediction on protein function in terms of physics calculation. The ab initio method of function predication based on atomic structure requires huge amounts of time and makes the computing process very long and inefficient. We still have to use it, however, because it is a physical calculation that exactly expresses the entire physical mechanism of the protein interactions. In order to reduce the computing time, the probability-based methods such as the Monte Carlo algorithm and molecular dynamics (MD)—infinite element method—are adopted, to get efficient computing algorithms, owing to their essence of employing partial information for calculation. But this is regarded as a kind of simulation and still cannot replace the ab initio method because of the essence of their stochastic mechanism. Other alternative methods for protein function predication are: comparing the known protein structure with the synthesized structure based on functional similarity; using machine learning algorithms to infer knowledge about protein function; stochastic-mechanism-based searching algorithms; and biological analysis in terms of evolution knowledge.

In order to know biochemical reactions of proteins in cells, we need equipments with high resolution of images on protein structures in order to observe the structural changes in biochemical reactions. This is a physical approach to investigate proteins in details with high reality, although the cost is still a problem.

To understand the whole cell function in biology, we need to study the systematic relationship among the molecules in cells, like the protein-to-protein network for example. Basically, two types of knowledge are needed. One is biology and the other is computing technology [13]. Some factors in the intersection of these two fields we have to consider are:

- Genome-wide information: the complete genome information used to understand the biological function of cells. Normally, genome-wide screening is used to investigate the gene's influence on cellular function.
- Interactome: the protein-to-protein interaction at the level of the genome-wide information.
- Molecular structure of proteins and "domains" (specific structural part) of the proteins that affect the biochemical reactions among proteins.
- Feasibility of biochemical reactions among proteins.
- Protein-to-protein networking (far away from being complete).

One of noticeable methods for protein-to-protein networking [13] is to integrate the knowledge of the graph of the pathway network [14] and corresponding structure features of molecular complexes (molecular compounds involved in the pathway network). The general knowledge of pathway and structural molecular complexes of proteins is integrated to give a hypothesis. The experimental verification of protein-to-protein networking is a benchwork to check the proposed hypothesis. The computer-based method is mainly aimed at using the obtained data and knowledge to infer the unknown function of proteins.

Also noticeable is the application of systems biology in the studies of cellular function. An example reported lately is systems-biology-based numerical calculation [15], which can be used for biological knowledge discovery based on wet experimenting. The information, including location of the molecular complexes, quantity of local concentration, and biochemical function of the enzyme on specific parts of molecular complexes, is the key measurements for understanding the behavior of proteins in cells, in which the location of the molecular complexes is obtained by imaging technology. The structural biology method is also an important reference for methods of biophysics and biochemistry in theory and experiment.

Similarity of proteins is one of the most commonly used measures of protein function predication. Similarity of two proteins is used for comparison of the relationship between the two proteins, which is a kind of local relationship. Similarity among many proteins can be used for quantitatively describing global relationships among proteins: protein similarity network [16]. The networking process goes toward to the direction of macro-scale. As a local feature of protein-to-protein interaction, the pattern of proteins (called the motif) helps us to understand the building block which is the element of complex protein-to-protein networking [17]. The motif is constructed by two proteins, three proteins, and four proteins for *Saccharomyces cerevisiae* [17]. The information from protein-to-protein is useful for medical practice. The proteomics and bioinformatics methods are integrated with the experimental approach for exploring new therapy of tumors where the informatics-based recognition is used to compensate the insufficiency of information caused by the technical limitation of the current experimental tools [18]. The research on protein function is desired to be used for the study on signal transduction and its link to the disease (e.g., the inflammation, immunity, and cancer [19]).

2.3 Cellular Structure from the Viewpoint of Molecular Biology

The complete genome sequences and corresponding proteome functions are the basis for us to get systematical understanding of the cellular signaling. At the cellular level, knowledge on different aspects of cells can be integrated in terms of systems biology. A cell is a biological system that contains molecules, biochemical reactions, information, and energy.

A cell that looks like a ball covers a lot of material [20]. The "skin" of the cell is a membrane that separates the internal spaces of the cell (intracellular compartments) with that of outside the cell. The cell is similar to a container. Although there are enormous molecules within the cell, a cellular nucleus containing chromosomes (in which genetic matter called DNA is included) is still the key for us to trace the biomolecular information flow within the cell.

Figure 2.3 is a cartoon-like schematic figure of the major eukaryote cell (of an animal) structure directly related to the topics we will talk about in later chapters. It is based on eukaryotes (a species whose cell has a nucleus). A complete picture could be found in textbooks such as *Molecular Biology of the Cell* by Bruce Alberts et al. [20]. We omit biological details in order to extract a concise conceptual structure of the cell for an informational description.

A chromosome, an object often observed by imaging technology, moves in the period of the cell division (meiosis) process, and influences biological functions such as reproduction (forming of the egg). The centromere is the central part of the chromosome, which is a mark for spatial information. As reported in [21], the telomeres (end part of the chromosome) move during meiosis. The featured part of a chromosome is useful information for cell study.

There are many proteins within the cell. They are the major components of cytoplasma. Within the cell, the organelles related to nanobiomachines and molecular information processes include microtubules, mitochondrion (in an animal cell), chloroplast (in a plant), and the Golgi apparatus.

Microtubule is a kind of organelle that exists in cells whose shape looks like a tube. The biological function of a microtubule is to make the cell move, which is important for the cell growth. With respect to nanobiomachine, microtubule is often used for molecular motors that carry physical media of molecular messages. The motor proteins move along the microtubules, which act as the "rail."

Mitochondrion is a kind of organelle in an animal cell that looks like a baby-silk-worm. Mitochondrion existing in animals is related to the energy flow in cells. The oxidative phosphorylation process occurs in mitochondrion.

Chloroplast is a kind of organelle in plant cells that has similar functions to mitochondrion in animal cell.

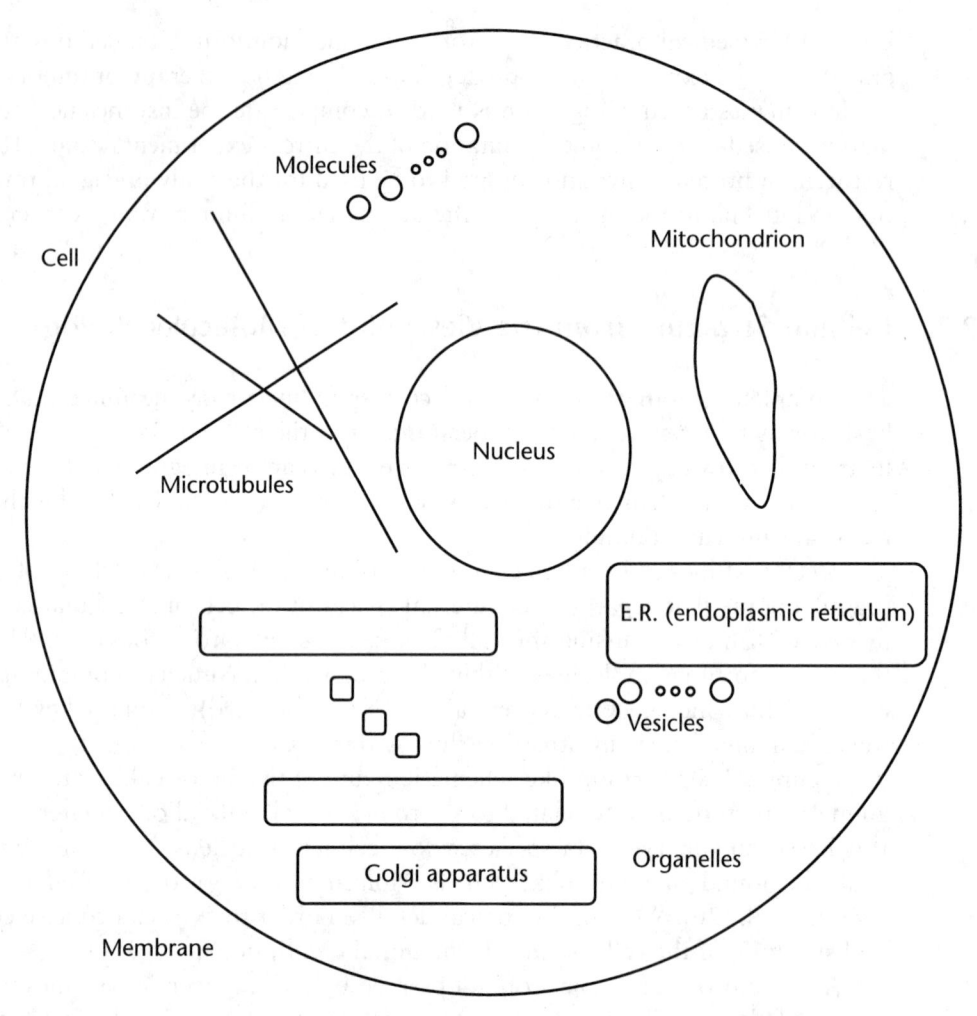

Figure 2.3 Cell structure in a cartoon-like brief.

Golgi apparatus is a kind of organelle responsible for molecule transportation between the membrane and nucleus. It controls vesicles moving around the endoplasmic reticulum and acts as post offices of intracell communication destined for the membrane. The nuclear receptor (NR) is on the nucleus.

The biological and molecular mechanism of the Golgi apparatus is still a topic at the frontier of molecular biology. The molecular information flow of the Golgi apparatus observed at the organelle level requires advanced technology with certain tools. An example is the method of immunofluorescence imaging, which has made a great step in imaging the Golgi apparatus for the observation of the Golgi apparatus activities. The molecular activities determine the working mechanism of the Golgi apparatus, which is crucial for molecular trafficking in cells. There are two possible models or hypotheses to describe molecules transported by the Golgi apparatus in a cell. A reported result [22] has confirmed one of the models—maturation model—with modification of the model by detailed structural information obtained from imaging technology.

2.4 Cell as a Nanobiomachine

The significance of a cell is not only in its behavior as a system, but also in its power on biological functions. The cell itself has a strong capability to process complicated molecular information that is encoded to represent complex biological functions. Under the regulation of cellular signaling, the molecules in cells are controlled to move among the organelles so that molecular flows are formed. The derived molecular flows with rich patterns are the causes of the biological functions of cells. The molecular signaling mechanism of controlling molecular movement in cells gives us hints for building molecular motors [23–29], for example, from the investigation on molecular mechanism of dynein dynamics [25] in cells, the molecular movement can be controlled under the activations of motor proteins. The radical function of motor proteins in cells is to transport cargo molecules, so it is feasible to use these cargo molecules to encode information and transport messages. The term "cargo molecule" refers to the molecules that are uploaded on molecular motors, are transported, and then downloaded at the destination. Briefly, the "cargo" is the cargo molecule, the "vehicle" is the molecular motor. The molecular motor is a kind of nanobiomachine that acts as a molecular vehicle. If we observe the cell from the viewpoint of mechanics and informatics, we will find that the cell behaves as a nanobiomachine, which is *adaptive, efficient,* and *robust*.

Adaptation: The environment that the cell is under is dynamic, but the cell can self-regulate the cellular signaling mechanism and molecular concentration to maintain cellular function.

Efficiency: When organs absorb nutrients, they will be transformed into the energy stored in cells. The energy is expressed in the molecular form such as ATP, GTP, and other molecules. The energy consumed by signaling processes in cells is much lower than the energy consumed in electronic computers. For example, the complexity of 1 cm × 1 cm transportation network realized by a molecular motor can reach an order of complexity of the whole cargo transportation network in Japan [30].

Robustness: Signal transduction is sustained by cellular pathway networks. With the exception of cancer as a negative example, the self-organizing function in cells is the outcome of biological evolution of cells, which adjusts the internal structure of cells so that the stable signaling processes are kept under the condition of the certain variation of molecular concentration in cells.

Controlling molecular flow is important for nanobiomachines. It consists of two aspects: controlling the information flow in moleware form and controlling the molecular flow in the biophysics/biochemistry mechanism.

Among the molecules used as the materials to build nanobiomachines, DNA molecules are the basic genetic material responsible for self-replication, self-reproduction, and self-repairing. Proteins in cells are crucial for a programmable nanobiomachine. The term "programmable" means "the cell as a collection of protein machines" [23] as Bruce Alberts [23] points out, in which three kinds of functions—motor protein, proofreading devices or "clocks," and assembly factors—are involved. GTP/ATP regulates the signaling proteins in cells and is directly concerned about the behavior of protein machines. Proteins are also related to the self-organizing mechanism of cells, which is important for the explanation of cellular functions.

The signaling molecules of second messengers in cells are responsible for the molecular information flow in cells. The Ca2+ is one of them, which is ubiquitous in cellular communication [20]. The kinase cascade is an information unit to connect the signaling mechanism of the molecular switch with the biochemical function affected by the molecular switch in a cell. Autophosphorylation has the biochemical function of regulating Ca2+ signals in cells and can be used for studying the phase locked–like phenomena in the cellular communication. The kinases take a big portion of the proteins. The distribution of molecular concentration offers information quantity about the molecular flow in cells. Based on this information, logic operators for information processing in nanobiomachines could be designed (see Figure 2.4).

Molecular information flow is reflected in the molecule transportation among organelles within a cell. The molecules called cargo molecules are attached on vesicles and transported from ER to the Golgi apparatus or networks and then to lysosomes in cells. This molecular transportation process is called vesicle trafficking in cells. The GTPase pathways are involved in the molecular transportation. It shows that the molecular switches of GTPases are related to the molecular message transportation. The information representation derived from the GTPase pathways is a binary form corresponding to the states of molecular switches. From the signal cascade, information about intra-cell communication thus can be coded, which is the basis of efforts towards "programmed" moleware communication in cells [31].

Medium within the cell is heterogeneous. Different molecules can be regarded as different data. Different signaling pathways can be modeled as different instructions.

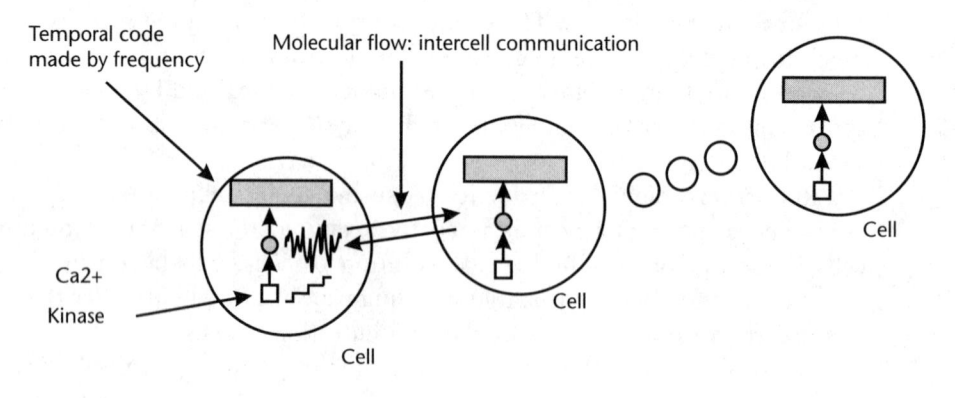

Figure 2.4 Brief relationship between temporal codes (Ca2+) and kinases in the cell.

In addition to the spatial organization of molecules in cells, temporal features can be inferred from the development process of cells as well. As reported by Y. Watanabe et al., the inactivation of Mei2 (a kinase) has the dephosphorylation function and activates the transition from mitosis to meiosis of the fission yeast *Schizosaccharomyces pombe* [32].

The functions of proteins are activated by the genetic mechanism of cell development. The molecular function is triggered at certain times according to the "clock/rhythm" of the cell, which is a kind of temporal feature. The second messenger acts as a ubiquitous signal within cell, which refers to the signaling molecules. The cellular communication process involves the signals to be sent, received, modified, and stored.

The molecular motor is a kind of biologically inspired nanobiomachine. The natural biological function of motor proteins contributes to the molecular movement in a cell division process. The locations of the organelles in cells are inevitable data for investigation on the molecular information and corresponding biological function in cell division. The name "chromosome" comes from its color by specific chemicals and can be observed by its colored shape/structure. It is natural to select chromosomes as the location reference because of the fact that chromosomes are easily observed. The location of centromere is important and often used in the observation of cell division. The telemere is a mark for the movement of chromosomes in cell division [21]. Locations of the information show the places of the molecular movement, which are the hints for the corresponding information flow in biochemical reactions. The dynamical information of signaling molecules requires high speed detection tools. The fluorescence imaging system is used for this purpose.

The biological signaling mechanism is the basis of designing the structure of nanobiomachines in the aspect of cellular communication involving molecular movement within the cell and the corresponding molecular messenger transmission. In the concept of ubiquitous communication, the spatial, temporal, and contextual features of cellular communication are the major research theme for the cellular nanobiomachine. The knowledge of information flow and architecture of the parallel computer is helpful for our understanding of the molecular information flow and corresponding operators in nanobiomachines.

The intracellular compartment structure and the molecular information flow of cellular communication are natural existing molecular mechanisms with spatial and temporal information. Two kinds of principal knowledge of the cell—*moleware mechanics* of the molecular transportation process in the cell and *molecular informatics* of the molecular signaling process in the cell—are helpful for constructing nanobiomachines.

2.4.1 Moleware Mechanics for Cellular Nanobiomachine: Molecules Carrying Messages

Three kinds of motor proteins—kinesin, dynein, and myosin—are used to realize the movement of molecular motors. They are called motor proteins because of their function of molecular motor which is to make molecules move and carry cargo molecules. The cargo molecules are the molecules that are transported within cells among different locations. Direct usage of molecular motor for molecular information processing is data flow by moving the molecular messengers. To control molecular information flow is an active way to guide a nanobiomachine to change the biological function of cells. Owing to the biological effect of molecular transportation in cells, the medical application of molecular motors is expected to deliver specific signaling molecules to those cells whose function is abnormal. Among motor proteins [24–29], dynein is related to degenerative disorders of motor neurons, such as a kind of neurodegenerative disease called motor neuron disease (MND) [29]. The locomotion of molecular motors comes from ATP hydrolysis.

2.4.2 Molecular Informatics for Cellular Nanobiomachine

The cellular communication provides a framework of massively parallel information processing in terms of signaling pathways in cells. In order to automatically operate the molecular signaling processes where a nanobiomachine is applied, we need to program the nanobiomachine, that is, to control the information flow in predefined order. This is actually the task to work out a computational description (formulated model) for the information flow in cells (see Figure 2.5). Here, the term "programming" refers to adaptively self-regulating the cellular behavior. In the sense of reconfiguring architecture (moleware structure) of a nanobiomachine, "programming" can be repeated. The function of programming is repeatable because the

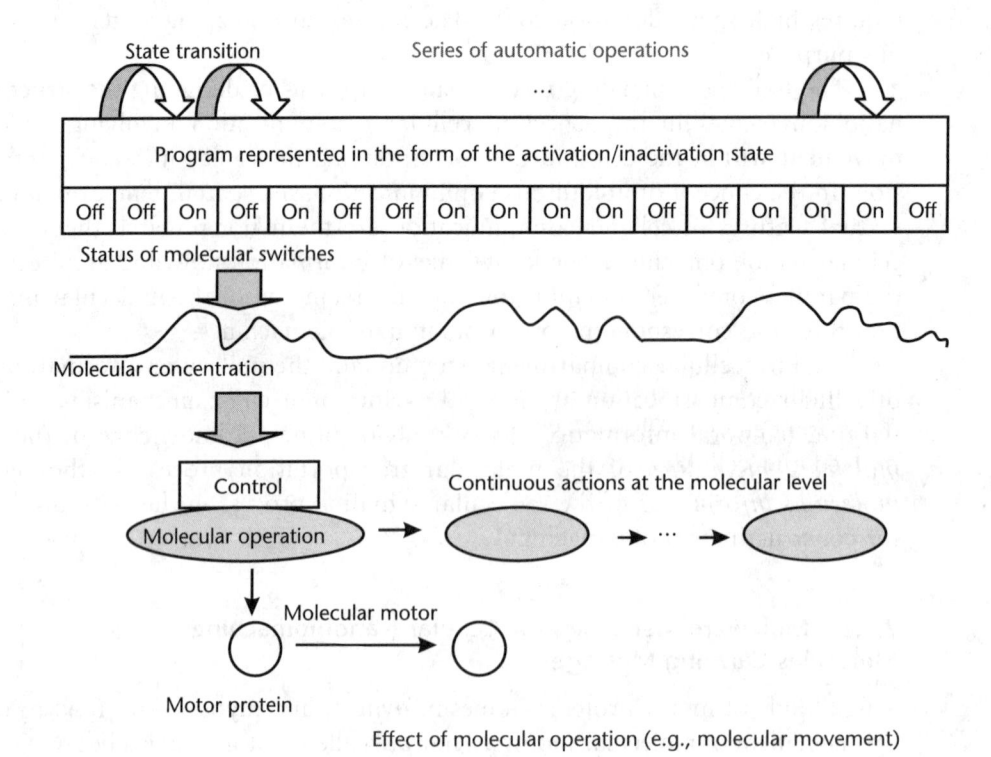

Figure 2.5 Schematic description of information flow from "moleware" program to molecular motor.

reconfiguration operation is repeated. By using those proteins that exist both in yeast and in plants, the cost of the experiment can be cut and the related molecular operation becomes more easier. M.-T. Chen and R. Weiss reported artificial cell-cell communication systems using yeast *Saccharomyces cerevisiae* and the molecules from the plant *Arabidopsis thaliana*. The knowledge about yeast and plant *Arabidopsis thaliana* is rich. The latter is a sample species in plants, which is similar to the *E. coli* in microbials. Basically, the biochemical level work and its bioinformatics support for building nanobiomachines should be based on the availability of signaling pathway structures and feasibility of the signal transduction mechanism.

2.5 Signal Transduction and Signaling Pathways of Cells

Within the cell, we have to make sure the molecules work together for certain biological functions. The information flow within the cell is complex and often modeled as a network. It can be summarized in terms of signal transduction or signaling pathways. The term "signal transduction" in cell biology has been popular for many years, but it is still important to explain the word "transduction." In electronics, a transducer is a device that transforms signals from one form (e.g., electricity) to another form (e.g., light). The term "signal transduction" in molecular biology refers to the signal cascades whose molecular flow takes the form of cellular pathways. Information about signal transduction and cellular pathways is rich. Signal transduction knowledge environment (STKE) [33] is one of the most excellent sources on signal transduction. KEGG is another information source on signal transduction. The term "signal transduction" in the discussion of this chapter, refers to cell-to-cell signaling and intracellular signaling, and the term "cellular pathway" refers to the molecular signaling mechanism in cells whose function is signal transduction. According to the concept of biochemistry, the biochemical forms of pathways are defined as a series of biochemical reactions in cells. Without confusion, we use the term cellular pathway to describe the molecular signaling process of cells in this section. Strictly speaking, cellular pathways are the pathways in cells, and there are two types of cellular pathways including metabolic pathways and signaling pathways (pathway for short). The function of the metabolic pathway is to sustain the energy circling in a cell. ATP is regarded as a medium storing energy. When ATP is transformed into ADP, the energy will be released to feed the cellular activity, which means that the function of the conversion process between ATP and ADP is to realize the energy transformation in cells. ATP is adenosine 5'-triphosphate, ADP is adenosine 5'-diphosphate. They are molecular complexes (compounds) existing in cells. There are three phosphates embedded in the molecular structure of ATP and two phosphates embedded in that of ADP. Different from metabolic pathways, the function of the signaling pathway is to regulate the molecular information flows within cells.

In informatics, the data structure of a pathway is represented by a directed graph showing input and output. The arrow shows the direction of the biochemical reaction. Pathway is tightly related to cell communication, which includes intercellular communication (cell-to-cell communication) and intracellular communication

(communication within the cell). In cell communication, molecules that represent messages move and activate series of biochemical reactions (a signaling cascade). In cell-to-cell communication, there are sender cells and receiver cells. Within the cell, the signals normally come from the membrane and activate a signaling cascade. The reactant of the signaling cascade finally interacts with the nucleus.

Pathway instances are named according to their biological functions in cells or key signaling molecules that have significant functions in cells. The molecules transfer messages in cell communication. The message is represented by molecules. At the material level, the material form of cell communication is realized by the molecular trafficking. The signaling molecule outside the cell is called the first messenger. The signaling molecule inside the cell is called the second messenger.

There are many pathway instances in cells discovered in plants, microbials, and animals. The following are the pathways that are expected to be used for biomolecular computing:

- Phosphorylation and dephosphorylation pathway;
- GTPase pathway;
- Oxidative phosphorylation pathway;
- MAKK pathway.

The pathways that have a close relation with signal transduction and medicine are:

- EGF/FGF pathway;
- Pathway for visual perception where kinases for phosphorylation are involved;
- Ca2+ signaling and neuron activity;
- NGF pathway;
- T cell antigen receptor (TCR) pathway.

Since the number of pathways is huge, it is impossible to discuss them all. We only give some remarks on some pathway functions in which cell communication is involved for a purpose of studying the signal transduction processes in cells here:

1. Cellular development in the sense of cell growth is the main function of the EGF pathway and the FGF pathway. The two pathways have common modules, that is, some of their subpathways are the same. The cellular function of the EGF/FGF pathways is related to diseases including diabetes, atherosclerosis, and cancers [33] where receptor tyrosine kinases (RTK) are involved. The feedback mechanism has been observed in these pathways. The existence of feedback is evidence to explain the dynamical features of the signaling processes in these pathways.
2. The function of the T cell antigen receptor (TCR) pathway is responsible for T cell signal transduction in which Lck—a Src-family protein tyrosine kinase (PTK)—is involved. From studying the function of Lck and corresponding pathways, the relation between kinase and immune function becomes clear.

3. Intracellular pathways interact with cellular nuclei. The nuclear receptor (NR) acts as a relay station of the cell communication between membrane-to-nucleus signaling and nucleus-to-membrane signaling.
4. A pathway called "killer pathway" has been reported [34]. Certain solutions to regulating pathways for medical purposes are hoped to be found from it.

Through cellular pathways, cell communications are realized in the form of transporting messages by different molecules. The media constructed by different molecules is heterogenous, which means different in structure and feature. This feature shows that signaling processes of different subpathways are asynchronous, thus a derived molecular information processing system is a distributed system. Here, we select two pathway instances to briefly explain their bioinformatics features that are useful for investigation of molecular information processing.

2.5.1 The Link Between the Signaling Pathway and Molecular Movement

The signaling pathways in cells are closely linked to the molecular movement. GTPase regulates cell movement of ameba. Kinases (whose function is for phosphorylation) involved in the intracellular pathways regulate the flagellum movement of bacteria. In order to explain how the proteins (including kinases) make cells move, the CheW-CheA-CheY pathway [20] is an example that shows the relation between motor proteins and kinases. The main information flow of the CheW-CheA-CheY pathway (where phosphorylation is involved) is outlined as the following:

$$CheW \to CheA \to ((or)CheZ \to) CheY$$

$$\to \text{the action of "molecular" motor}$$

First, protein CheW is activated by an external signal. Second, kinase CheA is activated by CheW. Third, the action of the "molecular" motor is activated by CheY that is activated by CheZ. There exist the processes of autophosphorylation and autodephosphorylation in the CheW-CheA-CheY pathway. The ChA and CheY have the functions of autophosphorylation and autodephosphorylation. Autophosphorylation refers to the phenomenon that kinase activates the phosphorylation process for itself. Autodephosphorylation refers to the phenomenon that phosphatase activates the dephosphorylation process for itself.

It can be concluded that in the CheW-CheA-CheY pathway, the molecular switches of phosphorylation/dephosphorylation are connected to the molecular motors.

2.5.2 The Links Between Signal Pathways and Neuron Function

The signaling pathways describe the molecular level of cellular activities, the neural function is at a little more macro-level than the molecular signals at cellular pathways. Research results in experimental biology and neuroscience [35, 36] to show the existence of links between the signaling pathways and neuron function.

In the cell, the function of signal transduction is realized by various types of signaling molecules in which the so-called second messengers are included. The second messenger Ca2+ in cellular pathways is linked to the neuron activities. The ions are important signals for models in neuroscience. In addition to Ca2+, the signaling molecules Mg2+ and Na2+ are also linked to the neuron activities. In the cell [20] the pulses of Ca2+ are regarded as kinds of temporal codes and are activated by kinase signals. The spatial distribution of molecular concentration of Ca2+ can be formulated by the stochastic model to describe the vibration features of Ca2+. Through the observation of Ca2+ signals, we can study adaptive behavior of signaling pathway networks. The phosphorylation (regulated by kinases) is also linked to the vision neurons. Matsumoto et al. has reported the effect of the phosphorylation process on visual perception [37].

The long-term depression (LTP) mechanism is an important cerebellar function of brain neurons leading to the learning and memory of the brain. By designing a computational bioinformatics model for signaling pathways, Kuroda et al. have successfully simulated the kinetic behavior of the phosphorylation process of the alpha-amino-3-hydroxy-5-methyl-4-isoxazolepropionic acid (AMPA) receptor, that leads to the function of LTP [35].

Doi et al. has studied the relation through bioinformatics simulation between dynamical Ca2+ signals and neuron signals—spikes—in cerebellar Purkinje cells. The research result obtained from this work is the basis of exploring the relation between the molecular mechanism of second messengers and cellular mechanism of neural systems [36].

The relations between signaling pathways and nanobiomachine/brains exist in nature. The bioinformatics model and simulation software are used to compensate the shortcomings of the current experiment tools.

2.6 Measurement and Detection in Material Science: Towards Manipulation of Biological Molecules

Without special tools in nanobiotechnology, we cannot observe the nanobioworld because it is so small. The object of measurement is the molecule, so the molecular feature decides the requirement of the moleware experiment for the related technology tools. Basically, the purpose of measurement of the molecules is to get their quantitative description, for example, concentration of the molecules.

Measurement for molecules helps bring about the development of various types of instruments in biochemistry laboratories. Concentration of molecules, which is one of the measurements, is a basic physical quantity in experiments of molecular biology. The molecular concentration is an analog form and its quantity is technically measured by biochemical equipment. The name of the molecule is symbolic. In genome sequencing, measurement is conducted by automatic machines and the result is a symbolic sequence of genome information of specific species. The symbolic sequence made by DNA molecules consists of A, T, C, and G. To detect the molecule (the existence of the molecule), we have to measure the quantity of the molecule, that is, its concentration and to judge whether or not the quantity reaches the required threshold. The symbolic sequence is also used as the representation form for RNA and protein molecules.

2.6 Measurement and Detection in Material Science

Detection of the molecules is for the verification of molecular functions. Microarray is used to detect DNA. The array consists of many tube-like small containers. When the sample is put into the array, the biochemical reactions will begin under the designed condition and operations on specific biochemical reactions designed in advance will be carried out on the array simultaneously. It is efficient and can be connected with an automated machine. The principle of RNA detection is similar to that of DNA. Detection of proteins is much more difficult than DNA. From protocol books for molecular biology experiments, methods for detecting proteins can be referenced.

Immunofluoroscence is often used to detect whether or not the proteins are bounded with specific molecules (e.g., phosphorylation or dephosphorylation of photo-proteins) in biochemistry experiments. Fluorescence resonance energy transfer (FRET) is a relatively new technology for detecting the fluorescence effect of proteins in cells. It can produce images of proteins that often have important biological function, with high resolution. In FRET technology, molecular probes are used for exactly detecting the fluorescence effect of proteins in spatial and temporal ways. Here, the word "probe" refers to the special molecules or molecular complex that can react with the target molecules or molecular complex to produce detectable signals. Light is one of the common forms of the generated signals. The protein connected with the molecular probe can be detected by certain physical effects, such as fluorescence (emitting color light). The microscope used here is with high resolution and can detect weak molecular signals (see Figure 2.6).

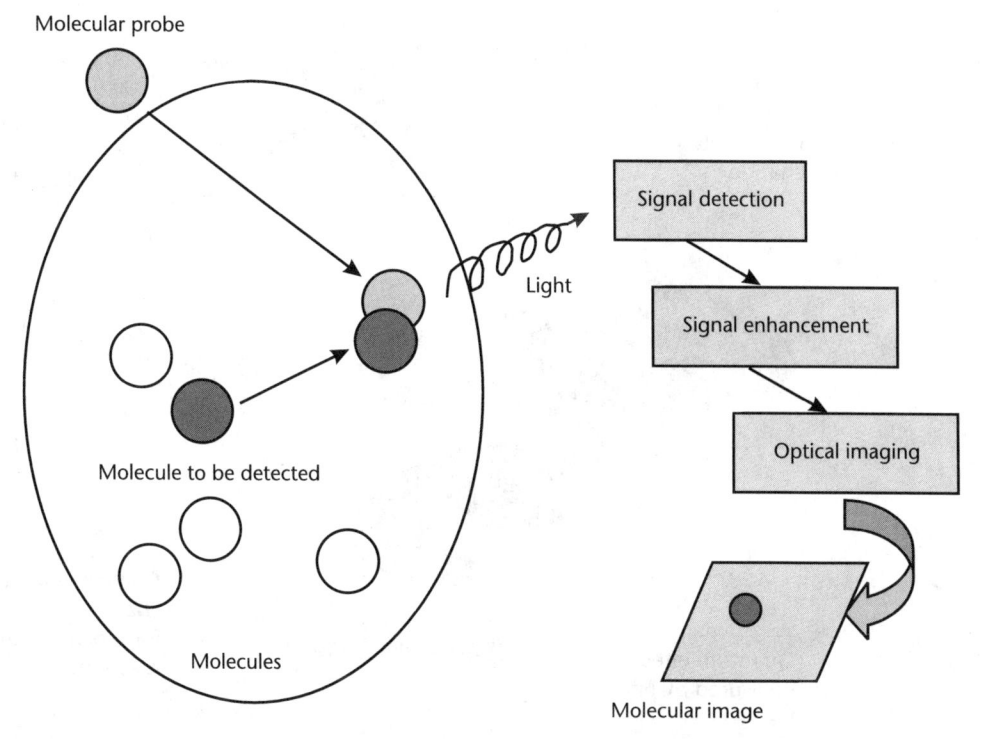

Figure 2.6 A microscope imaging process of molecules using molecular probe and fluorescence.

The history of the electron microscope is long. Two representatives of high resolution imaging technologies that are often used for the measurement and detection of the physical features of materials in the field of material science are atom force microscope (AFM) and nuclear magnetic resonance (NMR). In recent years, functional NMR (fNMR), an imaging method to measure the neuron activities of brain, is often used in brain science. Using it to get the images of the protein structures is relatively new. The principle of NMR in the application to the protein is to use NMR spectra to recognize the states of proteins in terms of the quantum effect caused by the atomic structure (see Figure 2.7). Different structures of protein result in different signals in NMR. Because the NMR image is with high resolution, location-dependent molecular-complex information of proteins thus becomes available. The spatial information of the protein structure is helpful to describe the conformational changes in biochemical reactions. The NMR detection method is also capable of detecting the states of protein continuously for the temporal measurement of the changes of protein, called protein dynamics.

The above-mentioned measurement and detection technologies are material-oriented and are being advanced by modern nanotechnology. Researchers are seeking new materials that can be used as molecular probes as well as new physical mechanisms of the imaging process for the better of molecular measurement and detection.

In the applications of molecular measurement and detection technology, the physical and biochemical features of the material—the object of measurement and detection—have to be taken into consideration. For example, the methods on

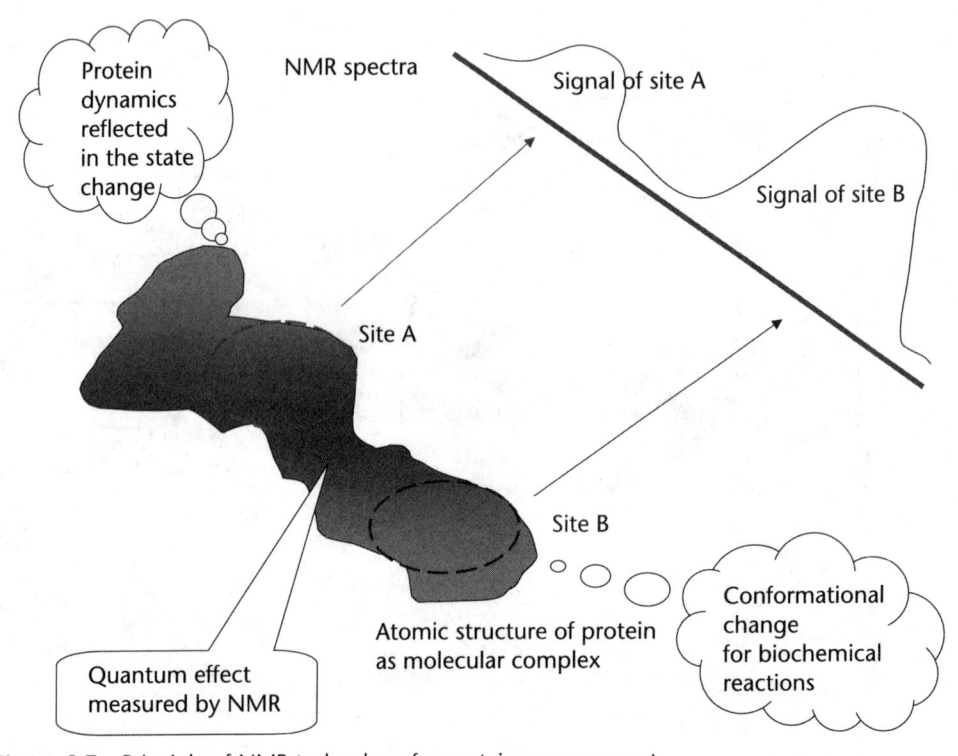

Figure 2.7 Principle of NMR technology for protein measurement.

observing the structures of carbon nanotube and porphyrin are different. AFM is efficient for observing the structure of carbon nanotubes, but it is not so helpful for the measurement of the porphyrin structure.

Molecular measurement and detection technology is for the manipulation of molecules, which is currently a hot topic in nanotechnology. The most common form of molecular manipulation is the self-assembly of molecules. The molecules are employed to build a supramolecular structure—the aggregate of molecules by the force of nature. It is better to grasp the meaning of "supra" from the definition of "supramolecular chemistry" [38]. It is clear that the self-assembly is a kind of biochemical process rather than an industrial one.

2.7 Pharmaceutical Nanobioinformatics

2.7.1 "Naïve" Thinking for Pharmaceutics

The nanobiosystem inherently links to medical applications. Pharmaceutics is the heart of medical industries. The task of applying nanobioinformatics to medicine is to discover the relation between the diseases and specific molecules from the set of DNA, RNA, and proteins, whose function is regarded as the reason for the diseases. A tendency that makes sense is integrating nanobioinformatics with medical science that includes pharmaceutics. Among the enormous numbers of projects on bioinformatics and its medical applications in the world, nuclear receptor syndrome X (NR-SX) project proposed by Kaminuma [39] is worth being mentioned. As a bold exploration to combine bioinformatics with pharmaceutics, the NR-SX project was the first one announced at the CBI community [40] and has been given a lot of attention.

Thinking about the origin of the word "pharmaceutics" that comes from the Greek origin meaning drug [41], we can reveal that pharmaceutics has a long history. The form of drug is various. The dosage is its most familiar image for us. The entire process of drug creation is to sieve the best solution from candidates through tests. As caduceus [42] (the symbol of medicine) implies, two factors—the needle and the snake—are the metaphor of drugs and diseases. The needle is the metaphor of medicine including drugs. The snake is the metaphor of disease. In a general meaning, the needle is not limited to drugs and covers many medical treatments including surgical operations. Actually, acupuncture in traditional Chinese medicine which uses needles to operate on the skin of humans perpendicularly according to a conceptual network theory of the human body existed more than 1,000 years ago. The image of caduceus is that the needle directly pins on the snake. If the molecular transportation can be realized by nanobiomachines to activate the biochemical reactions that have the medical functions, the nano-needle will be available to directly pin to the disease.

Sometimes the disease is regarded as the result caused by some things entering the body. The dosage of a drug is used to fight against this unexpected thing. The antivirus drug is a simple example. Many cases in clinical medicines, especially for chronic diseases, show that the disorder or unbalance in the body is the main reason for disease. The treatment is mostly focused on the systematic one: regulation of the biological activities within the body for certain order. The methodology of molecular informatics—to control the molecular information flow and the nanobiotechnology—to manipulate, transport, and control the molecules artificially synthesized or naturally existing in cells, is consistent with the principle of systems biology.

The medium of pharmaceutical nanobioinformatics could be the cells. The next generation of drugs could be the hybridizing conventional and unconventional paradigms (for example, the cellular signaling could be regulated in a defined order that looks like a program in a computer). The most obvious "unconventional" feature of nanobioinformatics applied for pharmaceutics is its methodology from the viewpoint of bioinformatics. In the human body, the drug causes a series of biochemical reactions of molecules. The idea of nanobioinformatics for drug application implies that we need to control the molecular information flow according to the bioinformatics knowledge through the means of nanobiotechnology.

If the earlier-mentioned artificial process of molecular transportation can be realized for cellular signaling by nanobiomachines, a revolutionary nanobiotherapy will emerge. The medical treatment nowadays is even dreamed of by nanobiorobots in science fiction. For the detailed operation of a nanobiomachine, to transport specific molecules to the desired places in a cell by nanobiomachines is a good idea. To develop a programmable nanobiomachine, the support of genomics and proteomics is needed, especially the bioinformatics on signal transduction. In signal transduction in cells, the cellular pathways are studied for the purpose of medical sciences [43] from which the pathways are regarded as the roads leading to possible treatment of diseases. This is natural because the technology of molecular biology has been embedded into almost every aspect of modern medicine [44]. The molecular level of cellular signaling is the key to the new generation of drugs/medicine [45–54] that has broad applications such as molecular codes of neurodegenerative diseases, the signaling mechanism connecting metalloproteinases, and stroke [52].

2.7.2 Molecular Information Flow as a Possible Solution Towards Potential Application of Nanobioinformation Processing Systems

Nanobiotechnology provides high possibility and rich chances for unconventional methods of pharmaceutics mainly based on the idea of controlling the molecular information flow of cells in terms of molecular biology and related nanotechnology. The success of genomics makes it easy to apply genomics to pharmaceutics. For example, the research on pharmacogenomics at Kyoto University [55] is to recognize the orphan G protein-coupled receptors (oGPCRs) for pharmaceutics. oGPCR is a kind pf protein receptor called orphan receptor. Orphan receptors are a class of receptors whose ligands were unknown [39]. With the advances of experimental technology of biochemistry, the number of identified ligands is increasing. The pharmarceutics-oriented technologies have been extending from the experiment level to the system level that needs knowledge of systems biology.

The state of the art of molecular biology requires us to study the cell in a systematic way that involves data base, experiment skills, knowledge about biological models, simulation software, and automatic equipment. The experiences gained from studies on different functions of cells provide us a more complete knowledge about the entire image of the cell. Through the information processing

2.7 Pharmaceutical Nanobioinformatics

tools of molecular bioinformatics, we hope to find medical solutions in terms of the systematic knowledge of signaling pathways (mechanism of molecular information flow in cells). Docking is an important work that connects the bioinformatics research and medical application. Docking in medicine refers to the work to find the interaction relationship between ligands and receptors. The grand challenge is the acquisition of sufficient data and information, that is, knowledge about cellular signaling. Development of new equipment that assist our biochemical experiments is imperative for obtaining data in proteomics. In essential proteome informatics, the basic idea of using the information of molecular flow in cells searching for the possible medical solution to the disease is to find the proteins that can be controlled and used to block the function of the proteins that cause the disease. We can easily grasp the meaning of this process from the case of Ras-induced cancer.

Normally, "Ras" refers to the protein kinase (notice the capital letter R), "ras" (the first letter is "r") refers to the gene that generates Ras protein. The Ras pathway is related to cancer [14]. The core therapy on Ras-induced cancer is to use the Ras inhibitor which is a signaling protein having the reverse effect compared with Ras. Ras activates the phosphorylation on photoproteins. The Ras inhibitor makes Ras ineffective so that the "harmful" phosphorylation will be curbed in cells (see Figure 2.8). The difficulty is finding the pathways that can activate the Ras inhibitor.

To study pathways of cells, we need not only the benchwork methods that include the powerful methods of structural biology for understanding the protein structure and function, but also the computational methods such as machine learning, pattern classification, signal estimation, and system identification. We notice that the gene information is very useful for drug discovery. Some mutation in gene

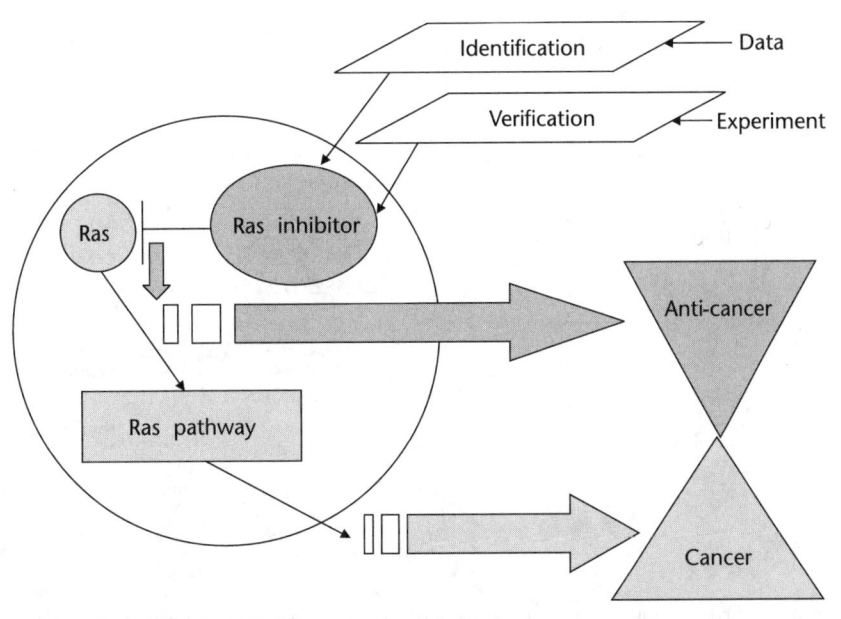

Figure 2.8 Activation and inhibition of Ras pathway.

causing cancer [14] is well known. For this reason, the gene therapy is the focus of application of nanobiotechnology in medicine. Now the progresses in proteomics make it possible to study the proteomic therapy directly, which can also be cooperated with the genomic therapy.

Kinase pathways (e.g., the MAKK in the MIF-activated pathways) determine the cell communication, which is the key to understand the cellular signaling mechanism of the living system where the cellular function decides the fate of the cellular life [56]. The significance of signaling pathways is obvious. In the NR-SX project [39], the work on pathways/networks for syndrome X (a name that summarizes various types of lifestyle diseases) is the central issue. Within the domain of signal transduction of cells, the biological functions of signaling molecules are closely related to the medical science. In order to understand the method of applying molecular bioinformatics to drug discovery, the intracellular signal transduction process of MIF-action reported by Morand et al. [56] is a good example for signal transduction process. Macrophage migration inhibitory factor (MIF) is a kind of cytokine. Cytokines are "soluble proteinaceous substances" [56] and act as the external cellular signals for signal transduction. The cytokine signaling mechanism is strongly connected to the immune system of the cell. There are three types of cytokines: lymphokines, interleukins, and chemokines. The names of the cytokines come from the biological function in cells involved.

As shown in Figure 2.9, MIF activates the intercellular signal cascades that generate the signals causing rheumatoid arthritis (RA) and atherosclerosis. The relationship between cytokines and kinases is tight. The well-known MAPK pathway is involved in the above signaling process. Since MAPK is a kinase, the phosphorylation

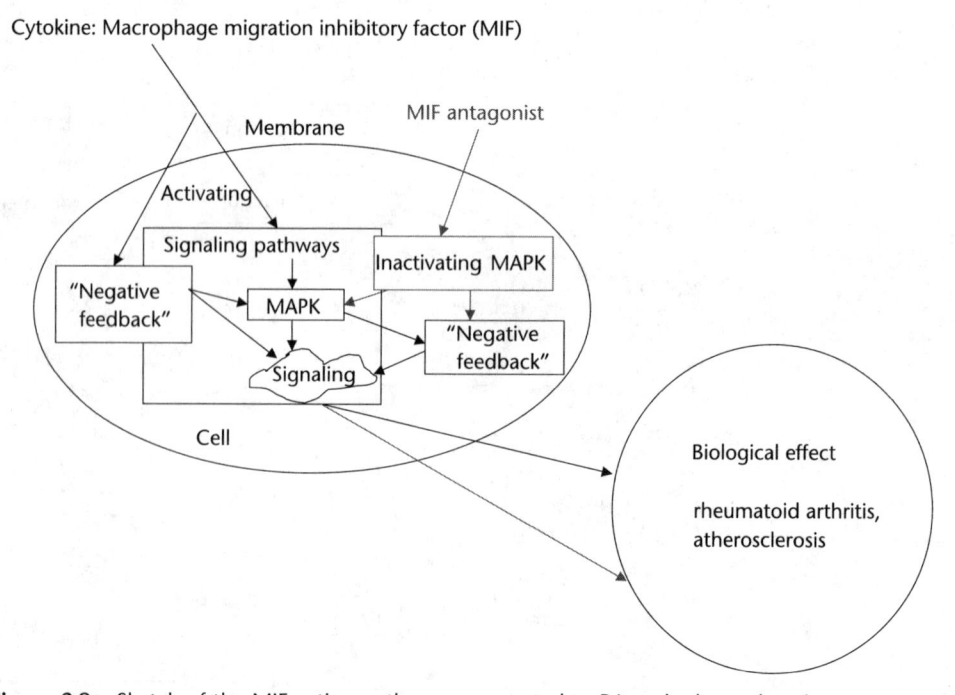

Figure 2.9 Sketch of the MIF-action pathways connected to RA and atherosclerosis.

processes occur in the signal cascade. The intracellular pathway connecting MIF and RA/atherosclerosis regarded as a complex system shows nonlinear behavior through negative feedback. The signaling mechanism of a drug for RA and atherosclerosis can regulate the MIF-activated pathway through MIF antagonists and negative regulation of MAPK. In the viewpoint of mathematics, the MIF-antagonist-activated signaling process is an inverse process of MIF-activated positive signaling. The mathematical multiple-input multiple-output (MIMO) model of cybernetics can be used to describe the dynamical behavior of the above signaling process.

From this discussion, it is concluded that the feasibility of generating the drug effect is dependent on the reliability of controlling the signal transduction towards the desired direction.

As an additional remark following this discussion, one of the breakthroughs for medical treatment could be expected from the integration of experimental observation of protein activation at the membrane and simulation of intracellular pathways of cells. The G-protein pathways and the phosphorylation pathways are connected with the membrane receptors. The intracellular pathways need to be simulated because certain parameters such as coefficients of biochemical reactions corresponding to the signaling pathways are not available by the current experimental tools. The simulation of the signal cascade can give us the clue of the logical relations among the functional proteins so that a global vision of the signaling pathway can be obtained. In the sense that proteomics is a field in which multiple disciplines are integrated, the experiment and simulation on intracellular communication are beneficial to revealing the real process of cell communication, where the normal functions and malfunctions of proteins in the living beings are studied in terms of the measurement of pathway quantification [57].

References

[1] http://www.sanger.ac.uk/HGP/overview.shtml.
[2] Sakharkar, K. R., and V. T. K. Chow, "Strategies for Genome Reduction in Microbial Genomes," *Genome Informatics*, Vol. 16, No. 2, 2005, pp. 69–75.
[3] Wu, H., et al., "Prediction of Functional Modules Based on Gene Distributions in Microbial Genomes," *Genome Informatics*, Vol. 16, No. 2, 2005, pp. 247–259.
[4] http://www.genome.jp/kegg/.
[5] http://www.hupo.org/.
[6] Potapov, A. P., et al., "Topology of Mammalian Transcription Networks," *Genome Informatics*, Vol. 16, No. 2, pp. 270–278.
[7] Krogan, N. J., et al., "Global Landscape of Protein Complexes in the Yeast Saccharomyces Cerevisiae," *Nature*, Vol. 440, March 2006, pp. 637–643.
[8] LaCount, D. J., et al., "A Protein Interaction Network of the Malaria Parasite Plasmodium Falciparum," *Nature*, Vol. 438, November 2005, pp. 103–107.
[9] Jones, R. B., et al., "A Quantitative Protein Interaction Network for the ErbB Receptors Using Protein Microarrays," *Nature*, Vol. 439, January 2006, pp. 168–174.
[10] Rual, J. F., et al., "Towards a Proteome-Scale Map of the Human Protein-Protein Interaction Network," *Nature*, Vol. 437, October 2005, pp. 1173–1178.
[11] Gershon, D., "Probing Real-Time Protein Interactions," *Nature*, Vol. 432, November 2004, p. 249.

[12] Zarrinpar, A., S. H. Park, and W. A. Lim, "Optimization of Specificity in a Cellular Protein Interaction Network by Negative Selection," *Nature*, Vol. 426, December 2003, pp. 676–680.

[13] Aloy, P., and R. B. Russell, "Structural Systems Biology: Modelling Protein Interactions," *Nature Reviews Molecular Cell Biology*, Vol. 7, pp. 188–197.

[14] Gomperts, B. D., I. M. Kramer, and P. E. R. Tatham, *Signal Transduction*, New York: Elsevier Inc., 2002 (Japanese translation, Medical Sciences International, Ltd. 2004).

[15] Gilchrist, M., et al., "Systems Biology Approaches Identify ATF3 as a Negative Regulator of Toll-Like Receptor 4," *Nature*, Vol. 441, May 2006, pp. 173–178.

[16] Weston, J., et al., "Protein Ranking: from Local to Global Structure in the Protein Similarity Network," *PNAS*, Vol. 101, No. 17, April 2004, pp. 6559–6563.

[17] Yeger-Lotem, E., et al., "Network Motifs in Integrated Cellular Networks of Transcription-Regulation and Protein-Protein Interaction," *PNAS*, Vol. 101, No. 16, April 2004, pp. 5934–5939.

[18] Oh, P., et al., "Subtractive Proteomic Mapping of the Endothelial Surface in Lung and Solid Tumours for Tissue-Specific Therapy," *Nature*, Vol. 429, June 2004, pp. 629–635.

[19] Karin, M., and F. R. Greten, "NF-κB: Linking Inflammation and Immunity to Cancer Development and Progression," *Nature Reviews Immunology*, Vol. 5, 2005, pp. 749–759.

[20] Alberts, B., et al., *Molecular Biology of the Cell*, 4th ed., New York: Garland Science, 2002.

[21] Chikashige, Y., et al., "Telomere-Led Premeiotic Chromosome Movement in Fission Yeast," *Science*, Vol. 8, No. 264, 1994, pp. 270–273.

[22] Matsuura-Tokita, K., et al., "Live Imaging of Yeast Golgi Cisternal Maturation," *Nature*, Vol. 441, June 2006, pp. 1007–1010.

[23] Alberts, B., "The Cell as a Collection of Protein Machines: Preparing the Next Generation of Molecular Biologists," *Cell*, Vol. 92, February 1998, pp. 291–294.

[24] Karcher, R. L., S. W. Deacon, and V. I. Gelfand, "Motor-Cargo Interactions: the Key to Transport Specificity," *Trends in Cell Biology*, Vol. 12, No. 1, January 2002, pp. 21–27.

[25] Hirokawa, N., "Kinesin and Dynein Superfamily Proteins and the Mechanism of Organelle Transport," *Science*, Vol. 279, January 1998, pp. 519–526.

[26] Hirokawa, N., and R. Takemura, "Molecular Motors in Neuronal Development, Intracellular Transport and Diseases," *Current Opinion in Neurobiology*, Vol. 14, 2004, pp. 564–573.

[27] Hirokawa, N., and R. Takemura, "Molecular Motors and Mechanisms of Directional Transport in Neurons," *Nature Reviews, Neuroscience, Advance online publication*, Vol. 15, February 2005.

[28] Vale, R. D., "The Molecular Motor Toolbox for Intracellular Transport," *Cell*, Vol. 112, February 2003, pp. 467–480.

[29] Hafezparast, M., et al., "Mutation in Dynein Link Motor Neuron Degeneration to Defects in Retrograde Transport," *Science*, Vol. 300, May 2003, pp. 808–812.

[30] http://www-karc.nict.go.jp/research/protein.html, in Japanese.

[31] Chen, M. T., and R. Weiss, "Artificial Cell-Cell Communication in Yeast Saccharomycess Cerevisiae Using Signaling Elements from Arabidopsis Thaliana," *Nature Biotechnology*, Vol. 23, 2005, pp. 1551–1555.

[32] Watanabe, Y., et al., "Phosphorylation of RNA-Binding Protein Controls Cell Cycle Switch from Mitotic to Meiotic in Fission Yeast," *Nature*, Vol. 386, March 1997, pp. 187–190.

[33] STKE (Signal Transduction Knowledge Environment), http://stke.sciencemag.org/.

[34] Vivier, E., J. A. Nunès, and F. Vély, "Natural Killer Cell Signaling Pathways," *Science*, Vol. 26, No. 306, November 2004, pp. 1517–1519.

[35] Kuroda, S., N. Schweighofer, and M. Kawato, "Exploration of Signal Transduction Pathways in Cerebellar Long-Term Depression by Kinetic Simulation," *J. Neuroscience*, 2001, Vol. 21, pp. 5693–5702.

[36] Doi, T., et al., "Inositol 1,4,5-Trisphosphatate-Dependent Ca^{2+} Threshold Dynamics Detect Spike Timing in Cerebellar Purkinje Cells," *The Journal of Neuroscience*, Vol. 25, No. 5, 2005, pp. 950–961.
[37] http://w3.ouhsc.edu/BIOCHEM/matsumoto.htm.
[38] http://en.wikipedia.org/wiki/Supramolecular_chemistry.
[39] Kaminuma, T., "Pathways and Networks of Nuclear Receptors and Modeling of Syndrome X," *Chem-Bio Informatics Journal*, Vol. 3, No. 3, 2003, pp. 130–156.
[40] http://www.cbi.or.jp.
[41] http://en.wikipedia.org/wiki/Pharmaceutics.
[42] http://en.wikipedia.org/wiki/Caduceus.
[43] Helmreich, E. J. M., *The Biochemistry of Cell Signaling*, New York: Oxford Univesity Press, 2001.
[44] *Nature Medicine*, http://www.nature.com/nm/index.html.
[45] Lesné, S., et al., "A Specific Amyloid-β Protein Assembly in the Brain Impairs Memory," *Nature*, Vol. 440, March 2006, pp. 352–357.
[46] Maudsley, S., and M. P. Mattson, "Protein Twists and Turns in Alzheimer Disease," *Nature Medicine*, Vol. 12, 2006, pp. 392–393.
[47] Cohen, P., and M. Goedert,"GSK3 Inhibitors: Development and Therapeutic Potential," *Nature Reviews Drug Discovery*, Vol. 3, 2004, pp. 479–487
[48] Freeman, M., "Pin-Pointing MAPK Signalling," *Nature Cell Biology*, Vol. 3, 2001, pp. E136–E137.
[49] Pulverer, B., "Pin-ning Down p53 Function," *Nature Cell Biology*, Vol. 4, 2002, p. E251.
[50] Wulf, G. M., et al., "Pin1 Is Overexpressed in Breast Cancer and Cooperates with Ras Signaling in Increasing the Transcriptional Activity of c-Jun Towards Cyclin D1, *The EMBO Journal*, Vol. 20, 2001, pp. 3459–3472.
[51] Ryo, A., et al., "Pin1 Regulates Turnover and Subcellular Localization of β-Catenin by Inhibiting Its Interaction with APC," *Nature Cell Biology*, Vol. 3, 2001, pp. 793–801.
[52] Zhao, B. Q., et al., "Role of Matrix Metalloproteinases in Delayed Cortical Responses After Stroke," *Nature Medicine*, Vol. 12, 2006, pp. 441–445.
[53] Ito, K., et al., "Reactive Oxygen Species Act Through p38 MAPK to Limit the Lifespan of Hematopoietic Stem Cells," *Nature Medicine*, Vol. 12, 2006, pp. 446–451.
[54] Takeda, H., et al., "Human Sebaceous Tumors Harbor Inactivating Mutations in LEF1," *Nature Medicine*, Vol. 12, 2006, pp. 395–397.
[55] http://www.bic.kyoto-u.ac.jp/COE/projects/tsujimoto_E.html.
[56] Morand, E. F., M. Leech, and J. Bernhagen, "MIF: A New Cytokine Link Between Rheumatoid Arthritis and Atherosclerosis," *Nature Reviews Drug Discovery*, Vol. 5, May 2006, pp. 399–411.
[57] Koshland, D. E., Jr., "The Era of Pathway Quantification," *Science*, Vol. 280, No. 5365, May 1998, pp. 852–853.

CHAPTER 3
Nanobiomachines for Information Processing and Communication: Exploring Fundamental Principles of NanobioICT

Rooted in material science and nanobiotechnology [1, 2], various types of nanobiomachines have been developed. The fundamental task of nanobiomachines is to realize molecular operations at the nano-level. Here, information processing and information theory of nanobiomachines are the two main fields of study that can provide us a functional testbed for exploring the fundamental principles of information and communications technology (nanobioICT) [3].

The information processing mechanism of nanobiosystems forms the basis of studies on nanobioICT systems, so it is natural to start from information processing approaches to ICT in the domain of nanobioscience. Among molecular information processing paradigms, biomolecular computing has an edge over other pursuits of nanobioICT (e.g., molecular communication) because it has been extensively studied over the past decade and many of the research results on algorithms have been accumulated. The informatics aspect of nanobioICT can be studied in terms of biomolecular computing with theoretical support from information theory. State-of-the-art technology in molecular biochemistry for cellular pathways [4–9] has made it possible to carry out molecular computation by employing the signaling mechanism of biological cells. From the viewpoint of steganography [10], biomolecular computing is promising for the design of encoder/decoders, one of the most important tasks in applications of information theory [11].

Looking back upon the short history of molecular computing, one can find good reasons for applying it to information theory (e.g., coding). Lipton [12] successfully used Adleman's DNA-strand-based molecular computing [13] to solve a contact network problem, which can be traced to Shannon's historic paper [14]. This work shows that molecular computing can be used to encode the information flow in communication or over a network. In this case, the molecules are the DNA strands. It is also notable that Karl-Heinz Zimmermann [15] applied molecular computing based on a DNA-strand sticker model to binary linear codes; however, there were errors in that approach. He analyzed the error-induced limitations of the DNA-strand-based coding schemes. Actually, DNA molecules have been investigated intensively for molecular coding. An overview by Mauri et al. on the theoretical aspects of DNA-strand codes can be found in [16].

As we know, living cells are types of nanobiomachines [1, 2, 4–9] that exist in nature. They are adaptive and can regulate and adjust themselves with their signaling pathways to maintain their robustness to a certain degree. Actually, molecular information processing, cell communication, and biochemical cellular functions are integrated in the cell, which is a biophysical form of a nanobiomachine. Furthermore, it can serve as a functional testbed for studying nanobioICT. For example, a biomolecular computing method [17] has been applied to encoding and decoding design, in which biomolecules for cellular signalling, such as GEFs/GAPs and kinases/phosphatases, are employed through an engineered-pathway mechanism (called kinase computing for short). Codes designed by this method can correct errors in the signaling mechanism of cellular nanobiomachines. This would provide a complementary method for codes in information theory if advanced nanobiotechnology [1, 2, 4–26] could be further integrated with the mainstream of information theory for future communication applications in the nanobioICT field.

In this chapter, we discuss the concepts and methodologies of nanobioICT, instances of coding, network structures for moleware communication, and the extension of nanobioICT to nanoscience, biotechnology, information technology and cognitive science (NBIC).

3.1 Mission of NanobioICT

Integrating information technology (IT) and communication technology is a promising direction, since the relation between IT and communication is so strong that it is nearly impossible to separate them. No one can doubt the significance of ICT, which influences society, economics, and many other aspects of human life [3]. The implementation of information processing systems mainly depends on computers and computing technology. Nowadays, IT technology is intricately connected with communication due to the wide proliferation of various networks. Consequently, the technological domain of ICT is very broad. In this new field, there are many open problems.

Nanobiotechnology originally emerged from the material sciences. Physics is one of the major pillars of this field. Information processing systems can be realized through different media in handling different signals based on the physical mechanisms of moleware. Moleware communication also includes various types of processes that take many forms. NanobioICT within the domain of the nanobioworld is the main topic of this chapter. The concept of nanobioICT can be easily understood as ICT that is implemented by biological technology at the nanolevel. The field of nanobioICT can be understood by considering the following major elements:

1. *Information theory of nanobioICT.* Information source, channel, encoding and decoding, and information capacity of nanobioICT systems are studied in terms of information theory.

2. *Information processing of nanobioICT.* Unconventional information processing paradigms of nanobioICT are explored to enhance the biological information processing capability at the nano-level. Models, algorithms, and applications have continuously emerged in recent years.
3. *Computing of nanobioICT.* Theoretical computer science and computing technology are two supporting tools for nanobioICT. This subfield includes nanobiocomputing, nanobiocomputers, and discrete mathematics for nanobiosystems. The field of biomolecular computing can be regarded as a generalization of DNA computing. Molecular electronics are included in this subfield as well, which is the basis of molecular circuits for electrical molecular computers. Parallel computing and networking also belong to the domain of nanobiocomputing technology.
4. *Communication of nanobioICT.* Communication processes of nanobioICT systems are studied by analyzing quantitative measurements in molecular biology and the signaling mechanism of nanobiosystems based on information theory. Here, network protocols, nanobiodevices, and nano-antenna are promising topics.
5. *Integrated systems of nanobioICT.* Elements from (1) to (4) are being integrated, and the kernel of this integration is the information processing mechanisms of nanobiosystems. Here, information representation and operation, nanobiocomputing function and the derived parallel distributed architecture, and nanobiocommunication processes work synergistically in the same system.

Nanomolecular-bioICT is another key term we need to keep in mind. According to our understanding, this concept refers to studying the molecular system of nanobioICT in the sense that a nanobioICT system is built at the molecular level.

NanobioICT is a new field, and the following items are important from the viewpoint of methodology:

- The objective of nanobioICT is to study the principle of nanobioICT and to build nanobioICT systems.
- The new idea of nanobioICT is that some future generation of information and communication technology will be realized by nanobiosystems having a different principle and manufacturing process from those of semiconductor technology. The biological information processing mechanism of naturally existing nanosystems consists of molecules, and it obeys the rules of biochemical reactions at the molecular level. The manufacturing technology for building nanobiosystems is based on the self-assembly process of molecules and self-organizing mechanisms of biomolecular systems.
- The research on nanobioICT is inspired by natural nanobiosystems, where our understanding of their description, operation, structure, and control is guided by the rules of system engineering. This approach is supported by the fact that naturally existing nanobiosystems can store and handle a huge amount of information within the nano-size space.

Molecular communication is a sub-field of nanobiocommunication. Researchers at UC Irvine [27] and at NICT-KARC [28] have pioneered the research in this field. T. Nakano et al. pointed out: "The class of molecular communication systems...consists of sender nanomachines, receiver nanomachines, carrier molecules, and the environment..." They imply that molecular carriers such as DNA, protein, and ion are controlled to deliver the information represented as molecules [29]. The proposal by S. Hiyama et al. [30] is on a molecular communication system directly applied to nanobiomachines.

Ron Weiss and Tom Knight have explored engineered communication at the molecular level [31], showing that it is feasible to employ the signaling mechanism of bacteria for basic communication. The sender-receiver structure is applicable to multiple cells where the programmability of the signaling process is possible [32]. Their molecular systems for communication among molecules employ signaling circuits and genetic circuits in cells. Cellular signaling also has other forms [33, 34].

The progress made in nanobioICT will have an impact on the ubiquitously communicative society through the development of nanobio-infrastructure. The indispensable components of nanobioICT include cell communication, molecular communication, and nanobiocommunication. Cell communication has been intensively and extensively studied as a significant domain in molecular biology [35]. Research results are rapidly filling in the blanks. By the term "moleware communication by nanobiotechnology," we indicate a molecular-computer-based communication system that is expected to be achieved through nanobiotechnology. According to the methodology of molecular communication [29, 30], intercell communication and intracell communication are regarded as having the same informatics process at the molecular level within a singular unified framework. This view is helpful in developing certain generic standards for molecular communication and related systems built by nanomachines. Two examples of molecular communication reported in [29, 30] are a (semi-)autonomous artificial molecular system and a natural molecular system. The former needs a nanomachine or molecular computer as the information processing mechanism so that the derived molecular communication processes can be (semi-)autonomous. The latter is assumed to be built on an existing cellular communication mechanism under external control.

Representative research pursuits in the field of nanobioICT include:

- Message transmission process realized by molecular motors;
- Codes of biomolecules;
- Controlled cell communication within cells and among cells;
- Communications of nanobiomachines;
- Biomolecular computing based on cellular signaling;
- Networking of signal transduction cascades.

The above efforts imply that information processing and communication are intrinsically integrated within nanobiosystems and that, furthermore, molecular information processing and cell communication can be integrated within the relevant bioinformatics mechanisms. Accordingly, the most suitable material for building a nanobioICT system is the biological cell, since it offers a functional "platform" that can satisfy nearly all of the requirements of building a nanobiosystem.

Now let us briefly discuss the expected future directions of nanobioICT.

In theory, we face the crucial tasks of universal code (length-variable code independent of the distribution of source or channel), information capacity, adaptive control, optimization of channels, and network protocol design.

In practice, implementation of nanobioICT systems is the most imperative task. To achieve this, one of the most obvious needs is to develop a technology to overcome the difficulties inherent in manufacturing, controlling, and testing nanobiomachines. Other expected challenges involve attaining excellent abilities in information compression, massively parallel and distributed information processing, and adaptive communication for unknown environments. Owing to the robustness of cellular signaling mechanisms, it seems reasonable to adopt them in error-correction schemes for nanobiosystems.

In the foreseeable future, four technologies need to be achieved for molecular nanobioICT systems:

1. Encoding/decoding algorithms;
2. Channel models;
3. Network protocol design;
4. MIMD structure and control.

The day of functional nanobioICT machines and nanobioICT systems will eventually come if these key technologies can be realized.

3.2 Information Theory of NanobioICT: Shannon Meets Feynman

In information theory, the basic mechanism of communication processes consists of three major parts: information resource, channel, and information receiver (Figure 3.1). The information theory for communication in nanobioICT should be a kind of moleware version of this communication process (Figure 3.2). The difference between moleware communication and the current electronics and electrical communication is obvious. The uniqueness of moleware communication systems can be seen in the moleware medium and the principle of molecular information processing. The molecular messages are "transmitted" in the form of biochemical reactions of the molecules. Here, the transmission process of molecular messages does not include molecular communication achieved through nanotechnology of *non*biomolecules.

The most basic difference between tele-communication and moleware communication is the medium employed. Moleware communication is realized by biochemical reactions. The informatics form of the moleware communication process is a molecular message that is detected biochemically. Molecular messages include the symbolic and digital information in the notations of molecules and a real number indicating concentration of molecules in the range of [0, 1]. These two kinds of signals—symbolic/digital and analog—are coupled tightly.

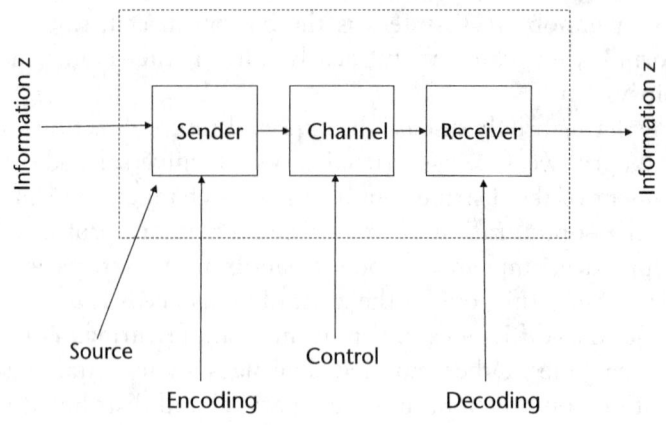

Figure 3.1 Description of communication process by information theory.

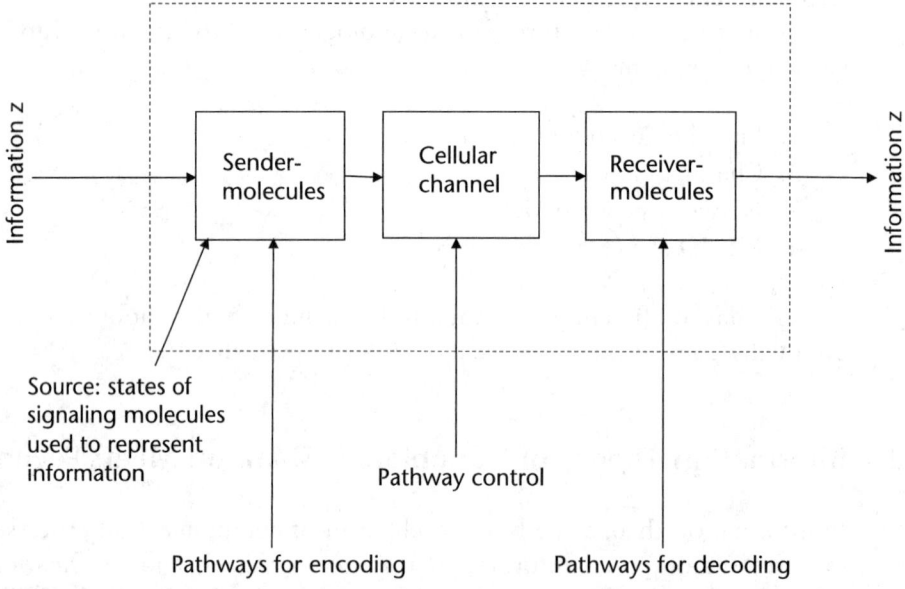

Figure 3.2 Description of nanobiocommunication process by information theory.

A binary form is available in the nanobioworld. In cells, the mechanism of setting the binary values is realized by cellular signaling. In comparison with the electrical operations in telecommunications on electrical messages with identical physical features, the molecular operations on molecular messages show different biochemical characteristics. This is because the implementation technologies are totally different.

As Feynman predicted, the nano-level architecture is a kind of powerful information storage medium. Based on the physical configuration of moleware systems, the informatics mechanism of moleware communication depends on information representation and operation. The signal transmission and translation require an information theory, that is, the theory of information encoding, transmission, and decoding. It is at this stage, where molecules are used for communication, that the molecular informatics extended from Shannon's information theory meets nanotechnology.

When biomolecules in cells are used to build moleware systems, the moleware structure of biomolecular information processing is reconfigurable and adaptive in terms of self-organizing operators at the nanobio-level. In inorganic-nanotechnology, the nanocircuit is made by self-assembly operations and the nanotechnology of molecular manufacturing. Thus it is necessary to develop tools that can measure nanobioICT systems by using information theory and communication technology. This is a challenging but promising task. In nano-sized space, molecular operations will contribute to the programmability and controllability of nano-level information processing with broad applications.

As shown in Figures 3.1 and 3.2, Shannon's information theory provides a theoretical framework for communication systems. In concept, the communication system includes the source, channel, and destination. In moleware communication systems, the source is a molecular signaling mechanism through which the information is encoded and sent out. The channel is the molecular medium where the biochemical reactions transport and/or transmit the molecular signals and messages. The destination (that is, the molecular receiver) is the signaling mechanism, whose state is activated by the received molecules in response to specific functions of cell communication.

Heterogeneity is one of the most important causes of complexity in moleware communication systems, which differs from the homogeneous semiconductor hardware used in electrical signaling transmission. The biophysics principle supports the information processing mechanism of moleware, where the conformal structure of molecules determines the interactions of the molecules. The dynamical behavior of molecular signaling processes can be observed in moleware systems. The information representation and operation is expressed in a way that we are not familiar with. Naturally, there arises a need to explore novel theory and technologies of ICT.

Molecules are connected to construct complexes that take the form of molecular compounds. Here, the fundamental task is to figure out how to represent information with moleware. The procedures for assigning informatics meanings to the molecules and designing efficient encoding schemes underpin moleware communication from the viewpoint of information source code. After we get the molecules to arrive at the destination, we have to decode them, that is, to translate the information represented by the molecules into an understandable form. This is the work of source codes. Molecular complexes have complex structures in three-dimensional space. The shapes and structures of molecular complexes lead to their complicated function in molecular biology. From the viewpoint of informatics, these structures can be mapped into certain symbols for coding. It is important to select the optimal code form from different mappings in order to quantitatively describe the maximum degree of information at minimum cost.

The channel in moleware communication is a chemical space in which biochemical reactions are used for molecular information processing. The implementation will take different forms. Increasing the quantity of information contained and transported by the channel is the goal of channel design and implementation, which are constrained by the molecular signaling mechanism in the channel. The road is indeed long from the analysis of the bioinformatics mechanism of cell communication to the engineered design of moleware communication. There are still many challenges in physical implementation of moleware communication systems.

3.3 Embryonic Approaches to NanobioICT

On one hand, we have cells in nature; on the other hand, we have various types of artificial nanobiosystems, ranging from molecular motors to controlled microbial communication systems. This means that work in cell communication has achieved a certain level of progress in understanding nanobiocommunication systems in nature, advancing the frontiers of molecular biology.

An analysis of cell communication based on experimental evidence provides a guideline for the synthesis of engineered communication technologies in nanobiosystems. Communication refers to the message transmission process in its original sense; cell communication refers to the processes that transmit molecular messages whose effects are biological functions in the cellular life within and among cells. The underlying mechanism supporting these communication processes is the network of signaling pathways. According to the principles of genetics and signal transduction in cells, biochemical reactions with complex interactions are activated adaptively and automatically within the cell. Pathway networks are ubiquitous. The molecules involved in the networks move within cells, determining the cellular functions without external controls on the concrete behavior of the biochemical reactions. Autonomous internal controls are perfectly realized by the cellular pathways. These outstanding works of art—pathway networks—are made by nature. What we must investigate are the design principles of these natural works of art. Lessons from experimental biology are emerging, such as the work by Markus Kollmann et al. on design principles of a bacterial signaling network [33]. Engineered communication [31] refers to the communication system that operates under engineered operation and control. So far, the most successful system of engineered communication was built by microbes [31].

Gaining a comprehensive picture of the organization of pathway networks would require a methodology encompassing systems science, cybernetics, information theory, and new emerging sciences. Cells will be handled by a system that involves synergic mechanisms at different levels and information flows among different biological objects. The counterpart to the systematic methodology for nanobiomachines is the methodology derived from systems biology, bioinformatics, and computational biology. At the nanobio-level, biocybernetics will be extended to nanobiocybernetics, based on the basic concept of feedback and various types of automatic control methods in mathematics and the related practical technologies. Robustness is one of the crucial factors in the analysis and synthesis of cellular pathway networks, and feedback is necessary for keeping the signaling mechanism stable. Accordingly, structural control should be investigated with models, algorithms, and techniques in the signal-processing field. Here, information theory is related to the informatics and communication aspects of nanobiosystems. The relationships among these disciplines are thus intertwined.

Logic description and logical relationships belong to the conceptual-level knowledge of complex networked nanobiosystems, such as cells. The natural and engineered pathways demand logical explanations of the cellular functions, owing to the current state of nanobiotechnology. Tu et al. discovered logical hints for the temporal biological function of yeast cells, specifically the temporal compartmentalization in the yeast metabolic cycle [34], which is also related to the robust nature of cellular systems. Multiple agent systems (MAS) may be used to model cellular signaling systems, where formal logic and other AI tools can be introduced in quantitative descriptions and possible engineering schemes for nanobiosystems.

The above methodology is reflected in the design of pathway structure for the collective behavior of cellular systems. Cellular systems work in a cooperative way to achieve cellular signaling, which is pivotal to the life of a living being. Abnormal signaling processes work under the same mechanism cause diseases. For example, cancer involves a complex signaling network that is not easily destroyed by external operations. Furthermore, neuron systems are strongly influenced by the cellular pathways. In the domain of molecular biology, the information flow among DNAs, RNAs, proteins, and other molecules in cells forms the basis of molecular bioinformatics. Cell communications are carried out by biomolecules, where the programs for these functions are the signaling pathways in cells. Those designing engineered mechanisms of moleware communication should ponder the biological cell communication in nature and learn something from it. The biochemical protocols and the network protocols for moleware communication should be identified and then implemented by engineered pathway architecture in nanobiotechnology.

In order to use proteins in the cellular pathways of GEFs/GAPs and kinases/phosphatases to represent information, the following notations are defined:

GTPase-x—GTPase representing x;
GTPase-y—GTPase representing y;
GTPase-z—GTPase representing z;
GEF-x—GEF activating GTPase-x;
GAP-y—GAP activating GTPase-y.

First, GEF-x and GAP-y are activated. In addition to GTP/GDP, the GTPases can be bound with the phosphorylated or dephosphorylated proteins. GEF-x and GAP-y set the GTP-bound state of GTPase-x and GDP-bound state of GTPase-y, while the combinatorial form of phosphorylation or dephosphorylation states of proteins is updated. GEF-x and GAP-y also set the state of GTPase-z as a GTP-bound state. In the code representation of the related signaling process, the updating operation can be formulated as follows:

{GTPase-x-GTP|GTPase-y-GDP|GTPase-z-???}
$\rightarrow$
{GTPase-x-GTP|GTPase-y-GDP|GTPase-z-GTP}
by
{GEP—x|GAP-y}(t),
and GEF-z and GAP-z are not specified

where "???" refers to either the GTP-bound or GDP-bound state.

In analyzing the phenomenon in which GEF-x and GAP-y set GTPase-z into a GTP-bound state, a biophysics-oriented bioinformatics model is presented to explain the mechanism. The codes of the GTPase complexes are 10x (x = 0 or 1), that is, 100 or 101 (before the operation) and 101 (after the operation). The interactive form of these three pathways is simulated in a complex network, where assumptions on basic protocols for signaling are introduced.

The ad hoc knowledge needed to develop novel methodologies for communication systems in nanobioICT comes from the characteristics of moleware media. The philosophy used in designing the components of moleware communication systems is thus reflected in the design tasks carried out between the levels of the physical medium and the information.

The main principles of the methodology used in nanobioICT can be summarized as follows:

1. Biochemical reaction is the physical form of the molecular communication process. Among cells, information is represented by molecules and transmitted by biochemical reactions. As opposed to the electrical signals in telecommunications, the molecular signals in moleware communication are represented by a series of biochemical reactions in analog values.

    At location Xa, molecules A0, A1,..., An are involved in the biochemical reactions. Here, there are biochemical reactions with input from Xa and output to Xb. When the molecule M involved in these reactions move in cells, they are created and consumed. At location Xb, molecules B0, B1,..., Bm are produced.

    It is nontrivial when the molecules at sender side Xa are different from those at receiver side Xb. Information is represented by the molecules, and the name of the molecules denotes the symbols. We can define these symbols as variables, and the chemical state of the molecules is defined by the values of the variables. During the biochemical reactions for molecular message transmission, the concentration value of the signaling molecules is represented in the analog form. Those molecules with a certain concentration above the detection threshold are measured. The waveform of the concentration versus time curves of the biochemical reactions is nonlinear and temporal. The related enzyme activation is dynamic. This makes the signaling mechanism completely different from that of an electrical communication device.

2. The process of detecting the signal and translating it into the information needed for communication relies on biochemical technology. The analog and symbolic or digital values exist in the same entity. Valid states of the signals are determined by the threshold. The signal represented by molecular concentration is regarded as the digital value of 1 when it is above the threshold and the digital value of 0 when it is below the threshold. Detection is made during a certain period because concentration is a temporal signal. In the current biotechnology, concentration is normally kept constant during a certain period of time that is sufficient for measurement. Amplitude and phase are two features of the concentration pattern used to recognize and identify the signals for information representation and transmission. The molecular signaling mechanism allows a real number and a digital number to be stored in the same physical medium.

3. The physical characteristics of biochemical reactions provide the basis for designing protocols in moleware communication. The response-answer function is fundamental for communication at the signaling level. Given two pathways for molecular signaling, pathway A sends molecular message C from molecule X to molecule Y, and pathway B sends molecular message D from molecule Y to molecule X. By detecting the concentration of X and Y, the response-answer process can be finished at the side of B.
4. The architecture of moleware communication is heterogeneous, where delays in signaling processes often occur because the values of coefficients of pathways and the activation degrees of enzymes are diverse. In order to obtain asynchronous signals from moleware communication, biochemical protocols should be established for asynchronous molecular mechanisms.
5. The signaling mechanism of moleware communication is a complex system. This is due to the existence of interactions among pathways. The molecular signal feedback and enzyme activation can efficiently regulate the states of molecules. By controlling molecular signals and engineered pathways, it is possible to achieve mutual signaling between sender-molecules and receiver-molecules, which is a prerequisite to designing protocols. At the top level of network protocols, the rules that define the basic steps of molecular communication at the abstract level have to be made according to the low level of biochemical reactions.

Here, we take an informatics-level example to explain the principle of moleware communication in which the phosphorylation/dephosphorylation and GTPase-hydrolysis/GTPase-dehydrolysis processes are used for information representation. The building blocks and the pathway network based on these processes are given in Figure 3.3.

A simple but basic communication protocol in informatics is discussed in the case where the molecular message is transmitted between X and Y, where X and Y are defined as the sender and the receiver, respectively. The protocols of signaling between X and Y handle the information processing task in algorithmic form (logic primitive):

Procedure X:
  Message-send (X):
    /* the signaling processes for sending molecular message by biochemical reactions are activated */
  Biochemical reaction (X, W, Y);
    /* the biochemical reaction (X, W, Y) refers to a series of biochemical reactions by cellular pathways that can set the state of Y, W refers to the molecules involved in the signaling process */
  Message-receive (Y):
    /* the state of Y is updated by the biochemical reaction (X, W, Y) */
Procedure Y:
  Message-send (Y):
    /* the signaling processes for sending molecular message by biochemical reactions are activated */

```
        ┌─────────────────────┐
        │   Phosphorylation   │
        └─────────────────────┘

        ┌─────────────────────┐
        │  Dephosphorylation  │
        └─────────────────────┘

        ┌─────────────────────┐
        │    GTP-hydrolysis   │
        └─────────────────────┘

        ┌─────────────────────┐
        │   GTP-dehydrolysis  │
        └─────────────────────┘
```

(a)

[Diagram (b): Kinase → GTPase–GTP (Activation state of GTPase); Protein 1 (p) ⇌ Protein 1 via Phosphorylation / Dephosphorylation; Protein 2 (p) ⇌ Protein 2 via Phosphorylation / Dephosphorylation]

(b)

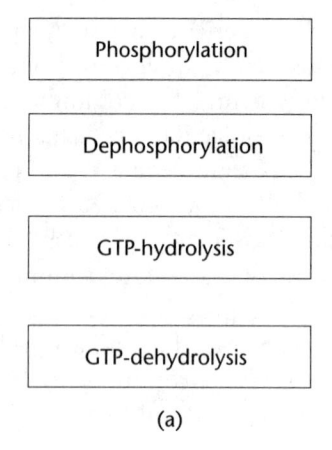

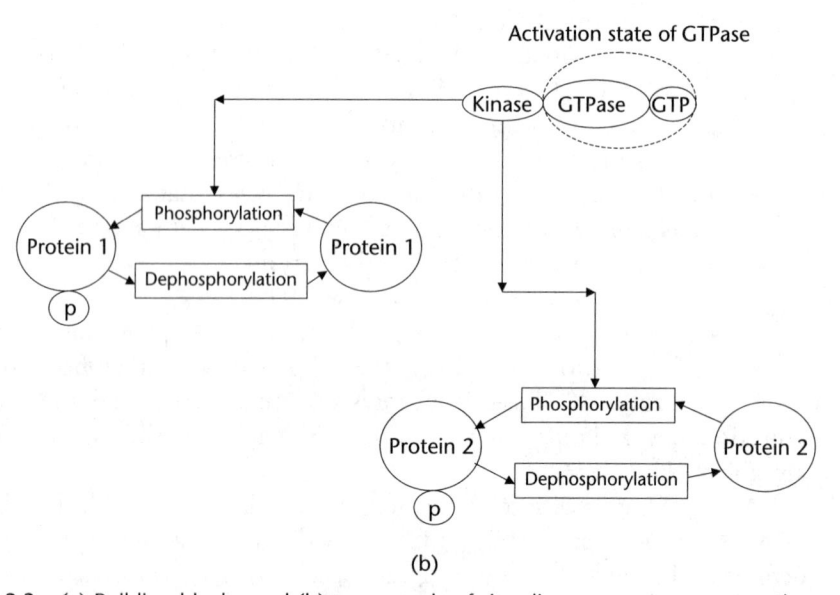

Figure 3.3 (a) Building blocks; and (b) an example of signaling processes among pathways.

Biochemical reaction (Y, W', X):
/* the biochemical reaction (Y, W', X) refers to a series of biochemical reactions by cellular pathways that can set the state of Y, W' refers to the molecules involved in the signaling process*/
Message-receive (X):
/* the state of X is updated by the biochemical reaction (Y, W', Y) */

In the sense of communication protocols, the two sides of moleware communication processes need to be harmonized in order to realize a basic communication function. A strategy recommended for the protocols is analysis by a train-schedule problem. Given two stations X and Y, from X to Y we can first take an express train to station M, midway between the two stations, and then proceed to station Y after transferring to a local train that stops at each station. The express train does not stop at station Y. The question becomes how to go to Y from X and then back to

3.3 Embryonic Approaches to NanobioICT

X. When one goes back to X from Y, there are two possible routes. One is the reverse route, and the other is to take a local train all the way back to X without using an express train. Under some conditions, the time to arrival at X by local train is even shorter than that of the route by express train from M. One of these conditions is that there is no express train available when one arrives at M by local train from Y, thus necessitating a long wait at M.

The rules that define the message transmission between X and Y are inspired by the above metaphor (see Figure 3.4).

For molecule X, a message is sent from X, and the period of the communication is divided into two stages: TTxm from Tx to Tm and TTmy from Tm to Ty. During TTxm, the biochemical reaction consumes molecule X, and the concentration of X decreases as time elapses; the concentration of molecule Y increases above a certain threshold due to the biochemical reactions. The signal of molecule Y, represented in measurable quantitative forms, shows whether the molecular messages have been received by Y. The corresponding pathways only enhance the signal of Y according to molecule X, without any inverse signaling function. This is a one-direction signaling process. The speed of the signaling is faster than that of two-direction signaling. During TTmy, the biochemical reactions correspond to two signaling processes: sending a message from X to Y and responding from Y to X. The signaling process from Y to X is set as a continuous process after TTmy, where the two-direction signaling processes involved are slower than when only a one-direction signaling process is activated.

The signaling operations are presented in the following algorithm that orients the informatics function.

Algorithm of sending messages from X to Y and responding from Y to X in parallel:

Concentration of X is initialized as a certain value Q above a threshold that is detectable for measurement.

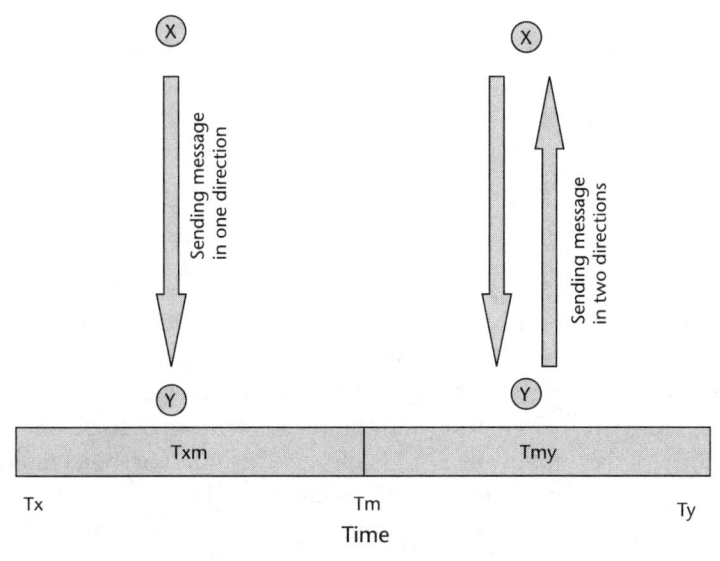

Figure 3.4 Protocols of molecular communication.

The biochemical reaction of signaling from X to Y is activated by signal X.

signaling:
    if concentration of X is lower than the threshold
    then goto ending
    if (T < Tm) and (the concentration of Y is bigger than the threshold)
        then
        pathway signaling (X->M->Y)
        The molecules are produced by biochemical reaction of signaling X to Y
            and moves to the location of Y
        else
    if (the concentration of Y is bigger than the threshold)
    then
    {
    [para]
    {
    pathway signaling (X->M->Y)
    The molecules are produced by biochemical reaction of signaling X to Y
    and moves to the location of Y
    }
  [para]
    {
    pathway signaling (Y->M->X)
    The molecule Y activates the inverse signaling process in biochemical
    reactions. The molecules that are produced by biochemical reactions of
    signaling Y to X moves to the location of X
    }
    concentration of X is enhanced by the signaling process from Y to X
  }
    detection of concentration of Y
    measurement of Y
    if the concentration of Y is lower than a certain threshold
    then
        goto ending
    else
        goto signaling
ending: end

where [para] refers to parallel information processing.

Measurement by information theory is the most rigorous way to describe the signaling process of the above communication quantitatively. The threshold-setting scheme allows the coefficients to vary over a certain range. The phases of the signals sent from X to Y and the response from Y to X vary consequently. The channel of the moleware is the cellular medium for biochemical reactions. The input signal to the channel is denoted as Sx. The output signal of the channel is denoted as Sy. The stochastic characteristics of the molecular channel require probability-based measurement.

The probability of X is denoted as P (Sx). In cellular signaling, even though the initial signals are set to a certain quantity at the beginning of the pathway signaling process, the output of the pathway will vary as a direct measurement of the quantitative description of the underlying channel.

The condition probability at Y is given as

P(Sy|Sx) = the probability of Sy under the condition of Sx

which refers to the occurrence of Sy according to the quantity of Sx. The influence of Sx upon Sy is exerted through the channel.

The entropy of Sy in Shannon information theory is

$$H(Sy) = -\Sigma P(Sy) * \log P(Sy)$$

where log() denotes that the logarithm base is 2.

We set the channel input as Sx, so the mutual information from the Sy side is the measurement of quantitative description of the encoding effect in the channel.

$$H(Sx, Sy) = \log P(Sx, Sy) \log [P(Sx, Sy)/ P(Sx) P(Sy)]$$

This measurement is used to quantitatively analyze the response of the channel to the input. From this measurement, the performance of moleware communication systems can be evaluated.

When the signaling molecules in the molecular channel are sufficient, Gaussian distribution and Bayesian theory are applicable to this case. The coefficients in the channel are the parameters of the Gaussian distribution. The multiple variant distributions result in

$$P(m_0, m_1, ..., m_n) = 1/ \text{variance} [\sqrt{(2\pi)}] * \exp[-(x - \text{mean})^2 / (2 \text{ variance}^2)]$$

where $m_0, m_1, ..., m_n$ are the coordinates of the variable, x is the vector of $m_0, m_1, ..., m_n$, and the mean and variance of x are vectors.

Here, we have the following rules:

1. IF the lower bound of mutual information H(X,Y) is obtained when the time of the signaling process from the start to Tm is maximum,
   THEN the one-direction route is dominant in the molecular communication process.
2. IF the upper bound of mutual information H(X,Y) is obtained when the time of the signaling process from Tm to the end is maximum,
   THEN the two-direction route is dominant in the moleware communication process.

The correlation degree ranges in the domain of H(X,Y) measurement. This measurement is used as a kind of semaphore for regulating the feedback mechanism of the response signaling process in molecular communication. This is because the mutual information of Sx and Sy reflects their quantitative relation in which we can observe the dependence of X on Y. The feedback control on the signaling sent from Y to X is dependent on the signal sent from X to Y:

enzymatic activation = appropriation to H(X,Y)

The speed of response from Y to X is quantitatively described by the previous formula. H(X, Y) shows the correlation degree of X and Y in statistics. If H(X,Y) is large, strong enzymatic control is necessary for the fast speed of response. If it's small, the speed is slow.

The underlying signal mechanism of molecular communication is dynamic if the fluctuation and noise have been taken into the consideration. The maximum likelihood (ML) estimation is appropriate for estimating the parameter of the phase of the curve of concentration versus time of the signaling molecules.

ML estimation: the parameter-vector M = argmax probability Pr(M)

where the probability takes the maximum value.

Since this parameter is defined by comparing the signals X->Y and Y->X, the information capacity of the underlying channel is estimated such that

Empirical information capacity = the bit number of the channel allowed
  = the peak number of the normal distribution in the channel
  = the number of the pathways corresponding to the channel
  = the number of the coefficient sets of the biochemical reactions of the pathways

The coefficients of the pathways are modeled as Gaussian distributions. Each of the Gaussian distributions is an independent identical distribution (i.i.d.). These pathways are connected in sequences, and the distributions are independent of each other. The feature measurement of each distribution is described by the coefficients.

Information capacity W = the bit number of the information transferred in the channel for the moleware communication processes of X->Y and Y->X, which is related to the information entropies of X->Y and Y->X in the sense of Shannon's definitions of information entropy and bit. The bounds of mutual information provide guidance for designing feedback mechanisms.

Let the coefficient sets of the channel for molecular communication be

$$\begin{matrix} M(0,0) & M(0,1) & \ldots & M(0,n-1) \\ M(1,0) & M(1,1) & \ldots & M(1,n-1) \\ & \ldots & & \\ M(n,0) & M(n,1) & \ldots & M(n-1,n-1) \end{matrix}$$

The detectable peaks for representing the signal transferred in the channel are translated into the code form of information contained in the channel. The peak signal is denoted as SS(M(i,j), t), where i and j = 0, 1,..., n–1. The speeds of their biochemical reactions are different. The occurrence of the peak is different for each peak signal. During the same period of time, some occur more often, others less often. The empirical statistical measures obtained from the corresponding histogram are used to design the codes that represent the information of the channel:

channel code (j) = SSc(0) SSc(1) ... SSc(n-1)

where SSc (j) = 1 if SS(M(i,j)) is in the peak; SSc (j) = 0 if SS(M(i,j)) is not in the peak. The set of channel codes is arranged as a matrix [channel (j)]. Here the channel code refers to the code that describes the signals in the moleware channel.

From this encoding process, we can infer that the code achieved is independent of the exact distribution of the channel, which means that it is not necessary for us to have complete information of the channel at hand. The automata that correspond to the state transitions among the states of the matrix given above are parallel-operable, that is, the current state of the signal of X-> Y and the transition rule designed in advance generates the next state of the signal Y->X. The automaton is oriented to formulate the molecular communication process as a computing process in computer science, to which signaling processing is applied for controlling molecular signals.

The moleware codes provide a means of applying signaling processing technology. For the channel of a moleware communication system, we need to encode the input of the system. The output is the result of the transmitted and decoded information in the channel. The quality of the encoded information is studied by measuring the moleware channel and the two sides (i.e., sender and receiver) of the communication system. By filtering the signals, the noise is suppressed. Feedback is the commonly used form of regulating the signals of channels. The characteristics of moleware communication compel us to consider a molecular dynamics mechanism that is different from the electrical channel of telecommunication systems. The signal processing and adaptive control are integrated. The concentration of a molecular signal varies temporally. The temporal codes of moleware channels are represented by the pathway structure and operated and updated by the corresponding pathway mechanisms. Varying with the temporal codes in the molecular message transmission, the reconfigurable pathway structure and operators are self-regulated within the pathway networks. The algorithm for detection of functional channel codes is formalized as follows:

```
for (i = 0 to maximum-steps-of-transmission)
{
   if (amplitude of the signaling protein is above the threshold)
   then
{
to measure the peak of the concentration.
   if (the peak is detected)
```

```
        then
        {
        the phase of the concentration is recoded
        }
    }
}
```

The quantities representing phase of the concentration curve versus time of the signaling molecules are used to encode the temporal codes of the dynamical signaling processes of the underlying moleware channel. The nodes that denote these quantities construct a graph that is updated in steps. The corresponding states of the graph are the measurements of the information capacity of the moleware channel. The distribution of these code values is independent of the activation states of enzymes that are involved in the corresponding channel. The corresponding biochemical reactions for representing the information encoded by these codes can be controlled. The molecular operators for writing/reading the information are related to the channel structure and protocol design of moleware communication.

NanobioICT is unconventional for three major reasons:

1. The entire nanobioICT system is a complex system, in which cooperation among the different molecular information processing units is prerequisite to realizing massively parallel molecular computing and communication. The interaction among molecules that upgrade the efficiency and capability of molecular information processing exists at the level of DNA, proteins, and cells. These signaling routes in molecular networks, resulting from the molecular interaction, satisfy the requirements in informatics and thus are used for the engineering schemes of nanobioICT systems. At the level above the molecular operators on the signaling molecules, the integration of the molecular relations is a condition of configuring the moleware architecture. Proper control of the signaling flow of the molecular network is a direct means to provide the expected medium condition for the molecular operations at the bottom level.
2. The biochemical characteristics of the molecular signaling are ad hoc in the sense that the concentration of the biochemical reaction is nonlinear. Unlike the digital circuits in electronics, the original form of the molecular concentration is analog. After the information representation mode of molecules is determined, the criterion for detection of molecular signals is:

    Whether the quantitative measurement of molecular information reaches the threshold according to the activation status of the molecules.

    Through signal processing technology, the molecular signals are analyzed in order to select the range of molecular signals that can get the optimal effect of information flow. Filtering is a well-known method in signal processing that is often used to modify signals and extract useful information from the original signals. In order to obtain a stable and unique signal from moleware, filtering the molecular signaling process is a promising approach. The nonlinearity of molecular signals generates the broad range of signal processing in molecular signaling. This is an important point for nanobioICT. In summary, molecular signaling is deeply rooted in molecular communication and computing.

3. The biological function of nanobioICT is adaptive and robust under uncertain environments. This merit of molecular biological systems is also rooted in the complicated logical structure of the molecular signaling mechanism in living biological systems. A cell demonstrates ample actions at the molecular level. As a follow-up to discovering the cellular pathways by experiment, making logical inferences about the data is a critical task in understanding a molecular network, and it is a normal course of study in nanobioICT research. The causal relations of biochemical reactions in cells give hints for pathway networking. In the case of nanobioICT constructed by molecular communication used in forming basic information processing architecture, the logical relations among the signaling molecules are located at the biochemical level of communication in experimental protocols. In nanobioICT systems, the informatics level of communication is constrained physically by the biochemical mechanism. As a result of mapping from the biochemical constraints to the informatics conditions, a protocol of moleware communication by molecular networking is adopted to define the global informatics specifications of a nanobioICT system. Chapter 7 gives details on forming logical inferences about cellular pathway networks with the aim of using the bioinformatics knowledge for molecular computing.

3.4 A Glance at Informatics of Moleware Communication

Cell communication was discovered by molecular biologists [35]. The biochemical level's signaling mechanism for cell communication is becoming clearer. The molecular bioinformatics forms of cell communication are pathways and the pathway networks constructed by the interactions of pathways. In order to move beyond biological cellular communication to computational cellular communication, we have to work on the informatics-level communication protocol based on the data structure of the pathway network. Computing technology itself can't answer the question of how to design protocols, but computable architecture makes it possible to program the cellular signaling in biochemical technology for engineered-like operation on cells. The informatics principle of computational cellular communication in nanobioICT systems is mainly based on the biophysics mechanism of molecular biochemistry.

The molecular structure of nanobioICT is reconfigurable, which means that the molecule number is changeable, that is, molecules appear or disappear during the biochemical reaction processes in cells. This is one of the unique aspects of nanobioICT from the viewpoint of an unconventional information-processing paradigm that can go beyond the domain of Shannon's information theory.

Along the biological information flow from exterior cell to internal cell, we can observe the bidirectional signaling process. In semiconductor technology, the circuits are fixed for the same signaling process, even in reconfigurable hardware. In contrast to this, the molecules in the biochemical reaction for the same signaling processes are dynamic. The signaling pathways from the start at the exterior cell's site to the end of the pathway at the target molecules' site show the one-directional type of biological information flow. The feedback also exists in the same processes. Feedback is the kernel concept in Nobert Weiner's cybernetics. In order to get the desired molecular cell communication effect, the feedback needs to be analyzed in order to control the molecular signaling process.

The scheme of introducing feedback into the computational model of engineered cell communication is unconventional. There are cross-talks among the molecules in pathway networks. Cross-talk refers to the phenomena or the mechanism of the signaling of molecules in the object pathway influenced by multiple molecules located in other pathways. In vitro, each protein can be studied under the condition of protein separation and its selected combination with control. But in vivo the cross-talk is complicated for observation and control. Decoupling methods in control theory deal with coupled systems in cybernetics. The cross-talk in biology is mapped into the coupling in cybernetics, so the informatics issue emerges as the analysis of the information structure of pathway networks in cells as described in [35]. In theory, the target of informatics research for nanobioICT is to gain the ability to efficiently control cell communication through systematic understanding of cell biology. The information processing of a molecular signaling process in which cross-talks are embedded behaves like a kind of networked computing.

The informatics-level discussion of moleware communication is presented in this section. The encoding/decoding at the molecular computing level is explained in Chapter 4, along with discussion on moleware communication by a molecular motor and its interface with molecular computing.

Previous works in the field of DNA computing, such as the one reported in [15], have shed much light on DNA encryption. We address this topic within the domain of DNA computing, since the information encoded by DNA molecules is mainly operated at the DNA molecular level, which does not involve DNA nanomachines. In this section, we focus on a molecular information-processing framework inspired by the naturally existing nanobiomachines. The information capacity of the cellular pathways that activate signal transduction is directly related to the functions of cells. Here, natural information representation arouses our interest in studying the information theory of the moleware communication process by designing encoder/decoder schemes in which the pathways are employed as the information processing units. The issue of information-source codes that correspond to the molecular codes made by molecular operations or signaling pathways is discussed in Chapter 4. This issue is related to the information processing mechanism of nanobiomachines and is regarded as an example of an artificial system designed according to what we have learned from nature. In order to emphasize the kernel of the communication process, here we discuss the channel code and the related adaptation mechanism of the proposed nanobioICT model based on the cellular signaling pathways.

As preliminary knowledge, a concise explanation of the information representation is presented here. Pathway control is the necessary condition of a molecular information processing system based on signaling pathways. Two technologies—moleware communication and moleware coding—are involved in this process. Moleware communication refers to the communication mechanism, process, and system that are designed by moleware. Moleware code refers to the code in which the physical medium of information storage is moleware. The molecular characteristics of moleware communication and moleware codes are what make the related theory and technology different from semiconductor hardware and software forms of communication and code.

The basis of designing molecular codes is information representation. In cells, two kinds of molecular switches—GTPase switches and kinase/phosphatase switches—are important for cellular signal transduction cascades and are employed for information representation.

Under the guidance of a GTPase switch, two states are generated:

GTP-bound for 1
GDP-bound for 0

Under the guidance of a kinase/phosphatase switch, two states are generated:

Phosphorylation for 1
Dephosphorylation for 0

Between Rho GEFs/GAPs for GDP/GTP switching and kinases/phosphatases for ADP/ATP switching, Rho GTPases [4, 5] are used to connect the upstream/downstream pathways in mammalian cells:

Rho (three isoforms: A, B, C), Rac (1, 2, 3), Cdc42, TC10, TCL, Chp (1, 2) RhoG, Rnd (1, 2, 3), RhoBTB (1, 2), RhoD, Rif, and TTF

The set of GTPases = {Ras, Rho, Rab, Arf, Ran} is the superfamily of Rho family GTPases.

For kinases and phosphatases, a 23-bit word can be designed by the set of kinases and phosphatases and other related target molecules [5] of

{PIP5-kinase, Rhophilin, Rhotekin, PKN, PRK2, citron, citron-kinase, Rho-kinase, MBS, MLC, p140mDia, p140Sra-1, Por1, PI3-kinase, S6-kinase, IQGAP, PAKs, MLK3, MEKK4, MRCKs, WASP, N-WASP, and Ack}

As an example of kinase/phosphatase pathways, a Rho-MBS-MLC pathway [5] is illustrated in Figure 3.5, where GTPase is interacted with the MBS pathway and MLC pathway.

In principle, based on the building blocks of the cross-talked GTPase/kinase/oxidation network of cellular signal transduction, combinatorial forms of pathway motifs can compose more complicated codes through pathway networks.

Considering the two molecular switches as information representation mechanisms, information compression becomes possible owing to the nonlinear mapping in moleware.

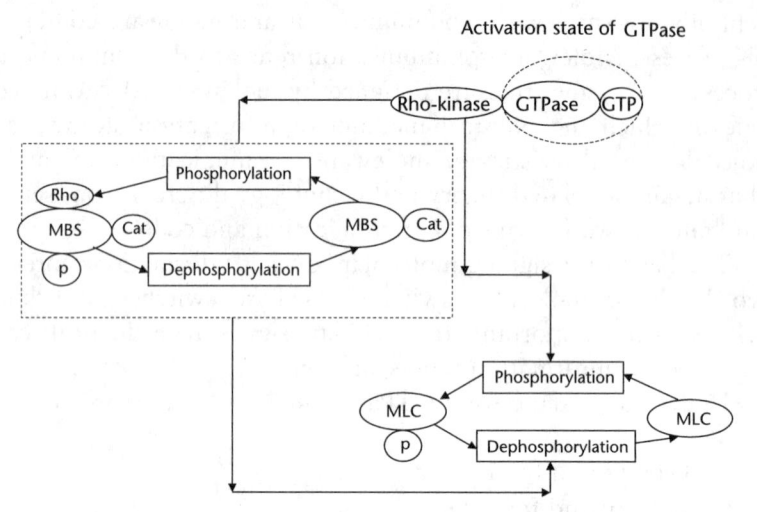

Figure 3.5 Example of Rho-MBS-MLC pathway for molecular information processing in molecular complexes.

Earlier we only discussed the molecular signaling process in a biological sense. In an information sense, we need to make certain abstractions to transform biological cell communication into informatics molecular communication. Based on the abstraction model, the derived models and their modifications need to meet the specific requirements of different applications. The basic abstract model of the encoding/decoding process is given as follows.

First, we use a GTPase complex to represent a binary word, and then we design a mapping from the GTPase word to an SPK word, where the SPK word space is much smaller than the GTPase word space. SPK refers to the signaling protein whose phosphorylation/dephosphorylation state is regulated by kinase/phosphatase in the cell. When the GTPase and SPK pathways interact, the molecular complexes consisting of GTPase complexes and SPK complexes with a structure in three-dimensional topology give rich features to express codes. The temporal features of signaling pathways on the codes are considered during the HeLa cellular mitosis process, where the GTPase complexes under the regulation of GEFs/GAPs are bound with the SPK complexes under the regulation of kinases/phosphatases. The cross-talks within and between the two types of pathways provide the basis of designing codes. The adaptation of the designed codes is reflected in the length-variable schemes of the codes that can be controlled internally by pathways or externally by enzymes. The information representation for the encoding/decoding process is shown in Figures 3.6 and 3.7.

In the spatial organization of cells, the signaling mechanism of oxidation also deserves study for its scientific value. We need to study the oxidative-phosphorylation pathway mechanism in cells. One of the abstract schemes for the description of the oxidative-phosphorylation states is convolution codes in the form of interacted molecules. A more generalized form is a kind of network coding in moleware. The labeling molecules and index molecules are used to access the molecular memory for molecular information processing. In the framework of cellular nanobiomachines, some engineered moleware schemes can be designated for coding, networking, and signaling. Extracted from the biomolecular form of coding

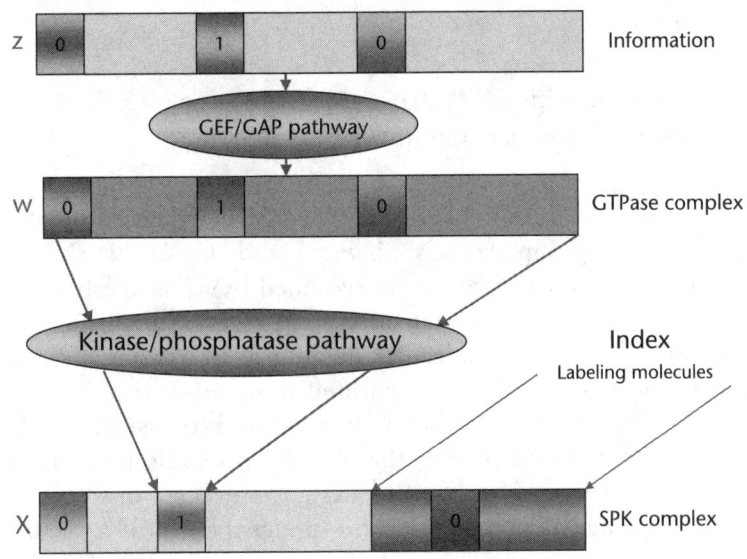

3.4 A Glance at Informatics of Moleware Communication 71

Figure 3.6 Information storage.

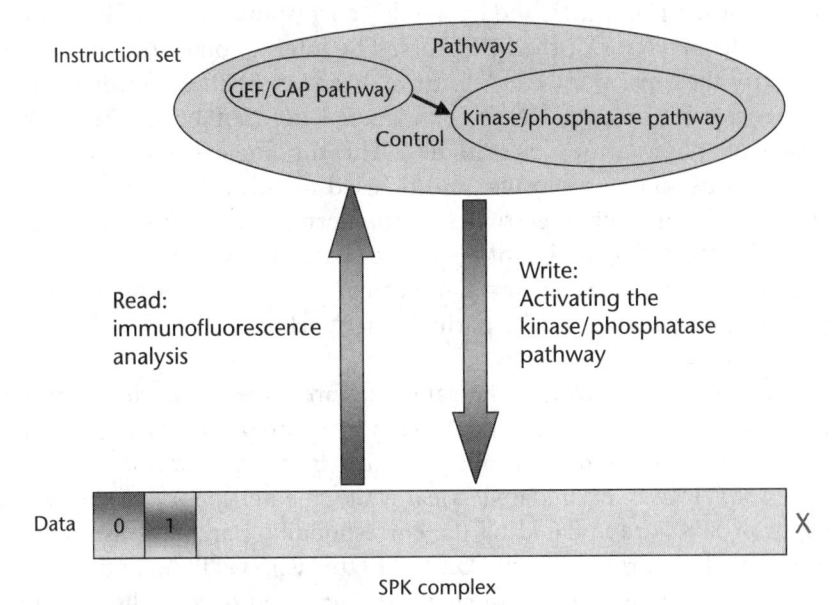

Figure 3.7 Schematic operators for instruction and data.

mentioned above, an abstract mechanism of molecular signal mapping is proposed as follows:

$\{X_i\ (i=0,1,...,n)\}$

is mapped to

$\{Y_j\ (j=0,1,...,m)\}$

under a network N_{xy} in which $E_w (w=0,1,...,W)$,

then

convolved with $\{Zl\ (l=0,1,...,L)\}$
or modified under condition of Q.

The operator of coding here can be constructed based on algebraic terms, for example, a semi-group. Then, we have to study how to work out the operator for this coding. The derived codes are designed by using the frequency of molecules that can lead to the generation of certain formal languages. The simple design scheme of coding at the molecular medium is to use one molecule for one bit of information at one time in a separated mode of an in vitro cell. In the spatial structure of pathways for parallel information processing, which is also the basis for trellis and network coding, the regulation function of different pathways on the same molecule can be used to design molecular operators for many-to-one mapping of states of signal molecules (under the condition of in vitro and multiple test tubes). The transition matrix (denoted as matrix $A[ij]$) representing these operable states is the key factor for understanding the parallelism of moleware operations. From state update in the corresponding signaling process, we can define the behavior of the operators. The interaction of the molecules corresponding to the operators can be formalized as abstract informatics forms. The sequences of the coded molecular complexes can be measured by histograms based on observations in statistics. The topological operators for coding with interactions among molecules can be used to encrypt the information in terms of stenography. The characteristics of the derived codes are generated by the pathway operators. The design of Nxy is discussed in the next section on network coding. The network structure of molecular code is designed by optimization and parameter estimation of the pathway network in terms of the physics of information technology.

As shown in Figure 3.8, the data structure is the bridge that connects the "chemical space" and the information space for nanobioICT. During the mitosis process, the GEFs/GAPs are dynamically activated by the pathways, and consequently the states of GTPases are updated. Meanwhile, the phosphorylation/dephosphorylation states of SPKs are updated by the corresponding pathways as well. As Oceguera-Yanez et al. reported [36], the Ect2 (GEF for Rho GTPase) and MgcRac GAP (Rho GAP) activate the Cdc42 during the mitosis process. The phosphorylation of MgcRac GAP is made by Aurora kinase B. This is only one of an enormous number of examples of combinatorial states of the signaling proteins under the regulation of the two types of cellular pathways. In order to understand the underlying informatics mechanism of cell communication during mitosis, we can learn from investigations based on information theory for the states of the signaling proteins regulated by the GEF/GAP and kinase/phosphatase pathways. These investigations show us the quantitative relations among these proteins under regulation of the corresponding pathways. The molecular states can be used to describe codes where the corresponding pathways are controlled. The information flow in the cell can be studied in a three-dimensional space, whose architecture is superior to that of a semiconductor. The GTPases and their cross-talks can be quantitatively measured in real time by the FRET-embedded, computer-controlled

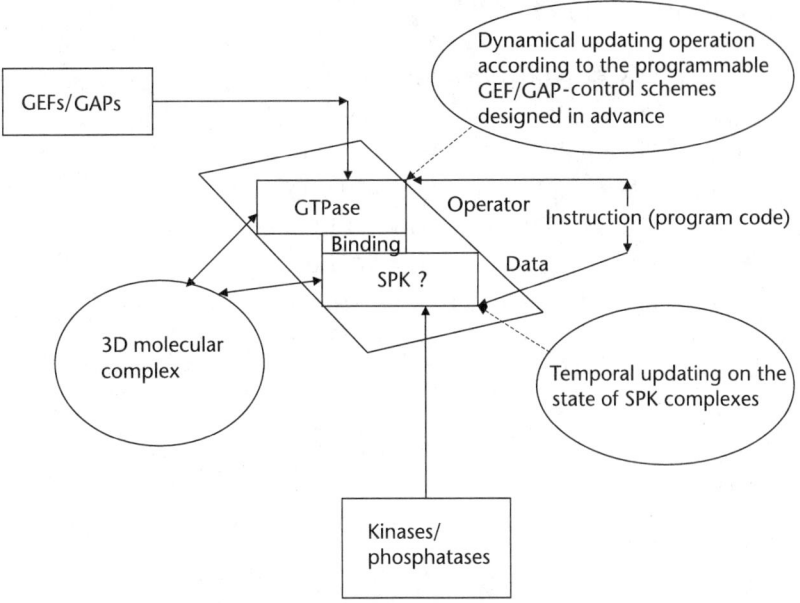

3.4 A Glance at Informatics of Moleware Communication

Figure 3.8 Relations among the three kinds of complexes.

3D molecular imaging system developed by the cell biology group headed by Y. Hiraoka at NICT-KARC [37].

The three-dimensional (3D) molecular imaging technology supports the idea of exploring the 3D structure for encoding/decoding information by GTPase pathways in cells.

From Figure 3.8, we can see that the central idea is to represent information by molecular complexes where GTPases and SPKs are employed with a binding structure. The GTPase complex and SPK complex encode information in binary form by their states as defined above. The binding material for these two kinds of molecular complexes is moleware. The two codes comprise combinatorial codes that involve two kinds of molecular switches of kinases/phosphatases and GEFs/GAPs. This data in moleware is programmable by the regulation operations of signaling pathways that are designated in a cellular environment. The instruction (program code for programming) in the sense of computing is made by three parts: GTPase complex for operator representation, SPK complex for data, and the binding molecular complex for connecting these two parts. The three-dimensional configuration of these molecular complexes is handled by biochemical technologies from which codes can be designed in a new way. The related data structure is graph and string. The coding process of phosphorylation/dephosphorylation is operated by the pathways. If the set of {Rho-GTPase, MBS, MLC} is used to design molecular codes, the encoding scheme will generate the molecular information by the Rho-MBS-MLC pathways.

The molecular complex is selected as Rho-GTPase-binding-MBS-MLC. The signaling molecules in this molecular complex have two states:

The states of Rho-GTPases: GTP-bound, GDP-bound
The states of MBS: MBS-p, MBS
The states of MLC: MLC-p, MLC

The code set becomes:

(Rho-GTPase-GDP)—-(MBS)—-(MLC)
(Rho-GTPase-GDP)—-(MBS)—-(MLC-p)
(Rho-GTPase-GDP)—-(MBS-p)—-(MLC)
(Rho-GTPase-GDP)—-(MBS-p)—-(MLC-p)
(Rho-GTPase-GTP)—-(MBS)—-(MLC)
(Rho-GTPase-GTP)—-(MBS)—-(MLC-p)
(Rho-GTPase-GTP)—-(MBS-p)—-(MLC)
(Rho-GTPase-GTP)—-(MBS-p)—-(MLC-p)

000
001
010
011
100
101
110
111

where the molecular state of bit2 activates the corresponding pathways of MBS and MLC, setting and updating the states of bit1 and bit0. Bit1 and bit0 represent data, and bit2 is the operation on data.

The algebraic terms in mathematics are the proper form to describe the informatics operators here. Interactions of signaling proteins are regulated mainly by cellular pathways. The conjugate word can be generated by this method as well.

$$G0G1...Gn1\text{-}X0X1...Xm$$

$$\rightarrow$$

$$X0X1...Xm\text{-}G'0G'1...Gn2$$

where X0, X1,..., Xm correspond to SPK, G0, G1,..., Gn1, and G'0, G'1,..., G'n2 correspond to GTPase, and m, n1, and n2 are integers.

The X part is kept to the same value while the G part is updated to G'. The algebraic operator of monoid is "introduced" to describe the word where the associate binary operation is assigned for the two molecular complexes, the SPK and GTPase complexes. From the viewpoints of molecular informatics and abstract mathematics, the operators provide the information-theoretical basis to design concrete algorithms for engineered communication. Based on the code representation and molecular operator, the automaton models and programming methods can be extended to molecular information processing.

With the reconfigurable pathway architecture of a nanobiomachine, a kind of self-regulation code can be designed with a self-adaptable signaling mechanism of reconfigurable pathways to achieve variable code-length in response to the dynamic environment. This is also independent of the channel and information source. Here, the parameters of the underlying channel are described by the coefficients of the biochemical reactions of the corresponding cellular pathways. The self-regulation of pathway networks makes it possible to carry out the transition of the information-representation structure for codes in a graph-rewriting manner.

Acquiring a huge amount of information storage with high intensity is one of the obvious advantages of using universal codes based on the signaling pathway structures.

In order to explain this, we use a conceptual example of coding an image by a pathway structure. The relations between the image in signal processing and the molecular mechanism in biomolecular communication, based on a concept of informatics, are given as follows:

pixel < - -> signaling molecule
grey-level/intensity value of pixel < - -> concentration of signaling
    molecules
data structure of resolution pyramid in homogeneous representation < - ->
        data structure of interaction network in heterogeneous description
compression space defined by content-independent information
    >> compression space constructed by biomolecular-computing-based
        communication method

The data structure from pathway units can be used to reduce the information quantity in compression processes (Figure 3.9).

The input to the atomic indivisible pathway is concentration; the feature pattern is the coefficient set for the network that condenses the information from a two-dimensional array of information. Here, the information compression problem is transformed into the identification of the network structure built from modeling and matching. The information processing in the pathway network is designed by a cellular pathway mechanism, which is quantitatively described by an Michaelis-Menten-equation-like model. At the informatics level, we can identify the pathway parameters by observing the distance between the pixel nodes and the concentration nodes under the minimum error criteria.

The above compression method is for information source code. In the case of moleware communication, the information encoded is transmitted through moleware channels. The biological feature of cell communication varies between the intracell channel and intercell channel cases. The protocols for the signaling processes influence the performance of moleware communication. Synchronizing the molecular signaling processes for accessing biomolecular memory and for the motion of biomolecular motors are helpful in controlling the speed of communication.

In a mobile wireless ad hoc network using sensor networks, the transmission modes of wireless LAN, the long-distance mode of WAN, and the broadcasting/digital mode of TV/HDTV can provide a high degree of realism in ubiquitous ICT systems. To develop the informatics algorithm for the network protocol, design skills must be applied in both the protocol layer and the physics layer of the transmission medium.

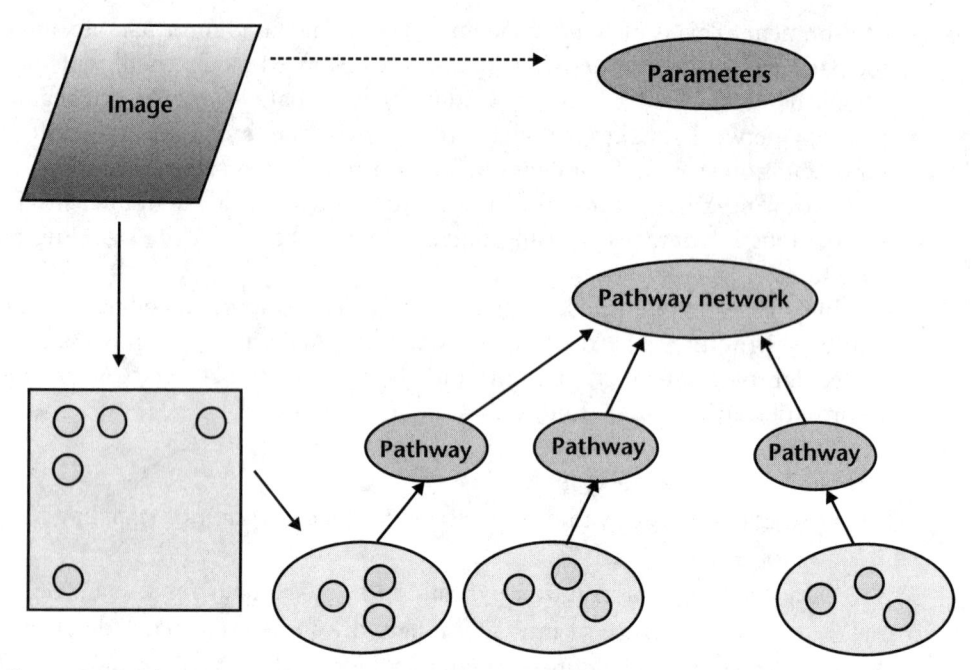

Figure 3.9 Schematic representation of the image compression by moleware communication.

Moleware systems in the nanobioworld behave much differently from telecommunication systems. But it is still possible to interface a moleware system and a telecommunication system. Within cells, we have an interconnection mechanism of signal transduction and the electron transfer that accompanies the signal transduction cascades. Electrons can be transferred to a molecular wire made by the coordination of porphyrins. This medium is capable of transmitting the optical signal for optical communication or connecting to the optical fiber communication in the physics layer. Engineered cellular signal transduction systems are also compatible with nanotube and moleware circuits by interfacing the electron transfer processes. In fact, even nano-antennas have been reported recently. Development of a completely automated moleware communication system that could cover all molecular signaling processes, from sender to receiver, would require the integration of different molecular devices to fuse different types of molecular signals.

3.5 An Informatics Form of a Molecular Viterbi Algorithm

The informatics mechanism of moleware communication is capable of parallel distributed information processing. It is natural to think of exploring the informatics form of a Viterbi algorithm by using a signaling pathway structure. The Viterbi algorithm (VA) was originally proposed by Andrew Viterbi, whose kernel idea was to find the route with the maximum probability—called the Viterbi path—based on observed data. It has broad applications that even include decoding in CDMA, GSM digital cellular, and 802.11 wireless LANs [38–40].

The information processing structure of VA is parallel and distributed. This is the natural link between VA in reality and a molecular system in theory. The motivation of exploring the moleware version of the Viterbi algorithm in informatics directly comes from the unique physical features of moleware communication and the O(M × N) complexity of VA for trellis codes. Up to now, VA has been successfully applied to the electrical medium, even though realizing a Viterbi algorithm by biomolecular computation is still a challenge in nanobioICT. The moleware structure of VA is given in Figure 3.9. Information processing is used in VA to estimate the path with the maximum probability based on a set of empirical data. The trellis structure is represented as a matrix X [n × m].

To fit the characteristics of molecular information, the procedure to select the node and to form the path at the level of informatics can be represented as follows:

min cost(route)

where

route = X(1,i-1)-X(2,i-2)-...-X(n,i-n);
i-1,i-2,...,i-n=1,2,...,m;
n is the column number;
m is the line number.

These operations are parallel in essence when all of the nodes are accessed at the same time. At each column X (j,i-j), one node is selected from the column {X (j,i-j)}. In the domain of information processing, VA is efficient.

One of the direct applications of VA is the VA decoder. The VA decoder uses the trellis structure to represent a network. The nodes of the network are assigned weights that are defined as a probability, that is, the occurrence of the codes' information bits in the decode application. The task of the VA is to find the path from the trellis structure that has the maximum probability. The inverse representation of this process is the minimum of the path cost, where the cost function corresponds to the probability. A smaller cost function indicates a higher value of the probability. The nodes of the path determine the decoding process. The decoded information has maximum probability, so the information obtained from the decoding is optimal.

In an electronic computer, VA takes the form of both hardware and software to calculate needed information. In the moleware form of VA, the concentration of molecules provides the analog quantity for determining the probability described above. The maximum probability estimation is equivalent to the biochemical reactions of the signaling processes that are generated by well-designed enzymes and the pathways regulated by these enzymes. After a certain detection threshold of the concentration is given, the biochemical reactions of the pathway are used as an estimation function that generates as output the selected signal corresponding to the solution of the optimal path.

More concretely, the moleware-VA-decoder uses the molecular information process to generate the encoded information. The constraint is embedded in the pathway structure in which only the required path among the signaling networks is activated according to the criteria of optimization. The molecules passing through the underlying pathways carry the information for coding, and they are read out right at the end of the pathway. The probability represented by concentration is also used for the decoding process.

In Figure 3.10, the trellis structure of VA is regarded as biochemical reactions among molecules. In the moleware medium, the molecular information processing mechanism of the VA processes becomes a form of biochemical reaction.

The molecular Viterbi decoder algorithm consists of three parts: trellis structure, path selection, and signaling for codes. The states of the selected molecules denote the code. The modification made is based on the molecular features of the biochemical reactions of the signaling pathways in cells. The branch metric is defined as the measurement of the cost between two nodes in the trellis structure. This branch metric is logical. The path metric is defined as the measurement of the complete route between an input node in the leftmost column and an output node in the rightmost column. This path metric is directly related to the global information.

The molecular-pathway trellis is a kind of structure whose signaling process is generated through pathways. The states in the pathways are designed and controlled to correspond to the states of the codes. Building an automaton requires the operations of detection, readout, and "judgment" in the form of an IF condition THEN action, where the condition is the set of signals detected according to certain thresholds and the action refers to the molecular operations on the signals. The automaton is controlled to transit the pathway trellis from the current state to the next state. One simple way to describe the data structure of the above process is to represent the state with molecules and to use a look-up table to describe the relationship between the input and output as the system's I/O constructed by biochemical reactions/pathways. In the primitive level of programming, the operators include {add, compare, select}.

We propose using the pathway units in cells for information representation and processing. This structure, which is the kernel of the Viterbi algorithm, takes

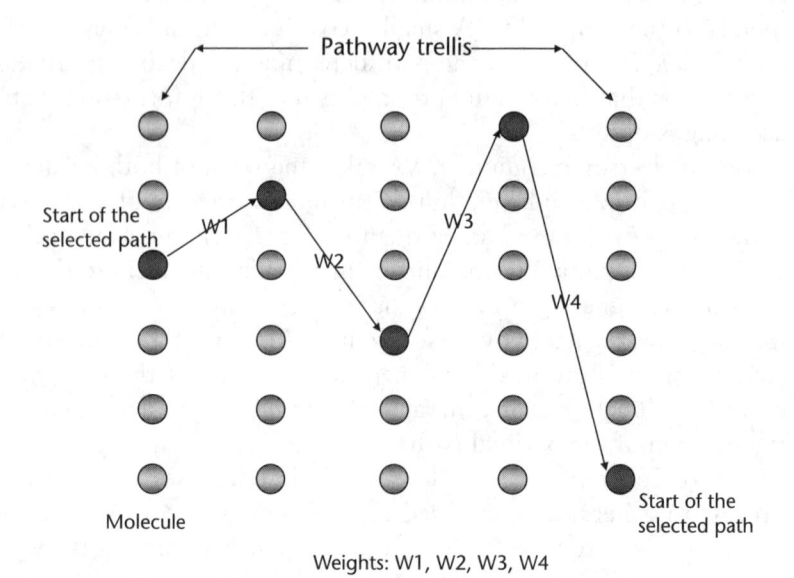

Figure 3.10 Moleware form of the trellis structure.

3.5 An Informatics Form of a Molecular Viterbi Algorithm

the form of a trellis. As Figure 3.10 shows, the path $W_1 \to W_2 \to W_3 \to W_4$ is selected through the related computing process designed by pathway structure. The weights in the Viterbi algorithm are represented by the coefficients of the biochemical reactions that correspond to the pathways. The minimal cost is obtained by selecting the fast reaction scheme. Here, the cost is represented by the speed of the corresponding pathways, and it is used to estimate the weight of the codes.

In mathematics, the signaling mechanism of the Viterbi algorithm is used to provide the "best" code in the sense of information theory. The physical meaning is what we aim to study. In biochemistry, the moleware features of the encoding/decoding process of the Viterbi algorithm form the basis of pathway design and testing.

The biomolecular computing process mainly handles the biochemical reaction that maps the Viterbi algorithm to a moleware form. The coded information in molecules is stored in molecule complexes. The informatics issue at the biochemistry level can be seen as determining what kind of cascades/pathway structure of a biochemical signal can be shifted to the informatics process of the Viterbi algorithm.

Consistency between the well-organized biochemical behavior in moleware and the informatics mechanism of molecular computing is crucial for the success of a moleware version of the Viterbi algorithm. This implies that it is necessary to find the configuration of the molecular computing algorithm that has an identical information process to that designated by mathematics. Semiconductor technology allows us to implement the Viterbi algorithm with hardware circuits and software tools. But wetware implementation remains an intriguing open question. We have to figure out the biomolecular-level mechanism of the Viterbi algorithm's computing process.

In Figure 3.10, the pathway trellis shows the working principle of the kernel process of the moleware VA. By judging various types of configurations, we seek the most feasible pathway structure while maintaining biochemical faithfulness. The informatics description of the VA algorithm is mapped to an operable biochemical reaction process in moleware. As the primitive form in programming and the predicate logic in operation, we adopt

$$y = \text{signal-pathway}(x)$$

where x and y are the input and the output of the signaling pathway, respectively.

Let $A(i,j)$ denote the nodes in the VA trellis, $i,j = 0, 1,..., n$. The vectors $\{A(0,0), A(1,0),..., A(n,0)\}$ and $\{A(0,n), A(1,n),..., A(n,n)\}$ are the input and output columns, respectively, of the node array.

The motifs of the signaling process corresponding to the array are oriented for the VA paths. The building-block-like motifs are dependent on the parameters of the coefficients of the biochemical reactions and an interaction-rating measure; these are the major specifications of the biochemical reactions. For node (i,j) in the path of the VA trellis, its signal is given in the logic form of

$$A(i+1, j+1) = \text{signal-pathway}(A(i,0), A(i,1),...,A(i,n))$$

where the input comes from the neighboring column to the left. Moreover, the mapping is many-to-one.

The corresponding biochemical reaction form is seen as

$$d/dt\ [A(i+1, j+1)]$$
$$= a(i,0)\ d/dt\ [A(i,0)] + a(i,1)\ d/dt\ [A(i,1)] + \ldots + a(i,n)\ d/dt[A(i,n)]$$

where $a(i,k)$ is the coefficient of the biochemical reactions ($k = 0,1,\ldots, n$).

The enzymes activate the corresponding pathways according to the path weight of the VA trellis. The selected pathway is the optimized path.

The trellis pathway structure is a moleware form of VA [38–40], in which the trellis structure is designed to optimize the path for the signaling process. The structure of biochemical reactions in the underlying pathway trellis is mapped into an informatics form that becomes a kind of application problem of network coding as well as the optimization process of the information flow. The scientific interest lies in searching and problem solving in the parameter space.

As an extension of the above structure, the molecular mechanism is used for the spatial and temporal coding in molecular information processing. The application of molecular coding is much broader than that of the molecular form of VA. The kernel of the idea of encoding and decoding information is to make the best use of the parallelism of molecular information processing. By using the interactions among molecules, the network structure of the signaling processes can enhance the information storage capacity of molecular codes.

3.6 Network Coding in Molecular Informatics

As we see from the above section in which the trellis—a kind of network structure—is discussed from the viewpoint of a moleware-based information processing algorithm, the molecular structure for information representation and the molecular operators for information processing are centered around a networking method. But the network structure discussed at the bioinformatics level of the molecular VA process is oriented to specific problems; the focus is not only on the structure itself. In order to study a generalized system of molecular information processing for communication, it is crucial to build a systematic method of codes by molecular networks. One of the theoretical issues related to molecular codes based on a molecular network structure is to apply network coding to the molecular-information processing systems. This kind of code is a moleware form of network coding [41, 42] (we need also notice the code concept in [43]).

Network coding is defined as a networking process with a network structure, as shown in Figure 1 of [41, p. 371]. Signal S is sent to two nodes, T and U. If T and U are separated, the signals of T and U are directly sent to Y and Z. The correlation of the signals T and U is not involved. After the signals of T and U are sent to node W, the signal of W is sent to X. The signal from X is then sent to Y and Z. The information is interacted and then separately extracted for individual outputs.

First, we study the network coding structure for the interaction of two molecular switches. Then, we extend the case of two molecular switches [35] to a greater number of multiple switches. The schematic representation of the network configuration is a graph whose nodes are connected by the links composed of the

3.6 Network Coding in Molecular Informatics

signaling routes. Multiple pathways are appropriate for this kind of structure. Figure 3.11 shows the network model for coding, where input is S and output is {Y1, Y2}.

Here, notations are explained by molecular informatics concepts:

S: the information source; molecule X is assigned to this node.

Two states of this node are "a" and "a'."

These two signals are used to activate different pathways if molecule X is assigned as an enzyme corresponding to the pathway.

K: kinase

The state of K is denoted as "b."

P: phosphatase.

The state of P is denoted as "b'."

Nodes "p" and "dp" correspond to molecules Y1 and Y2. The virtual links are used to denote the relation between kinase/phosphatase and phosphorylation/dephosphorylation. The related signaling processes are assigned by kinase/phosphatase pathways.

The activation signal to the above pathway is M.

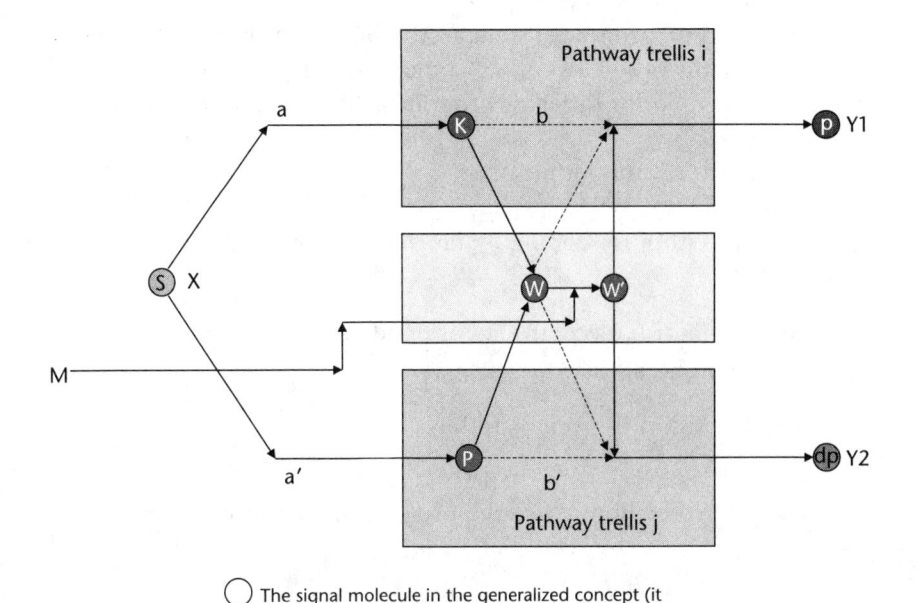

○ The signal molecule in the generalized concept (it can denote a vector for multiple molecules as well).

Figure 3.11 Network structure for moleware coding.

W: molecular switch.

W': molecular switch.

The signaling mechanism for the information flow from W to W' is a multiple-node-to-multiple-node switch structure. The informatics function on W and W' is defined as the setting and selection of the states of phosphorylation/dephosphorylation.

According to the schematic description of the network configuration shown in the above figure, the nodes are defined as molecules.

Data access:

data "q" are stored in the molecular complex/memory Q by a read/write operation, which is the basis of all machine/assembly languages, and updated by the molecular operator as the value q':

q -> q'.

Initialization: previous state of the molecular complexes.

At the K/P node, the kinase/phosphatase pathway for signaling is set to the binary status in bits. At the W node, the detection and selection operations are carried out according to the "program," the "instructions" are given in logic programming form, where the basic information representation is phosphorylation/dephosphorylation for 1/0.

The state of the switch at the W node and the related pathway unit is a kind of built-in structure/scheme.

At W', the "programs" update the binary information in the signaling molecules. The control signals for W' come from outside of the system. The same moleware can be used for different programs. This shows that the network code model is programmable.

Here, Y1/Y2 are outputs.

The algorithm for coding by the above network structures can be summarized as follows:

Algorithm of molecular network coding:
//for one operation //

The algorithm is given as follows:

{
For(j=0 to the number-of-the-program/instruction)
   {
read (instruction(j))
the instruction (set i1,i2,...,iL as the x1,x2,...xL)

```
        activating the signaling-molecules whose number
        /* number-of-the-signaling-molecules */
        according to the instruction (program code)
        to find the index q in molecular memory
        to activate the q indexed molecular complex
            in content-addressing mode.
        activate the pathway of kinase/phosphate
            /* it has the phosphorylation/dephosphorylation function on molecular
            switches/complexes */
        estimate the coefficients of the pathways
            /* according to the criteria of the stability of the pathways defined by the
            parameters */
            }
        for i=0 to the number-of-the-signaling-molecules
            {
        activate the pathway unit (i) through cellular regulation
        regulate the coefficients of the kinase/phosphates pathway
        set value of S(i) /* signaling */
        regulate the coefficients of the kinase/phosphates pathway
        set the state of K(i) and P(i)
        regulate the coefficients of the kinase/phosphates pathway
        set the state of W(i)
        regulate the coefficients of the kinase/phosphates pathway
        read(program/instruction(i,j)) from outside-signal;
            /* according to the read-information */
        to set the state of W'(i)
        to regulate pathways for Y1/Y2 by the control W'
            }
        the output-readout (molecular information in n bit)
        determine the coefficients that activate n number pathways
        }
```

At the informatics level, the channel model and channel design depend on the communication performance of nanobioICT systems, where it is necessary to estimate the parameters of the channel—especially for the MIMD mode of the related communication systems. The reaction-diffusion effect, discrete mode modeling, and approximation are considered methods of molecular communication and biomolecular computing, where the central task is to study how to represent information by biomolecules. Biophysics for bioreactions is related to the source and channel in code design. The dynamic procession of biochemical reactions in cells influences the characteristics of the channel. Of course, the signals are transmitted from sender to receiver. The performance of this communication should be studied using the medium-oriented model and in the principles of biophysics in molecular mechanisms.

In the nanobioworld, light and electronic forms offer alternatives to the chemical form of nanobiocommunication systems by biomolecular pathways in cells. A promising idea is the work using interface technology to connect organic molecular information processing to the inorganic molecular electronic circuits.

3.7 Quadruple Convergence

In today's world, the multidisciplines are "converging" quickly [44, 45]. The molecular network mechanism of cognitive organs, neurons, and signal transduction/cell communication is different from the networking developed in telecommunications. In order to study the life systems that systematically consist of different levels of biological information, we must integrate different biological tools. The convergence of nanoscience, biotechnology, information technology, and cognitive science ("NBIC") is a promising field that requires efforts from multiple disciplines [44]. "The convergence of nanoscience, biotechnology, information technology and cognitive science ("NBIC") provides immense opportunities for the improvement of..." [45]. The study of brain systems at the molecular level is one of the most important endeavors of NBIC, and communication is a factor that links different aspects of NBIC systems.

In the history of modern science, the cognitive and perceptual levels of human behavior have been studied in several fields. Cellular pathways such as G-protein have linkages with neuron growth in biochemistry [4]. The functions of the brain—thought, vision, audition, olfaction, and other biochemical sensing processes—have been intensively studied in cognitive science, physiology, and psychology. Nowadays, in the field of neuroscience, there is an emphasis on establishing a methodology of systematic neuroscience. The ATP and G-proteins are involved in many physiological mechanisms leading to cognitive function. The discoveries from molecular biology help us to gain a systematic understanding of the pathway networks.

From the viewpoint of complex systems, in nanobioICT systems, the levels vary in different mediums: cell, body, and brain in NBIC are located at the crucial crossroads. In order to study NBIC in terms of nanobiomachines, the following three key technologies are needed:

- *Signal transduction in cells.* The signaling pathway model of signal transduction directly describes the biochemical function at the molecular level. The neural signaling model of neuroscience quantitatively explains the high level of biological function of living beings. The most significant linkage is found in the biochemical reactions that connect the pathways of signal transduction and the neuron activities. For example, Kuroda et al.'s research on computer simulation [46] is beneficial for analyzing the cellular signaling model in which the LTP mechanism is involved.

  ATP is a kind of energy material in cells that acts as the fuel for nanobiomachines. The K-ATP channel, which refers to the signaling mechanism of activating the ATP-sensitive potassium, is an important factor for the degenerative mechanism of dopaminerigic (DA) mid-brain neurons. This shows the biomolecular flow between the ATP proteins in the cells and neurons in the brain [44].

- *Signaling in neurons*. The molecular mechanism of neural signaling takes a biochemical reaction form. The mapping from the pathway of biochemical reactions to the neuron structure of signaling explains the biological function of neuron cells.

  The corresponding biological functions of these pathways are related to the neuron signaling mechanism. The neuron signal models are normally oriented to signal processing, that is, the stimuli and response are quantitatively modeled for informatics purposes, which is also the basis of neural codes.

  Between the two types of codes, molecular and neural, a quantitative relationship exists in terms of biological function. The identification of this relation is a kind of data-driven process. The information space between the two sides is the framework of algorithm design.

  The scientific issue for the two kinds of codes is to identify the MIMO structure of the system whose input is the set of molecular codes and output is the set of neural codes.

  In the traditional methodology of biology, the exact location of neural signals can be precisely detected and thus a quantitative measurement can be obtained. The "geographical" information is almost completely understood for research purposes. The logical relation is normally studied by models based on computer calculation. Artificial neural networks were initially oriented to information processing paradigms. However, now the much broader field of neuroscience encompasses a systematic study of all aspects of the neural systems, even including the neurons of the brain. At the same time, a long tradition in molecular biology continues to study the logical relationship among the proteins in cells in order to reveal the dependence and independence among proteins and the underlying mechanism of the cellular signaling. Moreover, determining the exact location of proteins requires the advanced technology of molecular imaging [47].

- *Molecular imaging technology for tracing the information flow at the nanolevel*. As Mark R. Philips commented [48], "Until recently...the 'what' and 'when' of signaling...added the dimension of 'where' to the study of cellular signaling." This work was a milestone in nanobioICT study. The landmark FRET technology is crucial for observing the bio-information flow of GTPases in cells on line, in real time, and at the physical spatial locations. The knowledge obtained from comparison of molecular signals and neural signals can contribute to systematic neuroscience modeling. Meanwhile, at the neural level, samples of neural signals are helpful references for the comparative study of joint moleware-neuron modeling. The nanobiomachine synergistically works in the mode of parallel distributed biochemical information processing at the molecular level, in which the information flow shows a kind of MIMO structure. A preliminary task in studying this information flow is to observe and measure the molecular signal's exact location in the cell with sufficient speed. The proteins in cells are mobile and their movement is dynamic. The form of cell communication is the biochemistry basis of information flow in a nanobiomachine. The vector of the signaling molecule that is assigned to represent the information in a molecular processing element (PE) for nanobioICT systems can be obtained in the current technology:

Measurement of signaling molecule (physical location in three dimensions)

where the measurement takes different forms according to the detection technology adopted; for example, the fluorescence strength of the proteins is used in FRET.

Using an engineering methodology that satisfies the scope of the above conditions is the key to success in joint modeling. The concrete technologies involve biochemical detection and measurement, biomolecular signaling processing, bioinformatics theory, and experimental molecular biology. The informatics kernel is to identify the logical graph that connects the signaling pathway networks obtained from concentration data and the neuron structure determined by the neural signals. The bioelectrical signals of neuron cells, such as stimuli and response, are modeled in terms of a cybernetics-inspired system model. The protein-level factors need to be embedded in the model and recognized when hypothesis-verification schemes are used for the modeling procedure. The intercell communication is considered within the cell. The cellular pathway networks that have been reported are the basis of the neural-function-oriented pathway models. These models are being studied by imaging technology. The location of the activating signaling proteins and their influence on the neural signal at the cell are the objects of such modeling. These relations are described by a graph where the edges are bidirectional links. The signals from neuron to protein and the signals from protein to neuron correspond to the links, and these signals are also connected to the parameters of the neuron-level models and moleware-level models [49].

For example, vision is one of the important functions of biological cognitive systems where attention has been given to the photo-receptors and phosphorylation in invertebrate cells. In order to find the brain signaling that may give hints to neurodegenerative diseases, imaging technologies such as fMRI, NIRS, MEG, and PET are used to detect the firing activities of neurons in the brain. How to find the molecular bioinformatics relationship between images of neural activities in the brain and signals of molecules in cells is a promising theme in NBIC's converging technology.

A partial process of the cellular signaling can be observed in experiments, and modeling by computer simulation is being used to quantitatively describe the cell based on the current technology. However, complete knowledge of the whole brain behavior at the level of complex systems would depend on a systematic understanding of neuron functions and the entire set of cellular pathway networks. As the concept "cognome" in NBIC implies, many problems remain open. A deep understanding of the causal relations among the biological systems also requires a logical monf-del and reasoning technology, which is presented in Chapter 7. In the pathway network, the molecular objects of the graph that describe the pathway network are specific, and their number is not so big; accordingly, the graph can be reconstructed by the identification method. The topic of pathway network reconstruction deserves further study.[1] In such research, challenging tasks would include graph-based pathway reconstruction by identification, experimental verification of bench-work, and exploration of the robustness mechanism of fault-tolerance in nanobioICT systems.

References

[1] Ruggiero, C., et al., "The Nanobioworld," *IEEE Trans. on Nanobioscience*, Vol. 1, No.1, March 2002, pp. 1–3.

[2] Stix, G., "Little Big Science," *Scientific American*, Vol. 285, 2001, pp. 32–37.

[3] http://www.nict.go.jp/overview/about/vision.html.

[4] Etienne-Manneville, S., and A. Hall, "Rho GTPases in Cell Biology," *Nature*, Vol. 420, 2002, pp. 629–635.

[5] Kaibuchi, K., S. Kuroda, and M. Amano, "Regulation of the Cytoskeleton and Cell Adhesion by the Rho Family GTPases in Mammalian Cells," *Annu. Rev. Biochem*, Vol. 68, 1999, pp. 459–486.

[6] Kawano, Y., et al., "Phosphorylation of Myosin-Binding Subunit (MBS) of Myosin Phosphatase by Rho-kinase" In Vivo, *The Journal of Cell Biology*, Vol. 147, 1999, pp. 1023–1037.

[7] Hafen, E., "Kinase and Phosphatases—A Marriage Is Consummated," *Science*, Vol. 280, 1998, pp. 1212–1213.

[8] Helmreich, E. J. M., *The Biochemistry of Cell Signalling*, New York: Oxford University Press, 2001.

[9] Scott, J. D., and T. Pawson, "Cell Communication: The Inside Story," *Scientific American*, Vol. 282, 2000, pp. 54–61.

[10] Gershenfeld, N., *The Physics of Information Technology*, Cambridge, U.K.: Cambridge University Press, 2000.

[11] Blahut, R. E., *Principles and Practice of Information Theory*, Reading, MA: Addison-Wesley, 1987.

[12] Lipton, R. J., "DNA Solution of Hard Computational Problems," *Science*, Vol. 268, April 1995, pp. 542–545.

[13] Adleman, L. M., "Molecular Computation of Solutions to Combinatorial Problems," *Science*, Vol. 266, November 1994, pp. 1021–1024.

[14] Shannon, C., "A Mathematical Theory of Communication," *Bell System Technical Journal*, Vol. 27, July and October 1948, pp. 379–423 and 623–656.

[15] Zimmermann, K. H., "On Applying Molecular Computation to Binary Linear Codes," *IEEE Trans. on Information Theory*, Vol. 48, No. 2, February 2002, pp. 505–510.

[16] Mauri, G., and C. Ferretti, "Word Design for Molecular Computing: A Survey," J. Chen and J. Reif, (eds.) *DNA 9*, LNCS 2943, 2004, Springer-Verlag Berlin, Heidelberg, 2004, pp. 37–47.

[17] Liu, J. Q., and K. Shimohara, "A Biomolecular Computing Method Based on Rho Family GTPases," *IEEE Trans. on Nanobioscience*, Vol. 2, No. 2, June 2003, pp. 58–62.

[18] Liu, F., et al., "Direct Protein-Protein Coupling Enables Cross-Talk Between Dopamine D5 and γ-aminobutyric Acid A Receptors," *Nature*, Vol. 403, 2000, pp. 274–280.

---

1. Conrad's early vision of molecular computers gives us valuable hints. Figure 9 in [50] describes the information flow and some control operations in his design of a molecular computer for the future.

[19] Tarricone, C., et al., "The Structural Basis of Arfaptin-Mediated Cross-Talk Between Rac and Arf Signalling Pathways," *Nature*, Vol. 411, 2001, pp. 215–219.

[20] Digicaylioglu, M., and S. A. Lipton, "Erythropoietin-Mediated Neuroprotection Involves Cross-Talk Between Jak2 and NF-κB Signaling Cascades," *Nature*, Vol. 412, 2001, pp. 641–647.

[21] Katoh, H., and M. Negishi, "RhoG Activates Rac1 by Direct Interaction With the Dock180-Binding Protein Elmo," *Nature*, Vol. 424, 2003, pp. 461–464.

[22] Yoo, A. S., C. Bais, and I. Greenwald, "Crosstalk Between the EGFR and LIN-12/Notch Pathways in C. Elegans Vulval Development," *Science*, Vol. 303, 2004, pp. 663–666.

[23] Masip, L., et al., "An Engineered Pathway for the Formation of Protein Disulfide Bonds," *Science*, Vol. 303, 2004, pp. 1185–1189.

[24] Li, S., et al., "A Map of the Interactome Network of the Metazoan C. Elegans," *Science*, Vol. 303, January 2004, pp. 540–543.

[25] Han, J. D. J., et al., "Evidence for Dynamically Organized Modularity in the Yeast Protein-Protein Interaction Network," *Nature*, Vol. 430, July 2004, pp. 88–93.

[26] Mochizuki, N., et al., "Spatio-Temporal Images of Growth-Factor-Induced Activation of Ras and Rap1," *Nature*, Vol. 411, June 2001, pp. 1065–1068.

[27] http://netresearch.ics.uci.edu/nanotech/molecular_communication.html.

[28] http://www-karc.nict.go.jp.

[29] Nakano, T., et al., "Molecular Communication for Nanomachines Using Intercellular Calcium Signaling," *5th IEEE Conference on Nanotechnology*, July 2005, pp. 632–635.

[30] Hiyama, S., et al., "Molecular Communication," *Proc. of the 2005 NSTI Nanotechnology Conference*, 2005, http://netresearch.ics.uci.edu/nanotech/Nanotech-full-v2.1.pdf.

[31] Weiss, R., and T. F. Knight, Jr., "Engineered Communications for Microbial Robotics," in DNA6, *Lecture Notes in Computer Science*, Vol. 2054, A. Condon, and G. Rozenberg, (eds.), Berlin, Heidelberg: Springer-Verlag, 2001, pp. 1–16.

[32] Basu, S., et al., "A Synthetic Multicellular System for Programmed Pattern Formation," *Nature*, Vol. 434, 2005, pp. 1130–1134.

[33] Kollmann, M., et al., "Design Principles of a Bacterial Signaling Network," *Nature*, Vol. 438, December 2005, pp. 504–507.

[34] Tu, B. P., et al., "Logic of the Yeast Metabolic Cycle: Temporal Compartmentalization of Cellular Processes," *Nature*, Vol. 310, November 2005, pp. 1152–1158.

[35] Alberts, B., et al., *Cell*, 4th ed., New York: Garland, 2002.

[36] Ocequera-Yanez, F., et al., "Ect2 and MgcRacGAP Regulate the Activation and Function of Cdc42 in Mitosis," *J. Cell Biol.*, Vol. 168, 2005, pp. 221–232.

[37] http://www-karc.nict.go.jp/d332/CellMagic/index-J.html.

[38] Viterbi, A. J., "Error Bounds for Convolutional Codes and an Asymptotically Optimum Decoding Algorithm," *IEEE Trans. on Information Theory*, Vol. 13, No. 2, April 1967, pp. 260–269. (The Viterbi decoding algorithm is described in section IV.)

[39] Forney, G. D., "The Viterbi Algorithm," *Proceedings of the IEEE*, Vol. 61, No. 3, March 1973, pp. 268–278.

[40] Jarvinen, T., et al., "Systematic Approach for Path Metric Access in Viterbi Decoders," *IEEE Trans. on Communications*, Vol. 53, No. 5, May 2005, pp. 755–759.

[41] Li, S.-Y. R., R. W. Yeung, and N. Cai, "Linear Network Coding," *IEEE Trans. on Information Theory*, February 2003, pp. 371–381.

[42] Koetter, R., and M. Medard, "An Algebraic Approach to Network Coding," *IEEE Trans. on Networking*, Vol. 11, No. 5, October 2003, pp. 782–795.

[43] Berstel, J., *Theory of Codes*, New York: Academic Press, 1985.

[44] http://www.infocastinc.com/nbic/nbichome.htm.

[45] http://www.wtec.org/ConvergingTechnologies/Report/NBIC_report.pdf.

[46] Kuroda, S., N. Schweighofer, and M. Kawato, "Exploration of Signal Transduction Pathways in Cerebellar Long-Term Depression by Kinetic Simulation," *J. Neurosci.*, Vol. 21, No. 15, 2001, pp. 5693–5702.

[47] Liss, B., et al., "K-ATP Channels Promote the Differential Degeneration of Dopaminergic Midbrain Neurons," *Nature Neuroscience*, Vol. 8, No. 12, December 2005, pp. 1742–1751.

[48] Philips, M. R., "Imaging Signaling Transduction in Living Cells with Fluorescent Proteins," *Sci. STKE*, Vol. 2005, No. 314, December 2005.

[49] http://w3.ouhsc.edu/biochem/matsumoto.htm.

[50] Conrad, M., "On Design Principles for a Molecular Computer," *Communications of the ACM*, Vol. 28, No. 5, May 1985, pp. 464–480.

CHAPTER 4
Computing by Biomoleware: Diverse Methods from Diversified Materials

All electronic computers using semiconductors adopt the same standards of von Neumann architecture. Both fortunately and unfortunately, biomolecular computers tend to be different. The chemical features of biomolecules vary so much that various types of biomoleware systems have been set up for different purposes. The materials used for molecular computing are diverse, since the field of biomolecular computing is now only in its infancy. The physical biochemistry principle of moleware must be fully investigated to develop the methodology and guidance needed for molecular computing. After the fiftieth anniversary of the discovery of the DNA helix and a decade after Adleman's innovation of an experimental DNA computer, 2005 turned out to be "the year of physics," marking the hundredth anniversary of the publication of Einstein's first paper on his theory that reshaped our understanding of physics in the twentieth century [1]. From these three events, we should reflect more deeply on the notion of scientific inspiration. Currently, the precise measurement technology available for DNA molecular structure makes it possible for us to discover how DNA molecules are spatially located, which is the key to understanding any molecular interactions and not just those related to DNA. The creation of even a very preliminary form of a prototype experimental "DNA computer" by harnessing DNA molecules is indeed a daring enterprise, but such an advance would tell us that biomolecules can somehow contribute to computer manufacturing. The two distant "stars"—biology and the computer—have been bound together to form a harmonious pair in the constellation of biomolecular computing, in which the physical biochemistry is the driving heart. Stix once regarded Einstein as the first nanoscientist due to his work showing that the diameter of a sugar molecule can be calculated as 1 nm (one nanometer). With the advance of nanobiotechnology that makes centuries' dreams into reality and wild targets more reasonable in the nanobioworld, automated biochemical signaling processes at the nano level of moleware have become feasible. The nano-ship has been launched on the voyage to the future, at a time when we do not know how long the compass of Moore's law will continue to work for semiconductor technology. The counterpart to Moore's law in molecular biology is the Dickerson's formula [2] for protein structures. At least the star is shining when the compass is asleep. Physical biochemistry serves as a Polaris for guiding our direction.

The central idea of this book, that is, the holy grail of our exploration is to study how to make a biomolecular computer in the form of a nanobiomachine by using nanobiotechnology. Therefore, in Section 4.1, we start the discussion from the interface of nanobiomachines and biomolecular computers, which will bridge noncomputation and computation. Nowadays, various types of nanobiotechnology tools are available for us to design, test, and implement nanobiomachines. In addition to bottom-up nanotechnology and top-down biotechnology, mesoscopic paradigms and methodology are the key points we consider in this chapter. In nature, certain kinds of nanobiomachines existed even many years before the appearance of human beings. This evidence gives us encouragement and motivation and even provides us component materials for building artificial nanobiosystems. For example, the cell is one of the best nanobiomachines made by the blind watchmaker. For building computational systems based on a nanobiomachinery mechanism, it is necessary to extract efficient biochemical mechanisms for representing and storing information among moleware, which will be discussed in section 4.2. Sections 4.3 and 4.4 summarize molecular computing by nucleic acids and microbes, respectively. These two sections focus on the analysis of concrete molecular computing methods in terms of biomaterials, computing processes, and programming. With respect to exploring how to employ nanobiotechnology within the framework of nanobiomachines for computational tasks, the medium and structure employed are two important aspects of molecular computing, and these are studied in the general domain of unconventional information processing.

4.1 How to Build an Engineered Computational Nanobiosystem: Inspiration from Existing Nanobiomachines in Nature

4.1.1 Nanobioworld Becomes Observable with the Help of Innovative Measurement Technology: Schrödinger's Cat Is at the Door

The successful observations of single electrons and electron clouds have greatly extended our reach to the world of miniaturization [3–11], making "pervasive computing" capable of facing the challenges of in situ nanoprocessing [12]. Even though high-resolution STM imaging can display individual DNA molecules that can be easily recognized in the background of a helix [13], the nanobioworld remains a mysterious kingdom for human beings, mainly because of the limitations in detection technology. As we know, in the postindustrial society, information is absolutely important and seems to be ubiquitous. Analogous to the idea of atom and bit in the book *Being Digital* by Nicholas Negroponte [14], in which atom refers to real substance and bit to information, moleware is both the medium for storing information and the machine for handling information. The unity of biochemical signals and information in moleware is a feature peculiar to molecular computers, obviously different from electronic ones. Molecules are the basic units of information storage in nanobiomachines.

In this section, we discuss the methodology of how to build an engineered computational nanobiosystem inspired by existing nanobiomachines in nature. Although materials science and manufacturing technology nowadays can provide strong bottom-up tools such as scan tuneling microscope (STM) to artificially synthesize nano-level structures and top-down tools such as the so-called partial "human builder" to manufacture artificial materials as alternatives to human body parts, we prefer to study how nano-level mechanisms already existing in nature deliver signals, which may lead to such mechanisms being employed in building artificial systems. The challenge from the nanobioworld is not limited to the obvious phenomenon of its small size; its significance should be viewed as more or less a kind of "little big science" [15]. The scientific inspiration arising from recently unveiled principles and physical processes in "little stuff" can almost be viewed as giddy fantasy. This inspiration has spawned many pursuits: "computing beyond binary" by "phits" defined as phase digits where phase refers to electron spin; "The Einsteinian Pachinko Machine" [16]; wonderful carbon nanotube [17, 18]; cellular pathways of protein interactions in cells [19–23] for proteins, in which GTPases influence major aspects of cells and can be extracted even from plants [24]; the quantum dot from molecular electronics [25] to the quantum dot in biology; and various detection and observation tools such as mass spectrometry for proteins, fMRI, PET, MEG, NIRS, and fluorescence resonance energy transfer (FRET); and living cell imaging. Cellular nuclear receptors are important for using system biology to create drugs to cure chronic diseases such as diabetes, in which the proteome of human nucleoli is studied through in vivo fluorescent imaging [26] and adipokin (for example, visfatin [27, 28]). Among these new steps toward progress, single-molecular-level nanotechnolgy is at the frontier.

With the great efforts of material scientists, the energy of nano-structures will be released for powerful machines, as Kaushik Bhattacharya and Richard D. James point out in [29]. Also, with the semiconductor industry being eager to find technologically practical nano-size solutions to circuitry manufacturing [30], the nanotechnology for DNA structure assembly may provide more extended functions than we had expected [31].

As we know, the machine is one of the most commonly used artificial systems, and its history can be traced to the times before the industrial revolution of the 1800s and the information era of the 1900s. Even in the early stages of human civilization, material making with stone, bronze, and steel sustained the basis of mainstream society. Inspired by nature, artificial things have been continuously created over the long history of human beings. Nanomaterials show us nature-rooted features in the minimization of machines. The size of machines, as Feynman foresaw in the 1950s, will become extremely small, and the final detectable physical boundary around an atom can only be reached through new nanotechnology. Nanobiomachines are being built based on functional requirements for speed, energy consumption, and information capacity. Actually, nature is an excellent teacher for us. There will surely be revolutions in energy consumption for nanobiomachines, which is a crucial point for mechanical machines, and information granularity for computing as well as the information processing of computers. An extreme low cost of energy and a huge amount of high-speed information processing capacity can be achieved at nanoware level, which promises a more positive impact on society than the previous two industrial revolutions.

4.1.2 Seeking a Movable Nanobiomachine: Postman in Moleware

Steam locomotion in the industrial revolution opened an era of using massively greater power. This development has been augmented by various forms of energy developed by human beings, and consequently many kinds of machines are now running throughout the world. Even far away on Mars, *opportunity* and *spirit* rovers proved functional. Let us draw our vision back and look at the world in the reverse direction—the small size. Molecular motion is one of the most well-know features of nanobiomachines. Molecular motors can easily arouse a vivid image of nanobiomachines in our mind, but this is not the whole story about nanobiomachines. We have to study self-assembly capability for supra-molecular structures [32, 33] and energy-centered signaling mechanisms provided by edge-cutting nanotechnologies. In most current cases, molecular motors are made by modified biomaterials that are mainly microbes under design control [34–37].

Actually, a greater mystery behind the movement of moleware is the mechanism for delivering biochemical information and related messages in nanobiosystems. From the viewpoint of investigating biochemical signaling, it is essential to extract this information and communication mechanism.

The kernel of a nanobiomachine consists of three major parts: action, control, and information processing. The word nanobiotaxis, as we use it, refers to the movement of a nanobiosystem that can adaptively respond to signals from the outside environment and belongs to the action part. Here, we use the term nanobiotaxis rather than chemotaxis which means the movement often observed at the microbial level when the surrounding chemicals change so that movement only occurs in the domain of the nanobiosystem. Obviously, chemotaxis at the nanosize is a kind of nanobiotaxis that involves the aspects of light, bioelectricity, chemicals (with features in thermodynamics and mechanics), motion (in kinetics), and steering (in cybernetics). Molecular manufacturing technology can also synthesize DNA motors, to which DNA is fed as fuel.

A molecular motor at microbial level is a kind of nanodevice built on the principle of chemotaxis. Movable nanobiomachines, including molecular motors, not only run like a car but also transport chemicals to deliver messages in the form of chemical signals. They are really postmen in cells. When nanopostmen go somewhere, they transport proteins used as information in cells. For better understanding the function of postmen, we begin this topic from molecular motors. Actin filament, one of the materials used for constructing molecular motors, is an example often cited. Its biochemical features are frequently discussed in current topics, especially in structural biology. Three proteins—myosin, kinesin, and dynein—can motivate the activities of chemotaxis. Inspired by the kinetics of filament movement, one kind of molecular motor has been built using microbial material. For example, the molecular motor by dynein ATPase is in the form of a microtubule of *Chlamydomonas reinhardtii* [35] that can perform gymnastic action [36] (that is, when it moves the shape of its body it looks like a wave). In this case, dynein—one of the three kinds of proteins mentioned above—is used. The ATPase motor was highlighted physiology and medicine as well as by its 3D visualization in cells by Yoshida et al. [37]. GTPase can also control the chemotaxis in some bacteria.

4.1 How to Build an Engineered Computational Nanobiosystem

From the molecular motor we assume we can study how to design a communication scheme for integrating molecular computing and a nanobiomachine. The communication mechanism of a molecular motor and molecular computing is illustrated in Figure 4.1 as an example. The two locations corresponding to two processing elements (PEs) are denoted as A and B. At A, the molecular motor is first fed by ATP, and then the signaling molecules encoding the information for molecular computing are loaded on it. The molecular motor carries the signaling molecules to location B, where it delivers the molecules as messages that activate the reactions for molecular computing at B. Running the molecular motor requires motorprotein and energy (ATP or ATPase). In the PE unit, motor-proteins and energy can be controlled by reactions. At location B, molecular computing implemented by proteins can be programmed for specific sequences of reactions. The kinase computing presented in Chapter 6 controls and feeds this molecular motor. The phosphorylation and dephosphorylation processes are controlled for molecular computing and are programmed in the manner of biology to regulate the quantity of ATP/ADP. The kinases and phosphatases determine the phosphorylation and dephosphorylation states of the signaling molecules, respectively. The combinatorial forms of phosphorylation and dephosphorylation consist of binary words for representing data and the combinatorial forms of kinase and phosphatase consist of binary words for representing instruction code when phosphorylation and dephosphorylation are used to denote 1 and 0, respectively, and kinase and phosphatase are used to denote 1 and 0, respectively. The specially arranged signaling molecules transported by the molecular motor are part of the input to the molecular computer. Within an MIMD architecture, where the molecular motor runs between different PEs of molecular computing, the PEs of molecular computing can cooperate for the same computing task through exchanging messages using molecular motors. Multiple instructions are reflected in different signaling proteins and multiple data flow is formed through the motions of multiple molecular motors.

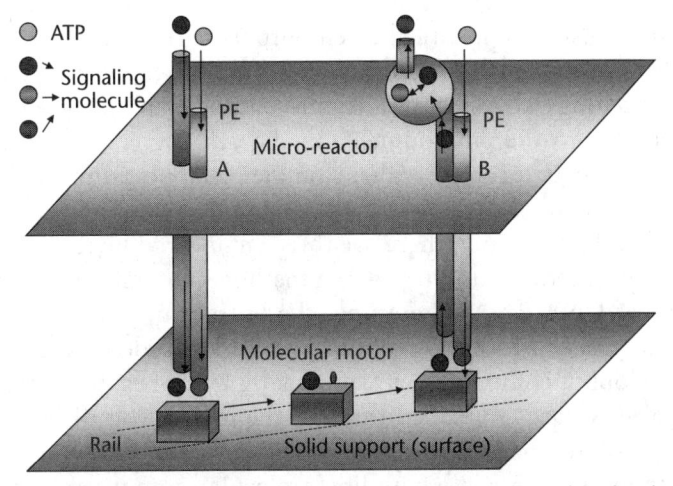

Figure 4.1 A diagram of communication between molecular motor and molecular computer.

The information flow for the molecular motors is:

Controlling kinases/phosphatases → ATP → motion

To succeed in the task of controlling the information flow as described above, we need the biochemical reactions of PEs, which are determined by the cellular communication in an assay of cell culture. This is because molecular biology tells us that cell communication is one of the key factors in biological living systems and that it is crucial for "cellmachines" ranging from microbes to multicellular organisms. In this sense, communications gave birth to related networks, and the network algorithms studied in information theory, whose application to the network moleware system is discussed in Chapter 3, are naturally oriented to nanobioICT. A biological environment is always dynamic for cells. Because chemical signals such as concentration change under dynamic environments, it is essential to design a kind of robust control scheme to sustain the stable states of nanobiomachines. The engineered method from nanotechnology is introduced in the framework of nanobiomachines based on naturally existing nanobio-processes in order to design a signaling mechanism for information processing and even computation.

4.1.3 Methodology Learned from the Cell and Beyond

Within the cell, various types of organic molecules are synergized to produce complex collective behaviors. Within the cell, various biochemical reactions occur. A pathway can be explained as the biochemical reactions of molecules in a cell where their input and output are clearly defined. All of the cellular pathways biologically interact within a complex network. More and more knowledge on these networks is being acquired from computational molecular biology and proteome bioinformatics. With the signaling network mechanism, certain types of architectures are expected to be developed for efficient nanobiocomputing processes owing to the advantage of the networks' parallelism. The most important difference between the biological life processes and biological computational processes is that the former may produce certain different results according the conditions, while the latter must have identical given outputs corresponding to the same input even if their environment has changed. This requires engineered technology for biological computing processes because the output of biological processes is nondeterministic due to the environment. In the cell, signaling pathways in the form of networks can be used to realize complex mapping through engineered pathway technology. A brief diagram for the description of this idea is given in Figure 4.2.

Generally speaking, there are three engineered methods—bottom-up method, mesoscopic method, and top-down method—in nanobiotechnology. In most paradigms of DNA/RNA computing, the bottom-up method is commonly adopted in the way of self-assembly of nucleic acids. Top-down methodology is beneficial for developing computational systems by microbes and organisms where signaling molecules are controlled through the interaction of DNAs and proteins. The work presented here requires mesoscopic methodology, since our purpose is to use nanobiomachines for building molecular computers. The mesoscopic strategy is at the heart of engineering designs that can compensate state description and

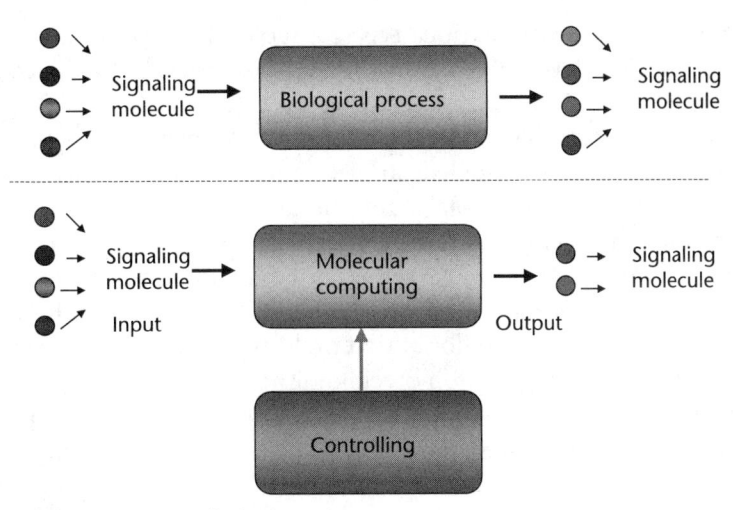

Figure 4.2 Diagram of controlled computing.

logic functioning, which is the main factor in the engineered nanobiomachines studied here.

It might be reasonable to consider molecular manufacturing an ad hoc approach to handling different tasks in the case of nanobiomachines made by DNA, RNA, protein, and other molecules; however, the biochemical reactions that are designed and controlled operate under the same laws of physical chemistry. The materials selected from macro-level living systems, such as microbial material, tend to be used in a top-down style. In contrast to this, a manufacturing process starting from a single molecule is a typical application of a bottom-up style. In the case of using a nanobiomachine for computing, the methodology becomes a double-sided style, that is, based on the framework of the nanobiosystem, the control schemes are carried out to regulate the signaling process toward the designated information flow. For example, in nanobiosignaling processes where an MDCK epithelial cell is used to obtain cellular signaling proteins in an assay, operations in vitro or in vivo for these proteins need to be designed according to the proteins' interaction attributes. Accordingly, multitest tubes are set up in parallel for these operations when we use single-protein detection technology for readout (e.g., computing by GTPase where we use FRET for readout). Currently, a tailor-made manner is common in the field of molecular computing; however, the final stage of the molecular computer's development should be automated machinery assisted by nanobiomachines.

4.2 Information Processing in Artificial Nanobiosystems: An Odyssey Beyond the Blind Watchmaker

Spitzer and Sejnowski said, "biochemical reactions within cells can be used for computation" [38]. The implementation of biochemical computational processes needs at least two component materials: molecules representing information and enzymes activating the corresponding reactions, prepared under the related condition of biochemistry. There are various types of molecules used to carry out computation at the molecular level [39–92]. In this section, we only discuss computing systems by biomolecules. Molecular electronics and quantum computing, which also use moleware at the molecular level, are two kinds of promising molecular information processing technologies; molecular electronics is briefly discussed in Chapter 8. As a foundation for the biomolecular computing explained in succeeding sections of this chapter, this section presents biochemical reactions in an artificial nanobiosystem as the main method of carrying out computation. In this way, information is represented by molecules and the computing process is realized by biochemical reactions. There can be no doubt that molecular information processing paradigms have been enhanced by nanobiotechnology.

In Adleman's pioneering experimental work [41], nodes and edges are represented by DNA molecular sequences, and the required results of computing are also represented by DNA molecular sequences. From a more generalized viewpoint, let us consider a certain mapping between one information set as the input and another information set as the output. When the input and the output are represented by molecules, it is possible to activate biochemical reactions equivalent to a mapping. Thus, an information processing procedure by a biochemical molecular system has been set up conceptually. In this case, information is explicitly represented for users. We know what each state of the molecules mean because the information is directly stored in the medium of molecules and the operations can be exerted directly on the molecules. This is in contrast to a conventional electronic computer, since in molecular computers, the instructions are "written" in the form of both biochemical reactions that correspond to operators and the states of the molecules that correspond to the data. Up to now, such tailor-made wet programs have been commonly used. Programming the molecules is defined as the task of assigning biochemical reactions to the molecules for computation.

As a first step, we need to define a table of the relationships among the molecules or molecular sequences used for representing biochemical reactions and corresponding information. These molecules or molecular sequences, selected according to their semantics, should be tested in benchwork. In electronic computers, a huge amount of time is used to translate or interpret high-level languages such as BASIC, fortran, pascal, C, and C++ into the assembly language, depending on the machine hardware or machine language, in the form of binary-bit electrical signals. This time spent for information transformation will be saved in molecular computing. The biochemical reactions can be realized in parallel where many molecules are placed in test tubes or assays and thus the biochemical parallel information processing can be carried out without any routers. As we know, routing and the related monitoring for networks are inevitable processes in conventional parallel computers. In molecular computing, the protocols of biochemical reactions are designed according to logical relations of related molecules. The basic molecular materials we use to build a nanobioinformation processing system obey the physical chemistry laws of molecular biology. Whenever two molecules are used for molecular computing in an artificial nanobiosystem, the capability of a biochemical reaction is judged according to the conformal characteristics of molecules in structural biology. In the case of DNA, two DNA sequences in a Watson-Crick complementary arrangement can be bound into double-strand DNA sequences by means of corresponding enzymes and biochemical reactions, since a hydrogen bond can bind two complementary DNA sequences. This is the basic mechanism of the structure of a DNA helix, and it is regarded as the fundamental physical chemistry of DNA computers. Many DNA molecules or DNA sequences can be placed in test tubes for ligase reactions at the same time. These molecules aggregate according to natural rules of hydrogen bonding. Although moleware for information processing basically obeys natural law through evolution, the biochemical molecular information in moleware and the information we want to represent in moleware are not identical in their essence. From the viewpoint of the biological concept that the metaphor of the blind watchmaker implies, computational nanobiosystems behave differently from natural existing nanobiosystems. Specifically an engineered-controlled mechanism, the biological mechanisms, or process unrelated to computation should be removed or inactivated. The advantages of moleware obtained from an information-processing nanobiosystem can be summarized as:

- Direct representation of information by molecules;
- Direct operation of information on molecules;
- Implicit parallelism in molecules.

The boundary between software and hardware in conventional computers does not exist in unconventional molecular computers. The moleware is the medium of information and the object of operations because the biochemical reactions are dependent on the molecules designated for information processing.

In any case, we have to figure out the potential moleware offered by nanobiotechnology and the possible processes we can harness for our designed targets. It is easy to understand that in cells there are many signaling pathways. The protein interaction network influences the most important aspects of cells in biological functions. The life of the cell is sustained by the cell itself, which works as a self-driven nanobiomachine. For the purpose of obtaining an engineered information processing mechanism in a nanobiosystem, we have to keep the desired parts and eliminate the unwanted parts of the biological materials in order to implement the biochemical information processing task for the target specified in advance. Controlling biochemical mechanisms through engineering will require going beyond the limitations of naturally existing nanobiosystems. The engineering work of achieving an information processing mechanism consists of three major parts:

1. Molecular memory built by molecular complex;
2. Molecular clock for moleware;
3. codes in moleware.

From the viewpoint of information processing, memory and code are necessary. If we design a synchronous nanobiosystem, we need a molecular clock. However, this is not necessary in designing an asynchronous system. Functions such as ALU and logic gates have been reported to be successfully realized by DNA computing systems. The kernel structure of the moleware for information processing appears as an aggregation of molecular complexes. In this sense, the entire system seems like a huge memory. The controls of information processing operate directly on molecules in handling the information. Consequently, the task of programming moleware is analogous to the reading and writing operation on memory or rewriting in mathematics. Computational moleware behaves in the same way for both biomolecular computers and molecular electronics circuits.

4.2.1 Molecular Complex as Memory—Memorizing Instead of Braining

A supra-molecular structure achieved by self-assembly is feasible for building nanodevices by DNA and other organic molecules. The size of molecular memory varies according to the materials used. In this section, molecular memory refers to the moleware built by biomolecules and used to store information. Baum [39] points out that a millimole of DNA sequences that is 200 base long can store 10^{20} words and can be used to make a DNA memory. This memory can surpass the capacity of the human brain which is estimated to be 10^{15} bytes. The DNA structures in 2D are used to expresses various types of combinatorial symbols of {A,T,C,G}. After defining specific mapping, four classes of formal languages of the Chomsky hierarchy—regular language set, context-free language set, context-sensitive language set, and recursively enumerable language set—corresponding to universal computation by a Turing machine can be generated [40]. The formal languages derived from molecules will be discussed in Chapter 5 from the viewpoint of theoretical computer science. In this section, we focus on the information issues of a molecular complex, and the functions of storing and accessing information are the main topics here. A molecular complex in biochemistry refers to a compound that consists of molecules and that can be sustained in certain structures under properly prepared conditions. In DNA computing, DNA complexes are used to make the DNA molecular memory. In this section, we use GEF complexes, GAP complexes, kinase/phosphatase complexes, GTPase complexes, and labeling-molecular complexes for designing molecular complexes.

On one side of the DNA world, the work on two-dimensional DNA structures is a milestone in the application of nanotechnology to DNA computing [86]. The rich features of the synthesized DNA structure [86, 87] based on the naturally occurring DNA helix make DNA computing even more powerful. In their hands, interesting topologically shaped spatial structures were created to realize unexpected molecular complex patterns for information representation, storage, and access. The different ways of accessing can provide different combinatorial forms of information, and the derived information capacity will differ as well.

After the spatial structure of the DNA helix is physically synthesized (see Figure 4.3), the interpretation of its information will depend only on the protocols in theory. Two strands of nucleic acid sequences are projected into two strings. As building blocks, subsequences extracted from the strings are combined to construct a one-strand string. Different protocols for these conceptual operations produce different interpretations of information.

As Figure 4.4 shows, the DNA memory built by the double helix structure has been demonstrated as a promising device for DNA computing. The corresponding structural manufacturing of the DNA sequences is mainly manual. In laboratories, building these DNA structures is labor-intensive. Recently, DNA recombinant technology has emerged as an aid for this task, but this way of writing information on

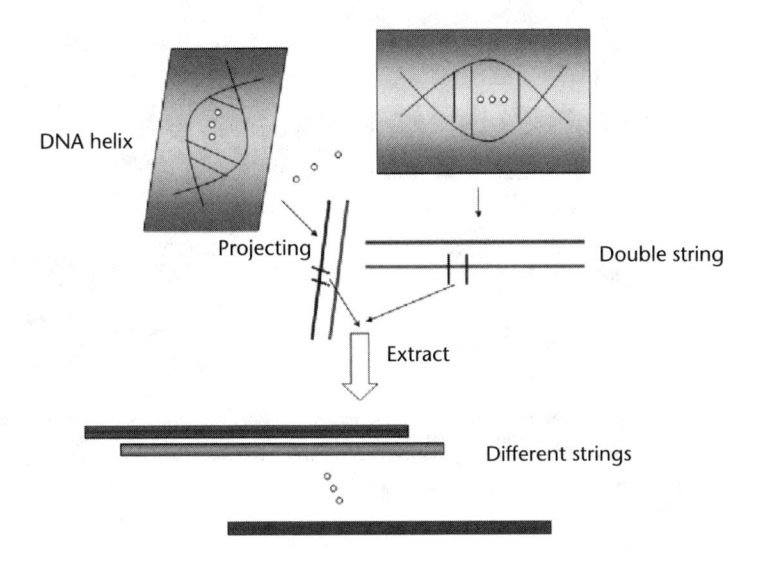

4.2 Information Processing in Artificial Nanobiosystems 101

Figure 4.3 Generation of strings from DNA helix structure.

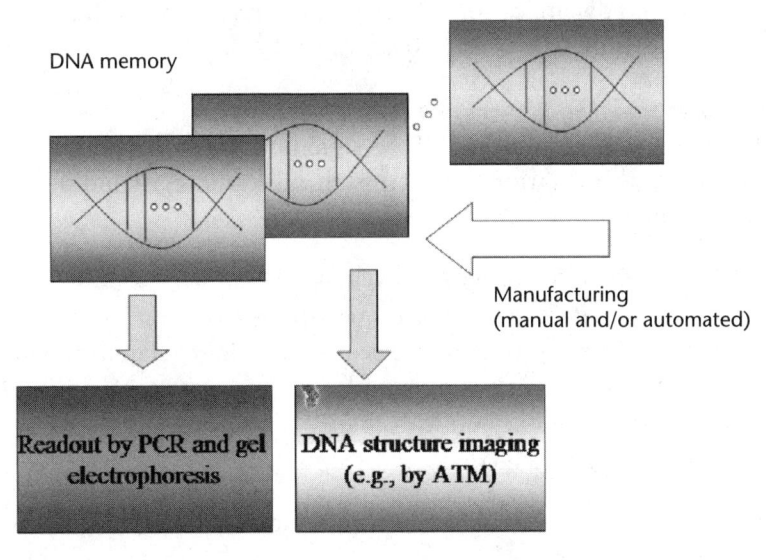

Figure 4.4 Operations related to DNA memory.

DNA molecules is not efficient. We hope to speed up the self-assembly processes by using certain machines. Biochemical readout and imaging are two major ways to read information from DNA molecules. Both of these approaches require technological support from devices. These processes become semiautomated when robot-like equipment assists them.

The two strands of DNA sequences are imagined as two strings in concept regardless of their topological structure. In the stick model, the two-strand form of the DNA sequence denotes 1 and the single-strand form of the DNA sequence denotes 0. The binary string is derived from the digits. If we designed a table to define the relationship between the DNA sequence and the alphabet, a word would be inferred (see Figure 4.5).

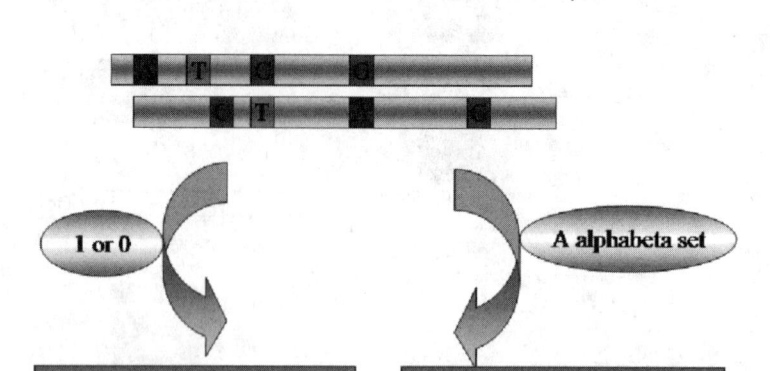

Figure 4.5 Information representation.

On the other side of the DNA world, bacteria such as *H. pylori*, *E. coli*, and others have a cyclic structure of the genome in the double-strand form. Modeling the genomic dynamics for DNA computing is also helpful for understanding the topological structure used to represent information.

Example: Signaling Protein Memory in Cell
To help our understanding of the signaling protein memory in a cell, we use the following abbreviations for terms in molecular cell biology in the subsequent discussions:

GEFs—guanine nucleotide exchange factors;
GAPs—GTPase-activating proteins;
SPKs—signaling-protein-molecules in cells regulated by kinases/phosphatases and their corresponding pathways;
GDI—guanine nucleotide exchange inhibitors;
p—phosphate.

The binary form is adopted for molecular memory built by GTPases and SPKs in cells.

As Figure 4.6 shows, the GTPase and SPK complexes store information in binary form [91]. Materials in vitro for these memories are expected to be an assay of cell cultures [45, 46] of epithelial Madin Darby Canine Kidney (MDCK) cells [46], which include ezrin/radixin/moesin family proteins (ERM), protein kinase C (PKC), botulinum for ATP, and tetramethylrhodamine B isothiocyanate (TRITC), prepared in advance [46]. Here, we use the set of GTPases that include {Ras, Rho, Rab, Arf, Ran} (that is, the superfamily of Rho family GTPases and the set of kinases/phosphatases).

One GTPase corresponds to one bit as shown in Figure 4.6(a). The state of the GTPase is determined by the corresponding pathways shown in Figure 4.6(b). As Figure 4.6(a) illustrates, the GTP-bound state of GTPase is represented as 1 in digit form, while the GDP-bound state is represented as 0 in digit form. These two states

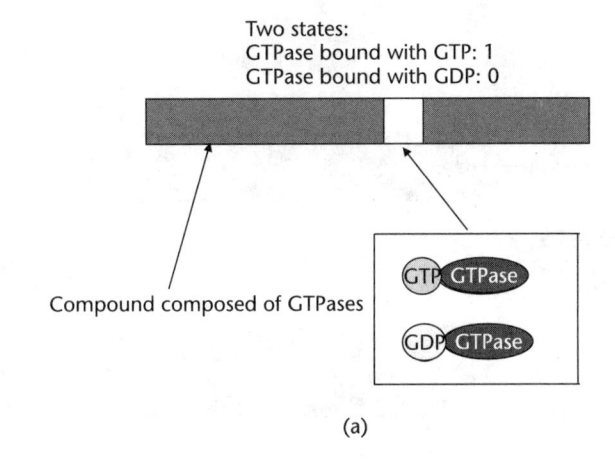

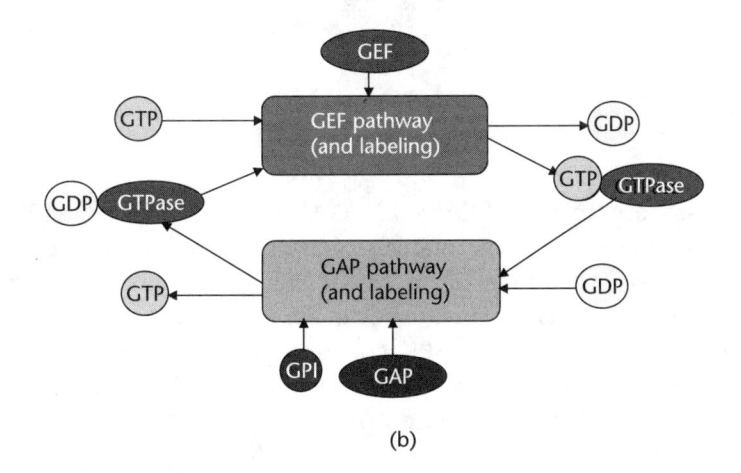

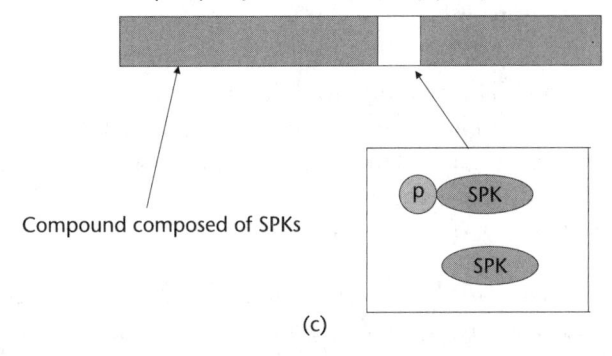

4.2 Information Processing in Artificial Nanobiosystems   103

Figure 4.6 Information representation and signaling pathways. (a) FTPase complex; (b) GEF/GAP pathways; (c) SPK complex; and (d) kinase/phopshatase pathways.

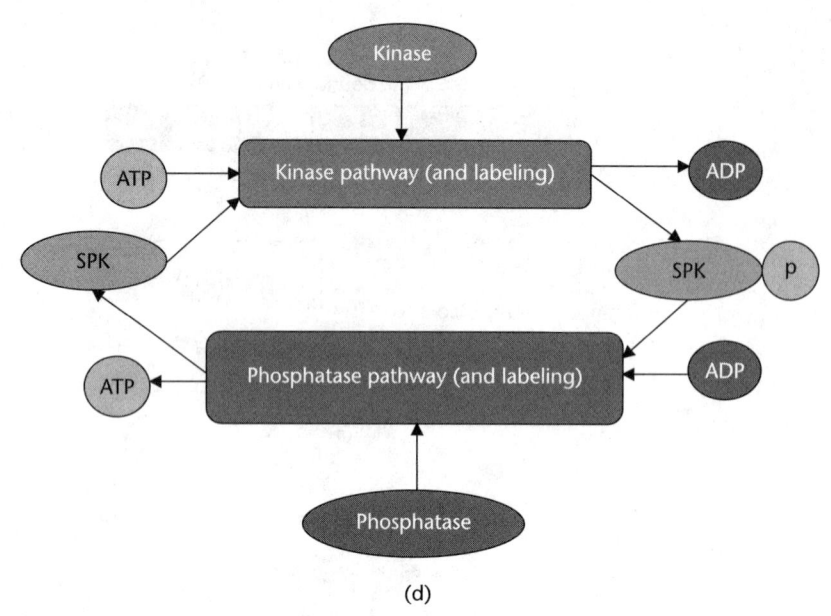

[Figure 4.6(d): Diagram showing Kinase pathway (and labeling) with ATP → ADP converting SPK to SPK-p, and Phosphatase pathway (and labeling) with ATP → ADP reversing the process.]

Figure 4.6 (continued.)

of GTPases are regulated by GEFs/GAPs through GEF/GAP pathways. The GTPase complexes form the material that represents the m-bit binary sequence. Similarly, GEF/GAP complexes represent the m-bit sequence. The signaling pathway used to write GTPase memory is given in Figure 4.6(b). Figure 4.6(c) shows the SPK memory, where one SPK corresponds to one bit. The phosphorylation state is 1 in digit form, while the dephosphorylation state is 0. The states of the phosphorylation/dephosphorylation of SPKs are regulated by kinase/phosphatase pathways, as shown in Figure 4.6(d). The SPK-complexes are the materials used for representing the n-bit binary sequences. The GEF/GAP and kinase/phosphatase switches are physically feasible for controlled pathways. In order to read out the information in SPK memory, immunofluorescence analysis will be used to detect the phosphorylation-dephosphorylation states on the phosphorylation assay. Also, in order to read out the information in GTPase memory, fluorescent resonance energy transfer (FRET) device will be used to detect the GTP-bound/GDP-bound states of GTPases. Transducers will convert the fluorescence signals to readable electrical signals. The readout tool for the output of kinase computing [56] is immunoanalysis. In biochemical technology, the term "readout" refers to the operation of transforming biochemical signals into readable forms (e.g., visible signals) by a certain kind of transducer.

We can define two kinds of binary words:

X—a binary word corresponding to an SPK molecular complex, $X = X_0X_1...X_n$, where X_i takes a binary value from $\{0,1\}$ ($i = 0, 1, ..., n, n \in N$). $X_i = 1$ for phosphorylation and $X_i = 0$ for dephosphorylation.

G—a binary word corresponding to a GEF/GAP complex, $G = G_0G_1...G_m$, where G_j takes the binary value form $\{0,1\}$ ($j = 0, 1, ..., m, m \in N$). $G_j = 1$ for GEF and $G_j = 0$ for GAP.

4.2.2 Molecular Clock—The Heart of Synchronous Moleware

In the cell, pathways are generally classified into two classes: metabolic pathways and signaling pathways. The energy for nanobiomachines is ATPase or GTPase in many cases. DNA motor and fuel are feasible in technology [90]. How to design a nanobiomachine for computing is a challenge for both engineering technology for manufacturing and biology theory at the mechanism level. We need a clock in synchronous architecture, where the signal of basic PE-like units can be efficiently controlled for computing. In the following, we discuss an example of a KaiC clock of phosphorylation/dephosphorylation and a signaling scheme for molecular computing based on it.

At the bacterial level, we can control the pathway for a KaiC clock of phosphorylation/dephosphorylation [43], where feedback is crucial for the biomolecular rhythm. As shown in Figure 4.7(a) based on the report in [93–95], the signal is zero when the KaiC is at the state of dephosphorylation, and then

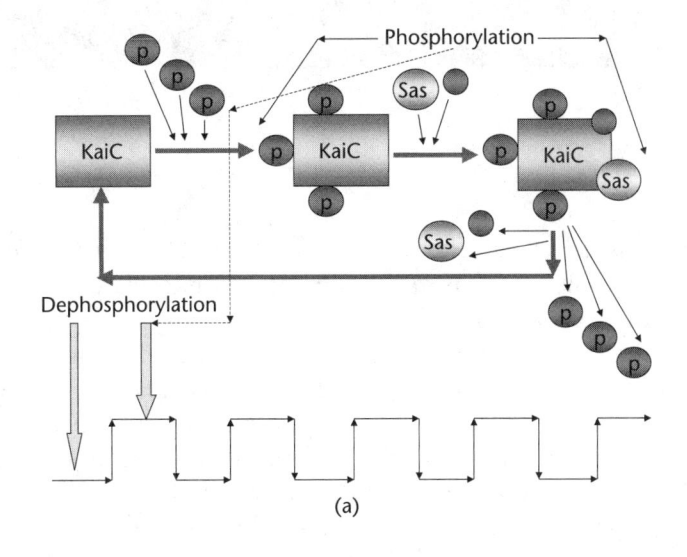

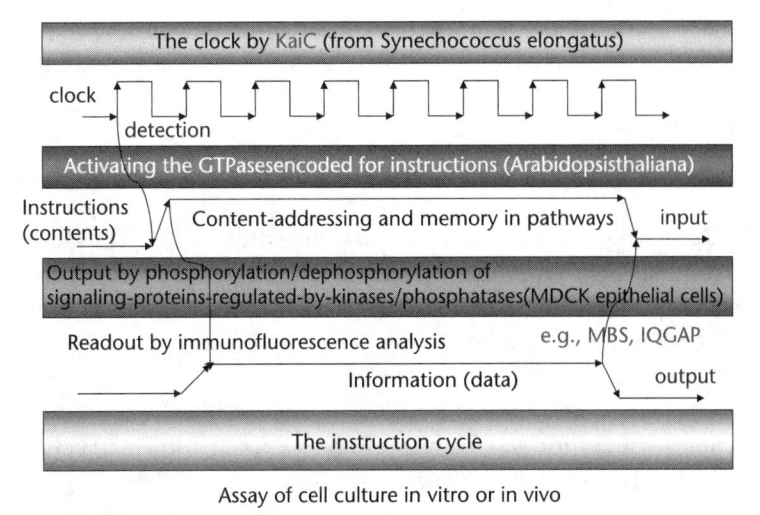

Figure 4.7 (a) KaiC clock of phosphorylation/dephosphorylation; (b) cycle of information processing in a molecular machine; and (c) diagram of computing unit.

input   Activating (words of GEFs/GAPs) = control/instruction

[Figure 4.7(c): Block diagram showing "The GTPases encoded for instructions (Arabidopsisthaliana)", "Pathway = memory", "activation", "detection", "Signaling-proteins-regulated-by-kinases/phosphatases (MDCK epithelial cells)", "operation", "data", "readout output", "Clock signal", "The clock by KaiC (from Synechococcus elongatus)", "Circadian rhythm control"]

(c)

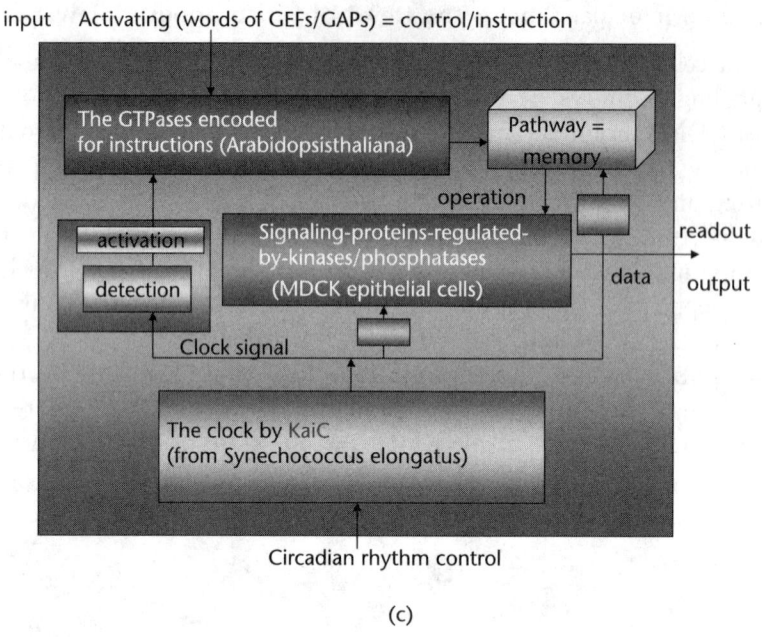

Figure 4.7   (continued.)

the kinase is activated and three phosphatases are attached to KaiC. The signal becomes one when KaiC is at the state of phosphorylation. Fortunately, this state activates the corresponding feedback pathway. Sas is attached to the KaiC so that the KaiC returns to the state of dephosphorylation denoting the signal of zero. This weakens the function of the feedback pathway because the strength of the feedback pathway is proportional to the concentration of KaiC.

In Figure 4.7(b), the KaiC clock is used as the basis of timing in the moleware. Phosphorylation and dephosphorylation are also used in kinase computing. The harmonic operations among signaling pathways can be easily understood in the above figure. The words of signaling proteins and kinase/phosphatases are given in binary form for the molecular memory.

The functions of each part of the clock in Figure 4.7(c) are given as follows:

(0) "Input":
Input = AcG word.

GEFs/GAPs representing the instruction activate the GTPase pathways, and the control function is assigned to them. The GEFs/GAPs can be prepared and activated in the assay of the cell culture in vitro or in vivo. The signaling enzyme proteins in cells can be used to activate the GEFs/GAPs according to the specific combinatorial forms of GEFs/GAPs.

(1) The pathways for generating GTPases encoded for instructions are expected to be built by *Arabidopsis thaliana*.

The input to (1) is AcG word.

This is used as the control/instruction. Here "instruction" is defined for a computer and refers to the machine instruction at the level of machine language. Control means the control of the information processing carried out in parts (2) and (3). This instruction is used to activate the related pathways in (2).

The bit information in words of GEFs/GAPs can also be assigned with different meanings/semantics according to the specific applications. We can design a kind of programming language for this molecular computer.

The output of (1) and the input to (2) is G word.

(2) Pathway working as memory:
This is a kind of content-addressing scheme. The related pathways exist in the assay of cell culture prepared in advance. The designed G word is used to activate those related pathways for generating the necessary kinases/phosphatases. The kinases/phosphatases are generated in word form as combinatorial forms of bit information. This pathway unit exerts the related operations on (3).
K word is the output of (2) and input to (3).

(3) The pathways for generating the signaling proteins regulated by kinases/phosphatases are expected to be built by MDCK epithelial cells.

After the operations exerted by (2), (3) is assigned with the related states of phosphorylation and dephosphorylation. The combinatorial forms of these molecules consist of the P word, where the phosphorylation is 1 and dephosphorylation is 0.

GEFs/GAPs, Rho-GTPase, kinases/phosphatases, and signaling proteins regulated by kinases/phosphatases can be prepared from mammalian cells (for example, MDCK epithelial cells). GTPase can be prepared from plant *Arabidopsis thaliana* with a low cost.

(4) The data is information encoded by the signaling proteins regulated by kinases/phosphatases (SPK). This can be read out by certain tools such as immunoanalysis.

(5) "Circadian rhythm control"
Circadian rhythm exists in the cyanobacterium *Synechococcus elongatus* [94]. Therefore, we can use a species made in the assay of cell culture in vitro. "Control" means maintaining this cycle of activity under a certain temperature condition.

The input here is the materials put into the test tubes/assays, and this is also the input of (6).

(6) "The clock by KaiC (from *Synechococcus elongatus*)"
The clock here means the circadian rhythm in cells. We can use the kind of mechanism existing in mammals, insects, plants, fungi, and cyanobacteria.

Here we give one example for this. The kernel mechanism of this clock is KaiC autophosphorylation-dephosphorylation in vitro. It works in vitro with an assay of cell culture at a certain temperature with feedback in the related signaling pathways, without the artificial control of a clock-gene and its expression (clock-protein).

(7) According to the clock signal, the related pathway units are activated in (1), (2), and (3). Bioluminescence imaging is used to detect the signals in (6). Then, this signal is used to activate the pathways in (1). As an example, Western blot analysis can be used to detect KaiC and Sas. Furthermore, signal can be displayed by densitometric analysis (software).

4.2.3 Moleware Coding in Nanobiomachine—A Solution from the Cell

As we have explained, molecular memory is physically designed in binary form for storing information in moleware and biochemical reaction is the way used to access the molecular memory. In order to understand the information capacity of molecular memory, we need to study the algorithms of coding by moleware.

In order to find an unconventional computing paradigm that is complementary to commonly used information theory, many efforts have been made. Among them, Lipton's work [52] used Adleman's DNA computing protocol [41] to solve the contact network problem, which can be traced to C. Shannon's pioneering paper [53]. This showed that DNA can be used to encode the information flow in communication or a network. It is also notable that Karl-Heinz Zimmermann [54] applied sticker-model-based DNA computing to binary linear codes, where double strands of DNA sequences mean 1 and a single strand of a DNA sequence means 0. However, errors were found in this approach. In experiment, the basic rule is to define codes of DNA sequences that can be distinguished by readout technology. This means that any two DNA sequences should have different information that can be detected by laboratory technology. From the viewpoint of information theory, the Hamming distance of any two DNA sequences should be larger than a threshold where Hamming distance is the sum of different bits in two DNA sequences. The quantitative difference of two DNA sequences is often reflected in the weight marker of gel electrophoresis. If the Hamming distance is smaller than the threshold, the weight quantity cannot be detected precisely. Therefore, Hamming distance is the most important criterion for designing DNA codes. An overview on the theoretical aspects of DNA strand codes by Mauri et al. can be found in [55]. In physics, DNA-based coding methods should represent the constraints on the fidelity of complementary DNA sequences and hydrogen bond biochemical features for ligase reaction. The concrete combinatorial forms of {A, T, C, G} are designed empirically according to the effect of their biochemical reactions in laboratories. Otherwise, we use physical calculations based on free energy and other factors.

From the viewpoint of steganography [50], it is promising for us to apply biomolecules in designing encoder/decoders, which is one of the most important tasks in applications of information theory [51]. In addition to DNA codes, other biomolecules can also be used to design codes. Cells can maintain their robustness by using pathways of signaling proteins [42, 44, 47–49, 57–65] such as GTPase and kinases/phosphatases. We have already discussed the molecular memory of GTPase and SPK; in this subsection, we will discuss an example of a corresponding coding algorithm, which is designed based on GTPase memory and SPK memory.

4.2 Information Processing in Artificial Nanobiosystems

Let z be the information vector $z \in F^k{}_2$ [54], which is encoded to code X. Code X is represented by a codeword that consists of $\{Xi\}$ where $Xi = 0$ or 1 ($i = 0, 1, ..., n \in N$). The notations we use in the latter discussion are given as follows:

$i = 0, ..., n, n \in N$.
$j = 0, ..., m, m \in N$.
$q = 0, ..., Q, Q \in N$.
$l = 0, ..., L, L \in N$.
$i' = 0, ..., P, P \in N$.

z—an information vector to be encoded, $z \in F^k{}_2$.

z-set—the set of all information vectors. The number of elements in this set is Q.

$z(q)$—the q-th element in the set of information entity z.

$z(q, j)$— the jth bit of $z(q)$. $z(q, j) = 0$ or 1.

X—a codeword called the X codeword. This corresponds to an SPK molecular complex, $X = X_0 X_1 ... X_n$, where X_i takes a binary value from $\{0,1\}$ ($i = 0, 1, ..., n, n \in N$). Here, it is $L \times (P+Y)$-bit long ($Y \in N$). $X_i = 1$ for phosphorylation and $X_i = 0$ for dephosphorylation.

G—a codeword called the G codeword. This corresponds to a GEF/GAP complex, $G = G_0 G_1 ... G_m$, where G_j takes the binary value form $\{0,1\}$ ($j = 0, 1, ..., m, m \in N\}$. $G_j = 1$ for GEF and $G_j = 0$ for GAP.

$G_j(z)$—the j-th bit of the G codeword corresponding to z.

$G_j(z(q))$—the j-th bit of the G codeword corresponding to $z(q)$.

G'—a codeword, called the GTPase codeword, $G' = G'_0 G'_1 ... G'_m$. ($j = 0, 1, ..., m, m \in N$. $G'_j = 1$ for GTP-bound and $G'_j = 0$ for GDP-bound.

$G'_j(z)$—the jth bit of $G'_j(z)$.

$G'_j(z(q))$—the jth bit of $G'_j(z(q))$.

G'-set$Q$—the set of the GTPase-encoded codewords.

$G'(z(q))$—the q-th element in G'-set$Q$ corresponding to $Z(q)$. $G'(z(q)) = G'_0(z(q)) \, G'_1(z(q))...G'_n(z(q))$.

$\Psi_{i'}$—this denotes the i'-th kinase or the i'-th phosphatase, Ψ denotes kinase or phosphatase, and i' is the index in the corresponding library.

$SPK(i')$—this denotes the i'-th SPK.

K—the number of cross-talked signaling pathways.

$X_{i'}$—the i'-th-bit of the X codeword, which corresponds to the SPK complex with a phosphorylation/dephosphorylation state.

X'—the X codeword with P-bit length.

$X^*(l)$—the l-th part of the X codeword. This is $(P+Y)$-bit long.

label(l)—this denotes the l-th element of the set of words that corresponds to molecular complexes of labeling molecules. This is Y-bit long ($Y \in N$).

The kernel of the encoding/decoding process is the signaling pathways in cells, which consists of two groups of pathways, which are the encoding and decoding pathways with reversed information flow by reversible switches of GEFs/GAPs and kinases/phosphatases (see Figure 4.8). In mammalian cells, the relevant techniques

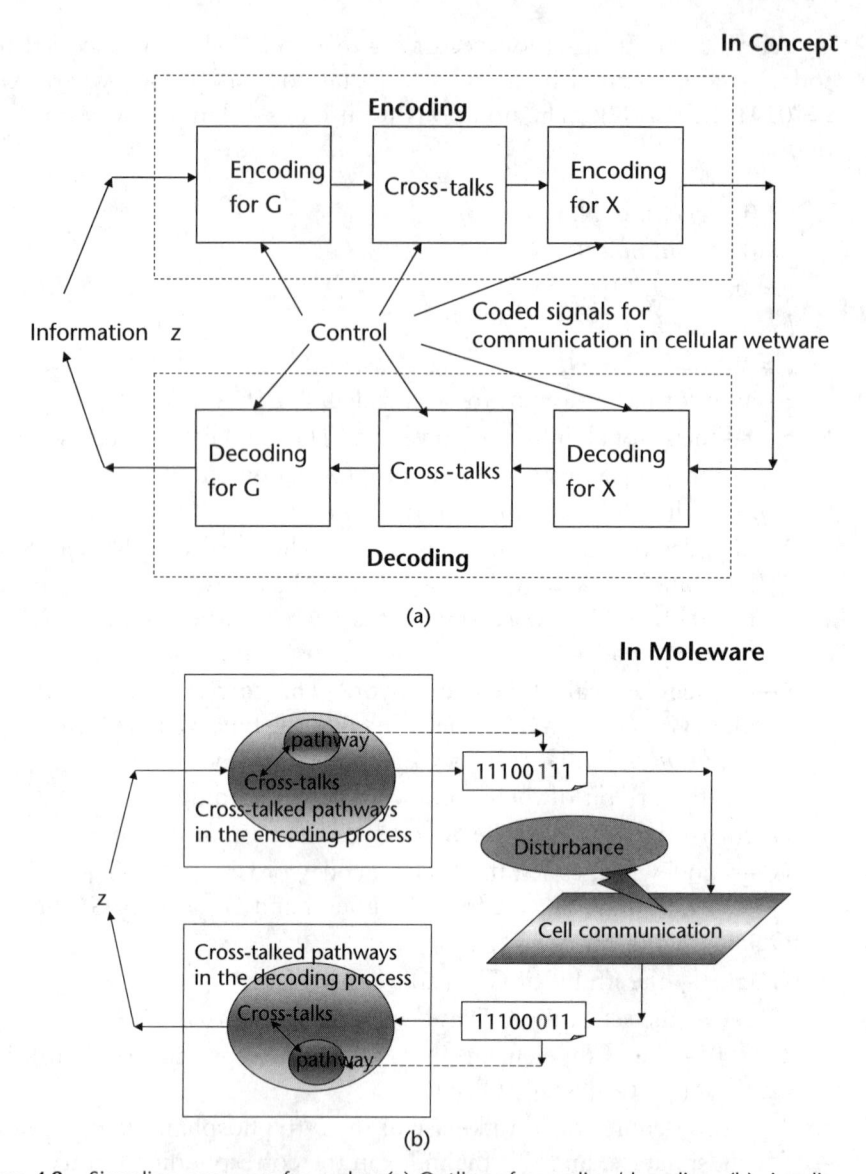

Figure 4.8 Signaling processes for codes. (a) Outline of encoding/decoding; (b) signaling pathway mechanism for correcting errors; and (c) conceptual diagram of biochemical process for codes.

are already mature and practical in laboratories of molecular biology and biochemistry [45, 46]. These techniques can guarantee the necessary accuracy of computation by biochemical reactions in living MDCK cells.

Algorithm 1: Encoding by GTPases and kinases/phosphatases.
Input: GEFs/GAPs: G.
Output: Molecular complexes of SPKs: X.
Main Body of the Algorithm:
1. Encoding_algorithm (z-set, Q, m, K, L, P, n)
2. Generation_of_G-codeword (z-set, Q, m)

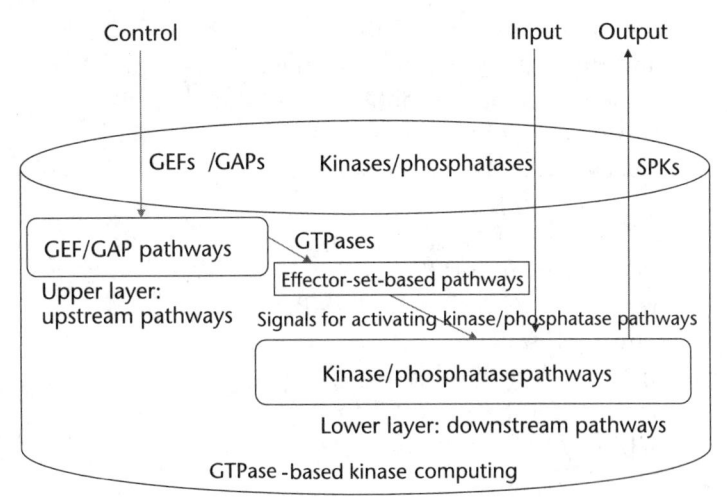

4.2 Information Processing in Artificial Nanobiosystems

[Diagram: GTPase-based kinase computing cylinder with Control, Input, Output arrows. Contains GEFs/GAPs, Kinases/phosphatases, SPKs at top. GEF/GAP pathways → GTPases → Effector-set-based pathways → Signals for activating kinase/phosphatase pathways → Kinase/phosphatase pathways. Upper layer: upstream pathways; Lower layer: downstream pathways.]

SPKs: Signaling proteins regulated by kinase/phosphatase pathways

(c)

Figure 4.8 (continued.)

$\quad$ (parallel) for $q \leftarrow 0$ to Q do
$\quad\quad$ (parallel) for $j \leftarrow 0$ to m do
$\quad\quad\quad G'_j(z(q)) \leftarrow$ GTPase_pathway
$\quad\quad\quad\quad$_activation($GTPase, G_j(z(q)), z(q,j)$)
$\quad\quad$ return $\{G'(z(q))\}$
$\quad$ return $\{G'\text{-set}Q\}$
3. Generation_of_X-codeword (L, P, K)
$\quad$ (parallel) for $l \leftarrow 0$ to L do
$\quad\quad$ (parallel) for $i' \leftarrow 0$ to P do
$\quad\quad\quad$ cross_talked_SPK_pathway
$\quad\quad\quad\quad$_activation($SPK(i'), K, G'(z(i'+l \times P))$)
$\quad\quad\quad \Psi_{i'} \leftarrow$ access_of_library ($_{i'}$)
$\quad\quad\quad X_{i'} \leftarrow$ kinase/phosphatase
$\quad\quad\quad\quad$_pathway_activation($SPK(i')), \Psi_{i'}$))
$\quad\quad$ return $\{X'\}$
$\quad\quad X^*(l)$ (labeling (X', label(l))
$\quad$ return $\{X\}$

where $L \times P = Q$ and "(parallel)" refer to the parallel operations, for example, "(parallel) for $q \leftarrow 0$ to Q do" means carrying out q operations in parallel.

Algorithm 2: Decoding by GTPases and kinase/phosphatases.
Input: The molecular complexes of SPKs: X.
Output: GEFs/GAPs: G.
Main Body of the Algorithm:
1. Decoding_Algorithm (z-set, Q, m, K, L, P, n)
2. decoding_of_X-codeword (L, P, K)
$\quad$ (parallel) for $l \leftarrow 0$ to L do
$\quad$ (parallel) for $i' \leftarrow 0$ to P do

```
        Ψ_{i'} ← access_of_library (i')
        complementary_kinase/phosphatase
            _pathway_activation(X_{i'})
        if (match()=F) then
        {feedback_pathway_activation(S_0,S_1,...,S_k)}
            G'(i'+l×P)←
        cross_talked_SPK_pathway
            _activation(SPK_{i'}, K)
        return {G'=G'(l×P)...G'(l×P+P)}
        return {G'-setQ}
3. decoding_of_G-codeword (z-set, Q, m)
        (parallel) for q ← 0 to Q do
        (parallel) for j ← 0 to m do
            G_j(z)←
        complementary-GTPase
        _pathway_activation(GTPase)
            if (Gj(z(q)) =1) then
                z(q, j) = 1
            if (Gj(z(q)) = 0) then
                z(q, j) = 0
            return {z(q)}
        return {z-set}.
```

Here, we use {Rho-GTPase, Rac-GTPase, Cdc42-GTPase} to encode the G' codeword, {MBS, MLC} to encode the X codeword, and Rho-protein as the labeling molecule. Provided that we have the set of {GTP-bound Rho-GTPase, GDP-bound Rac-GTPase, GTP-bound Cdc42-GTPase}, the set of {MBS with the phosphorylation state, MLC with the dephosphorylation state}, and the set of {Rho-protein with the activation state}, the X codeword is inferred as 11 from G-codeword 101. If an error occurs in MBS, that is, the state of MBS becomes dephosphorylation, the related cross-talked pathways and complementary pathways recover the original codeword as 11.

The operators for codes are given as follows:

For encoding part:
    kinase/phosphatase_pathway_activation ($SPK(i')$,$\Psi_{i'}$)—this activates the kinase/phosphatase pathway to determine the state of $X_{i'}$ (the state of phosphorylation or dephosphorylation).
    GTPase_pathway_activation (GTPase, $G_j(z(q))$, $z(q,j)$)—this determines the state of $G'_j(z(q))$, that is, the state of GTP-activation or GDP-activation of GTPases, according to $G_j(z(q))$ and $z(q,j)$, where $G_j(z(q))$ corresponds to $z(q,j)$.
    cross_talked_SPK_pathway_activation ($SPK(i')$,K,$G'(i'+l×P)$)—this activates the cross-talked signaling pathways whose number is K.
    access_of_library (i')—(this denotes the kinases/phosphatases according to their index in a designed library produces $\Psi_{i'}$.)

4.2 Information Processing in Artificial Nanobiosystems

labeling $(X', \text{label}(l))$—this merges X' with $\text{label}(l)$. The merged molecular complex is $X^*(l)$.

For decoding part:

The output of complementary_kinase/phosphatase_pathway_activation$(X_{i'})$ is $SPK_{i'}$ dependent on $X_{i'}$.

match()—this detects whether or not the product of

complementatry_kinase/phosphatase_pathway_activation$(X_{i'})$

matches SPKs that are identical to the pathway

kinase/phosphatase_pathway_activation$(SPK(i'), \Psi_{i'})$

in the decoding process.

If they are identical, it produces T in logic; otherwise, it produces F in logic. "feedback_pathway activation $(S_0, S_1,..., S_k)$" produces $SPK(i')$ if match() is F. $S_0, S_1, ..., S_k$ are necessary conditions of activation for the related pathways to restore the signals.

The pathway

Complementary_Cross_talked_SPK_pathway_activation$(K, G'(i'+l \times P)))$

produces $SPK(i')$.

The pathway

complementary_GTPase_pathway_activation $(GTPase)$:

produces $G'_j(z(q))$.

That is,

$G'_j(z(q)) = 1$, for GTP-bound state,
$G'_j(z(q)) = 0$, for GDP-bound state,

$G_j(z(q)) = 1$, for GEF,
$G_j(z(q)) = 0$, for GAP.

As Figure 4.8 shows, the encoding and decoding processes use the reversible mechanism of signal transduction in cells, and the encoding and decoding can be done by way of a complementary pathway structure:

complementary_kinase/phosphatase_pathway_activation$(X_{i'})$

in comparison with

kinase/phosphatase_pathway_activation($SPK(i')$,$\Psi_{i'}$),

and

complementary_GTPase_pathway_activation($GTPase$)

in comparison with

GTPase_pathway_activation($GTPase,G_j(z(q)),z(q,j)$).

In the encoding process, the following pathway is employed:

cross_talked_SPK_pathway_activation($SPK(i')$,K,$G'(z(i'+l\times P))$).

In the decoding process, the following pathway is employed:

cross_talked_SPK_pathway_activation($SPK_{i'}$, K).

The function of "memorizing" codes is reflected in the following part of the decoding algorithm:

if (match()=F) then
 {feedback_pathway_activation($S_0, S_1, ..., S_k$)}
 $G'(i'+l\times P)$(cross_talked_SPK_pathway_activation($SPK_{i'}$, K).

When an X code with errors is input to the decoding process, the pathways denoted by $S_0, S_1, ..., S_k$ will be activated by the condition of (match()=F). The "feedback_pathways" set the SPK_i to the states determined in the encoding process.
 Then the pathway

Cross_talked_SPK_pathway_activation($SPK_{i'}$, K)

will set the G' codeword to the states determined in the encoding process. It is obvious that the complementary pathways for mapping from G' to G will be free of errors.

4.3 Computing by Nucleic Acids

Along the arrow of time representing billions of years of evolution, some of the most amazing gifts from nature have been the biological information-processing machines that autonomously and adaptively deal with the information of biochemical reactions in living systems. These are expressed as DNA, RNA, and proteins in cells, where the pathway does the most important work. In the cell, transcription from DNA to RNA and translation from RNA to proteins are two remarkable biochemical processes for handling information crucial to life, whether simple forms of life such as bacteria or complex species such as human beings. Currently, with the development of genome sequencing projects, huge volumes of genomic information on human beings, rice, *H. pylori*, *E. coli*, *Arabidopsis thaliana*, and others in a rapidly increasing list have become accessible.

Due to the reduced costs of DNA sequences and DNA operations (e.g., vector, plasmid, related enzymes), DNA molecules have become one of the most popular media for molecular computers as a pragmatic technology. In the field of DNA computing, different methods have been developed with different materials for different purposes; consequently, design strategies are typically devised in an ad hoc way for the convenience of biochemical implementation of specific algorithms or computing models. In light of this situation, we discuss Adleman-type DNA computing in Section 4.3.1, RNA computing in Section 4.3.2, surface-based DNA computing in Section 4.3.3, and finally nanobiotechnology for DNA computing in Section 4.3.4. Although the materials used are different in Adleman-type DNA computing and RNA computing, their strategies are similar. However, surface-based DNA computing has a different strategy from those two methods. Since DNA computing systems reported in [88] can be considered a kind of automata, their method is discussed in Chapter 5.

4.3.1 DNA Computing

The pioneers who dreamed of moleware as the driving force of future information processing include Feynman, Drexler, Conrad, and Head. The available technology of their time did not prevent the blueprints from going beyond the empirical limitations.

4.3.1.1 One Week in the Lab: A Stride in History

About ten years ago, Leonard Adleman successfully realized a "DNA computer" [41, 66, 67] in test tubes. The experiment took seven days in the laboratory. Such a long time was mainly due to manual operations. Later, other research results reported the feasibility of experimentally implementing autonomous devices for DNA computing [88, 89]. New nanobiotechnolgy for DNA manufacturing continues to emerge, and thus we can reasonably expect the development of a completely automated molecular computer, especially when we recall the flight of a few meters accomplished by the Wright brothers' original aircraft about 100 years ago. This development positions the dream of a computer built by molecules within the realm of reality and thus also opens up a new era of computing technology, which in a more generalized sense implies a new kind of unconventional information-processing paradigm.

Although the material used in the experiments [41, 66–68, 88, 89] was DNA molecules, the significance of their work in this field is profound. It was later found that RNA, proteins/amino acids, kinases/phosphatases, membrane structure, and even cells can also be used to make the dream machine called the molecular computer. Through the sound benchwork reported in [88, 89], it is clearly and loudly announced to the world that molecules can be used to express, store and process information, where the computing tasks are carried out by biochemical reactions in the form of transformation from one kind of specific message to another. Here, the message is stored in DNA sequences, and the transformation is reflected in temporal processes of reactions, where some molecules are consumed and others are produced, as in writing letters on a blackboard.

The blackboard is the test tube containing the liquids, and the chalk is the DNA placed in the test tubes. Since the emergence of the first-realized DNA computer in a biochemical experiment by Adleman, several other successful experimental systems have been developed, including Adleman's method, surface-based DNA computing, and others. Most of these systems are normally formed in liquid (for example, in the form of in vitro). The advantage of the in vitro approach is that DNA molecules can be arranged to perform functional computation simultaneously, in parallel, free of the burdens of allocation and communication often imposed on parallel electronic computers made of hardware. Naturally, DNA assembly requires other technological manipulations such as how to handle hairpins and errors. Full implementation of an actual DNA computer still needs much more work. Manually handling the DNA is gradually being replaced by machine-based tools; for instance, electrophoresis will eventually be used as an alternative to DNA sequencing devices for automation, speed, and accuracy.

Thanks to the unique structure of helix DNA, we find that phosphate-plus-sugar as backbone and base pairs as message-storage, formed by the hydrogen-bond, play important roles and can be used to realize the operation of DNA computing. The hydrogen bond combines two DNA sequences into double-stranded DNA in the form of Watson-Crick complementarity (more detailed background on terms and explanations is given in Chapter 2). Under biochemical conditions, this biochemical constraint is employed to combine two strands into a single double strand. The biochemical technology based on free-energy calculations for particular experimental conditions, robust DNA sequences found through empirical experiments (trial and error), and computing support is studied to reduce the errors due to unwanted bonding of DNA sequences. In this subsection, we mainly observe the biochemical characteristics of DNA. In DNA computing, it is common practice to use artificially manufactured DNA sequences by arranging them into a kind of computing process based on the designed experimental steps of biochemical reactions.

In contrast to the artificial process described above, the natural processes of cells, in which messages of nucleic acids are transmitted to cellular functional units for evolutionary heredity, mainly start from DNA to proteins as the central dogma implies. We note that the non-Mendel heredity and cell communication (second messenger) also have similar functions. The DNA recombinant nanotechnology and supramolecular structural manufacturing technology for self-assembly of 2D topological structures of DNAs are inevitable tools for DNA machines in action. The DNA can work as the fuel [90], in the form of information-processing units, and also as the drive of the DNA machine. The automation technology for DNA is a promising technical support toward realizing a completely functional DNA computer as the final target.

By focusing on the characteristics of molecular computing as the moleware configuration, we attempt to analyze molecular computing processes from the following three viewpoints in methodology.

State Description of Biomachines
Imagine a DNA computing process as a machine in our brain in which the state of the moleware is the basic unit of measurement according to the computing tasks. The

state of moleware can be represented in the form of symbolic sequences consisting of {A, T, C, G}, which can be interpreted as binary words or symbolic words or mapped into other forms of information in the case of DNA computing.

With the finite state machine of DNA computing biochemically realized by Benenson et al. [88, 89], a continuous state transition process can be carried out in a stepwise manner by certain biochemical mechanisms such as self-assembly or self-organization. According to [96], "Rivka Adar, Yaakov Benenson, Zvi Livneh, Tamar Paz-Elizur and Ehud Shapiro of the Weizmann Institute of Science (Israel) have created the smallest biological computing device. A microlitre of salt solution containing 3 trillion self-contained DNA computing devices can perform 66 billion operations per second." This was the smallest DNA computing device in the world at that time and thus regarded as one of the best examples. This approach seems natural for automata-based representation of DNA computing.

Informational Structure of Computing Process
For computing tasks, the space of the molecules used for computing can be quantitatively modeled, and this space is often rich in nonlinearity in the sense of complexity. This phenomenon in informatics is based in essence on the fact that DNA computing is an unconventional computing paradigm. In applying DNA computing to NP problem solving, the problem of space complexity is the main factor to be considered. This is also a crucial factor in programming moleware as well as designing and testing algorithms with real-world problems.

Logic Operators for Moleware
Biochemically faithful logic operators are employed to realize the wetware-level molecular operations used for computing. The predicate form of logical programming is used to describe the corresponding biochemical reactions in various forms, and through this scheme, the advantages of formalization and theoretical analysis of molecular computers can be exploited to improve performance. This ability is based on rigorous computer science theory arising from axioms, theorems, and algorithms. It is thus very useful for building an operating system and compilers for molecular computers according to rigorous logic protocols.

The word "molecular computer" is used here in its generalized definition: any molecular computing system in the form of theory, simulation and implementation, even including partial implementation. The narrowly defined concept of a molecular computer refers to a completely functional computer built by molecules that can perform any computing task with full automation and no need of manual operations.

4.3.1.2 The Progression of Adleman's Classic Experiment

Here, we trace the route of Adleman's computing system from the biological stage to the informatics stage and then to the logic form.

The Biostate Description
The biochemical reaction by the self-assembly mechanism of the hydrogen bond is used to realize the state transition from the initial state set to the solution-contained ones, that is,

from the set of vertex-encoding 20-mer DNA sequences ∪ the set of edge-encoding 20-mer DNA sequences, under the conditions that
the edges and vertices satisfy the Hamiltonian graph, and
there exists an exception for V0 and V6 in the graph with 10-mer sequences (mer refers to the number of molecules),

to the set of generated chains in double-strand DNA sequences, where other unwanted sequences and other materials in the test tube may exist. This set contains the solution for readout if the HPP problem has a solution.

A random table of DNA sequences is designed in advance, and the vertices are denoted by 20-mer DNA sequences. The term "mer" refers to the number of nucleotides. The 20-mer DNA sequence for edge is designed in two parts: the first 10-mer part is designed as the complementary form of the sequence representing the first vertex, and the second 10-mer part represents the second vertex as shown in Figure 4.9 [92].

The two directions of the DNA sequences denoted as 3' and 5' are used to correspond to the edges in a directed graph in order to make the physical representation of each edge in a DNA sequence unique. The biochemical process of Adleman's experiment is given in Figure 4.10.

In Adleman's experiment, manual work is necessary to bring the solution form of a DNA sequence into the range of human vision through the amplification of polymerase chain reaction (PCR) and electrophoresis on gel (agarase). Here, two primers are used as clippers for the two ends of the chains that denote vertex 0 and vertex 6 as shown in Figure 4.11. The weight marks on the glass tell us the number of vertices and which vertex is along the chains in a spectrometric way. The information processing ability of DNA computing has been improved by the recent achievements in DNA computing exemplified by the systems of Braich et al. [68] and Benenson et al. [88, 89], which provided the greater efficiency and less exhaustive manual work that Adleman envisioned.

An informatics-based analysis

In the above example, DNA computing solved an HPP problem. In the performance of this computing process, space complexity is a crucial process, and this issue is examined in the following.

For HPP

Representation for an edge

| Vertex i | Vertex j |
| Edge i → j |

Representation for a vertex

| Vertex i |
| Complementary form of the above strand |

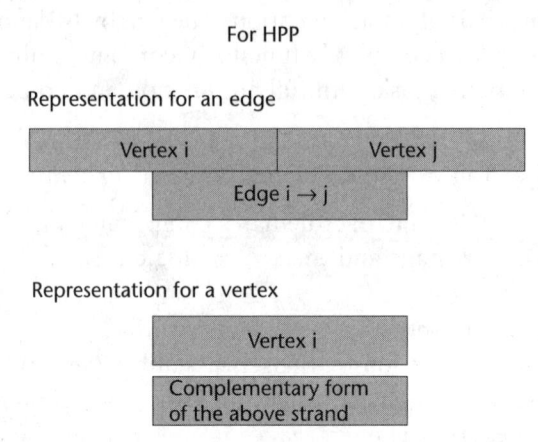

Figure 4.9 Description for HPP.

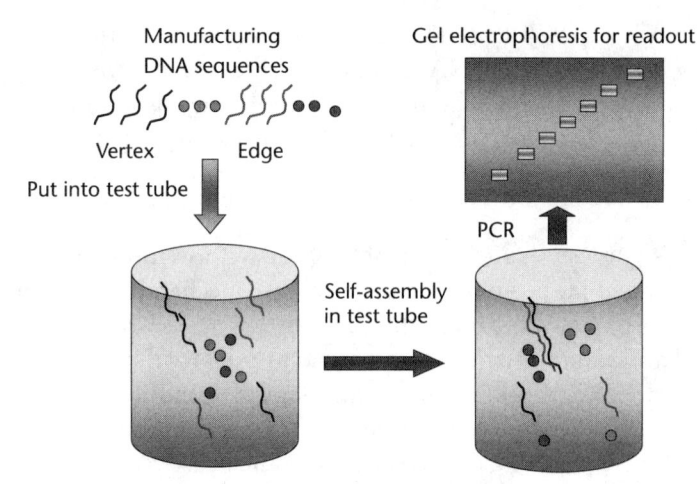

4.3 Computing by Nucleic Acids

Figure 4.10 Outline of Adleman's DNA computing process.

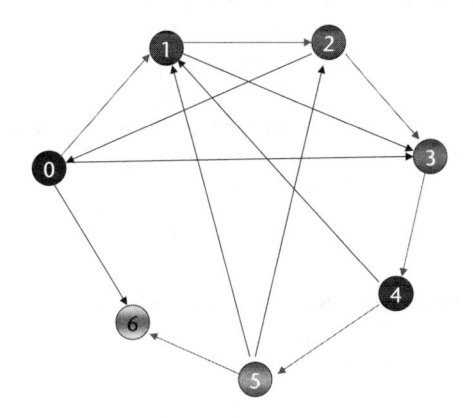

Figure 4.11 Hamiltonian path problem.

Let

Nv be the number of vertices, also denoted as n.
Ne be the number of edges, also denoted as m.
Nc be the number of randomly generated chains, also denoted as k.
Ns be the number of solutions, also denoted as L.

As an experimental setting, the number of molecules in the initial space is n + m. Then, the total number of merged chains is

$$\sum_{i=1}^{m} P_m^i \tag{4.1}$$

This is called brute-force searching. Heuristics techniques also have been introduced in DNA computing to improve the efficiency of the algorithm used for 3-SAT problem solving [97]. Even though the Schoening algorithm for SAT problem solving can be applied to DNA computing, the space complexity still needs to be improved owing to the huge number of molecules used in DNA computing.

The space complexity is defined according to the entire space with the molecular numbers n and m for the initial materials, and $\sum_{i=1}^{m} P_m^i$ is to be manufactured by Adleman's "oracle," which is capable of the self-assembly of a hydrogen bond.

It is a good idea to have the biochemistry mechanism act as the "oracle" instead of us. Unfortunately, however, this process requires ad hoc skills and is vulnerable to errors. This kind of intrinsic error is due to the fact that the hydrogen bond principle often does not work as exactly as a logical 1 and 0 operation. This error causes a bottleneck in this type of experimental DNA computing technique. The time complexity of the computing equals the time of the ligase reactions plus the readout time.

If we neglect readout time, the time of the DNA computing algorithm needed for 3-SAT problem solving is proportional to the time of the biochemical ligase reactions in physical processes. Equivalently, this is linearly dependent on the number of the vertices and edges. Consequently, it is obvious that

$$m \leq P_n^2 \tag{4.2}$$

The space complexity can be estimated as O(n) by computation theory.

Logic Operators
The logic operators are given as follows:

   put-into-test-tube (the vertex set, the edge set),

   ligase-reactions-in-parallel (the vertex set, the edge set)
   // it generates the chains //

   readout (the chain set | the set of ending vertexes, the length of the chain, the wanted-vertex-list).

All of these concepts correspond to biochemically faithful processes in physical test tubes.

The ligase reaction does not spontaneously occur in a physical sense. It is a kind of manual/machine-guided process, so it is listed here as a logical operator.

4.3.1.3 Adleman-Lipton DNA Computing for 3-SAT Problem Solving

Adleman's historic experiment showing DNA molecules in test tubes can be used to solve the Hamiltonian path problem, which is regarded as one of the most difficult problems and called NP complete.

In the following, we introduce an algorithm for satisfiability (3-SAT) problem solving by Lipton [52], based on Adleman's experimental protocols.

Let the set of variables be $\{X_i\}$ ($i = 0, 1, \ldots, n-1, n \in N$). The constraint (that is, the condition called the clause) is represented as the union set of $\{\Psi_j\}$

($j = 0, 1, ..., m-1$), where $\Psi_j = \Psi_{j1} \vee \Psi_{j2} \vee \Psi_{j3}$. Ψ_{jl} ($l = 1, 2, 3, j = 0, 1, ..., m-1, m \in N$) takes either a positive form of the variable X_i or a negative form of the variable $\neg X_i$. The three variables Ψ_{j1}, Ψ_{j2}, and Ψ_{j3} in the clause are called the literals. The goal of a 3-SAT problem is to find solutions that consist of n variables from $\{Xi\}$ and that satisfy the condition of the clauses given above.

A DNA computing algorithm based on the Lipton encoding scheme [52] for 3-SAT problem solving and Adleman's experiment protocol consists of the following major steps:

- Step 1: Initialization // three variables are encoded as DNA sequences in which the Watson-Crick complementary form is used to distinguish values of T or F in logic. These correspond to the vertices in a graph.
- Step 2: Ligation reaction for the library of all candidates // The combinatorial forms of these variables are generated by the method in (1) and the library in (8).
- Step 3: Selection of the sequences according to the constraint of the clauses.
  For j = 0 to k do
   { // clause //
     for i=0 to 2 do
       {
       DNA-sequence(testtube'(i))<- extract (literal-i, testtube)
       }
     testtube<- testtube'(0) + testtube(1) + testtube(2)
   }

The operation for one clause in a 3-SAT problem is given in Figure 4.12 and the overall extraction process is given in Figure 4.13.

The principle discovered by Adleman's brave work clearly showed that molecules can be used to build future computers based on an unconventional computing paradigm as opposed to that of electronic computers. Nevertheless, new technology is still needed to handle the bottleneck of the DNA molecular number used, the so-called space complexity involving a limitation of scale with 70–80 variables for 3-SAT computation [98], as well as errors.

After 10 years of molecular computation, it is natural to ask whether the Adleman-Lipton algorithm for 3-SAT computing as an objectively existing law is actually applicable to any type of molecular computer.

4.3.2 RNA Computing

The year 2000 is considered a historic year in this field because DNA computing was extended beyond surface-based DNA computing into RNA computing. Dirk Faulhammer et al. also reported their experiment on molecular computing by RNA [69] and applied it to solving the Knight problem.

The direct benefit of using RNA as the material for molecular computers is the promise of opening up molecular computing to practical and efficient RNA technology, which has been developed for years. Furthermore, the degree of automating the implementation of molecular computers can be expected to

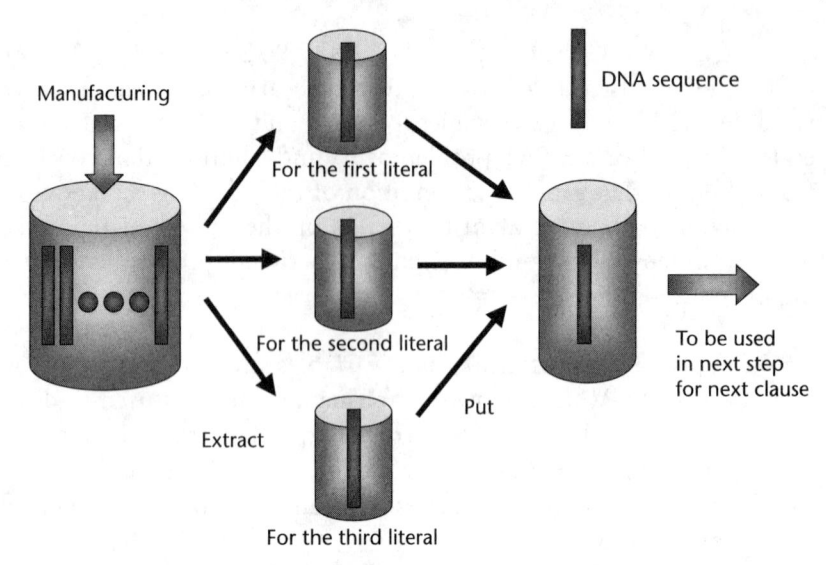

Figure 4.12 Operation of DNA computing for one clause in 3-SAT problem.

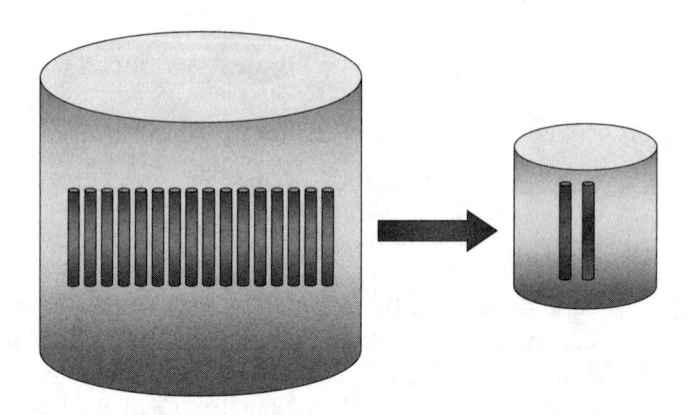

Figure 4.13 Extraction strategy of DNA computing for 3-SAT solving.

increase. If DNA computing and RNA computing were integrated, perhaps the living cells could contribute their central dogma to a guide mechanism for a totally automated computing machine.

In the experiment, the algorithm is designed based on the framework of operating candidate RNA sequences in a pool according to certain constraints. In the case of applying an RNA computing algorithm to solving the Knight problem, the problem is summarized as:

> Assume an n × n lattice, where n = 3 and the solution refers to the location of a chess knight in the lattice that satisfies the condition of no two knights being aligned. The rules of the chess game tell us that if knights are aligned, they will attack each other. In the field of artificial intelligence, searching algorithms have been developed by

many strategies. The constraint for the combinatorial forms of the locations is represented by AND and OR, and it can be regarded as a modified and derived form of the logical constraint problem as expressed in Skohoimers's method.

These are the main steps used in RNA computing:

Step 0: Prepare a DNA library and an RNA library.
Step 1: Select RNA sequences that satisfy the condition on locations.
Step 2: Read out the remaining RNA sequences in PCR.

Biochemical Features
Each location of the chess pieces is represented by an RNA sequence that corresponds to a certain DNA sequence in the related library. The destroyed sequences correspond to the removal of sequences designated as not-wanted. The set of solution locations is represented by an RNA sequence readout as the final result. Here, 10-mer sequences are designed and prepared in test tubes.

Problem Description
The problem space is $2^{n \times n}$, so the number of chess locations will grow exponentially when size n increases. The parallel RNA operations in the library of RNA sequences are efficient for exploring the parallelism of molecular computing. The time complexity is linearly dependent on the size of the Knight Problem. The scalability mainly depends on the technology of manufacturing RNAs, DNAs, and other related libraries.

Logical Programming
For operating in a library, the major logical operators are:

  set(library)
  destroy(bit, sequence)
  readout(sequence, primers)

These operations are run in parallel in the test tubes.

4.3.3 Surface-Based DNA Computing

Surface-based DNA computing was proposed by Liu et al. [70]. Compared with Adleman-type DNA computing, this method follows a different strategy for 3-SAT problem solving (see Figure 4.14).

The complementary form of the clause is used to remove the unwanted solutions from the pool of DNA sequences representing the combinatorial forms of variables. The details of this approach are given next.

4.3.3.1 Word Design

Encoding can be described as shown in Figure 4.14. Those parts that consist of a molecular computer are used for biochemical reactions. These schemes are

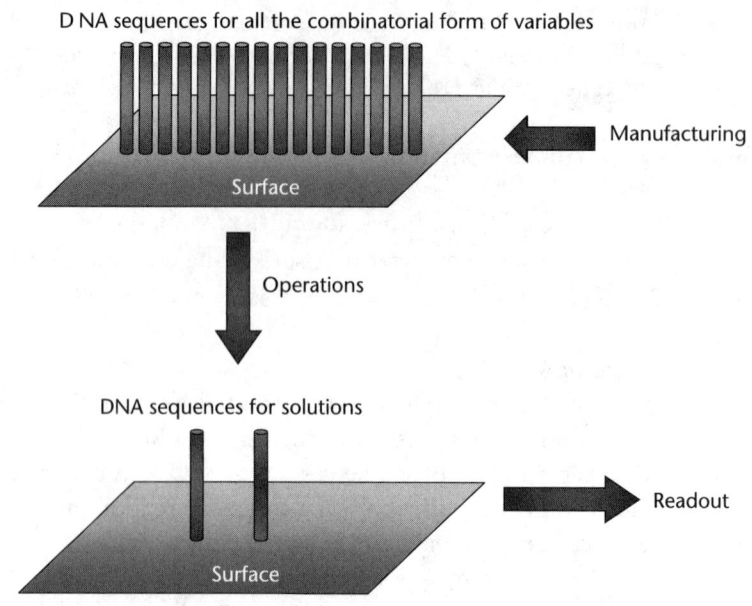

Figure 4.14 Basic strategy of surface-based DNA computing.

designed to achieve a low error rate, efficiency (e.g., speed) of biochemical reactions, and a high information capacity (e.g., coding ability).

4.3.3.2 Bioconditions

The complementary encoding scheme is implemented by ad hoc programming at the machine-language level, so it is not suitable for general-purpose computers. With the pool understood as the population, the candidates are selected during several cycles (generations). This kind of framework is similar to that of evolutionary computation, but the operators are completely different. The "hybridization" operation is often used to bind two DNA molecular sequences. It is also necessary to prepare a surface made from gold material.

4.3.3.3 Readout

The "readout" operators are for the I/O made by PCR. The "hybridization" operation here is used to attach the DNA sequences for final solutions to an addressed array. The detection of its fluorescent intensity can be realized on the support surface.

4.3.3.4 Operations

The set of bio-operations is {MAKE, ATTACH, MARK, DESTROY, UNMARK, READOUT}. MAKE here refers to making the DNA sequence, by encoding them with certain information. This is done for certain application problems and is commonly applied to biomolecular computing. ATTACH is developed especially for surface-based DNA computing in order to fix the DNA sequences on a surface made from gold material rather than letting them remain in the floating state in a test tube. This is an important improvement because it provides a practical form

4.3 Computing by Nucleic Acids

Word Design

Encoding

The encoded content

These parts are for biochemical reactions

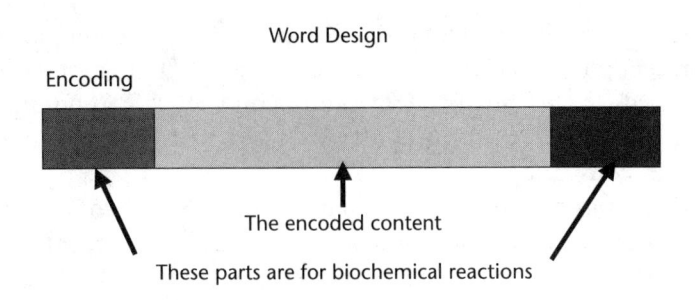

These schemes are designed to achieve a low error rate, efficiency (e.g., speed) of biochemical reactions, and high information capacity (e.g., coding ability).

Figure 4.15 Word design.

of I/O. MARK refers to arranging the DNA sequences representing the candidates to be "tested" by the DNA sequences that represent the clauses, where the satisfactory sequences are bound with the clause into double-strand forms. Unfitted ones are kept in the single-strand form. Accordingly, the desired sequences are marked. This is the principal operation in surface-based operations, and it plays a central role in the entire self-assembly process. DESTROY is used to wash out the unwanted single strands, which are similar to the unreacted DNA strands in test-tube form. UNMARK is used to restore the selected double strand into a single-strand form, where the DNA sequences for candidates are useful and the bound clause strands now become useless.

Word design in sequence-making and encoding (see Figure 4.15) also directly influences the performance of the related computing processes.

4.3.4 Nanobiotechnology for DNA Computing

From our findings reported in the previous sections, we can conclude that engineered technology for DNA manipulation is imperative for the development of DNA computers. The direct product of manufacturing spatial structures of DNA and/or RNA is molecular memory in the form of a DNA complex or an RNA complex. This makes us wonder how the technology will work for operators of much more complex DNA/RNA structures. In addition to finite automata of DNA computing, the computational capability of DNA computing should be understood as a theoretically equivalent concept to conventional computing, that is, the conceptual operators based on the engineered structure of molecules can carry out the computation of a Turing machine.

Several works have advanced the use of engineered methods to build a DNA/RNA [74] computer from the viewpoints of accessibility and programmability: the work on protein arrays and nanowires built according to DNA templates by Hao Yan et al. [73], the polymer assembly based on translation from DNA by Shiping Liao and Nadrian C. Seeman [72], and the Tectosquare RNA structure, which can be manufactured through certain designs, by Arkadiusz Chworos et al. [71]. The Tectosquare adopts jigsaw-like building blocks that can construct complex patterns of molecular compounds. In the sense of molecular memory, they are also molecular complexes just like an RNA complex used for RNA memory. We can go a bit further by exploiting the DNA nanomachine for possible computing. The recently reported translocases are types of DNA motors in cells. The linkage between the two fields of DNA motors and DNA computing lies in DNA nanotechnology, that is, the central idea we continue to emphasize: using a nanobiomachine for computing.

In hairpin-based DNA computing for 3-SAT problem solving, the hairpin structure is used to connect and disconnect the states repeatedly to realize the state transition in an autonomous way. The speed of biochemical reactions is mainly dependent on enzymic activities. The biomolecules cannot move quickly from one place to another without a motor mechanism, due to the diffusion principle in a liquid environment. A molecular motor can speed up the information delivery among separately located DNA sequences. Basically, DNA computing processes are constructed based on the operations of DNA sequences used for switching between the single-strand and double-strand forms. The DNA motors that move along the DNA are beneficial for frequently combining DNA strands in order to achieve fast communication among DNA-PEs.

Figures 4.16 and 4.17 show that a biochemical process can be used for communication, that is, to transport the information needed in PEs.

One of the new discoveries related to molecular motors is FtsK [75], a protein purified from *Escherichia coli*. This motor, called translocase, moves along or bonds with single DNA molecules. It can move in two directions. This attribute is beneficial for information exchange between two locations of PEs where DNA memory and a DNA computing unit work together in computing operations. Both short and long distances are reachable, and the motor's movement can use coordinates of the DNA sequence, thus providing the capability to access the DNA memory within the framework of DNA computer architecture. The speed of this motor is 5 kilobases per second. We can estimate that the communication speed of two PEs in a DNA computer can reach 250 bits per second if we consider a 20-mer DNA sequence as 1 bit (mer means the number of DNA molecules in the DNA

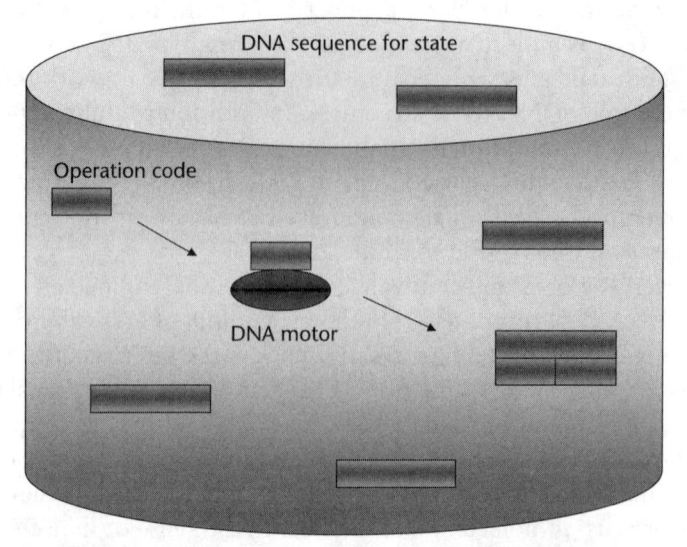

Figure 4.16 Communication between DNA motor and DNA computing.

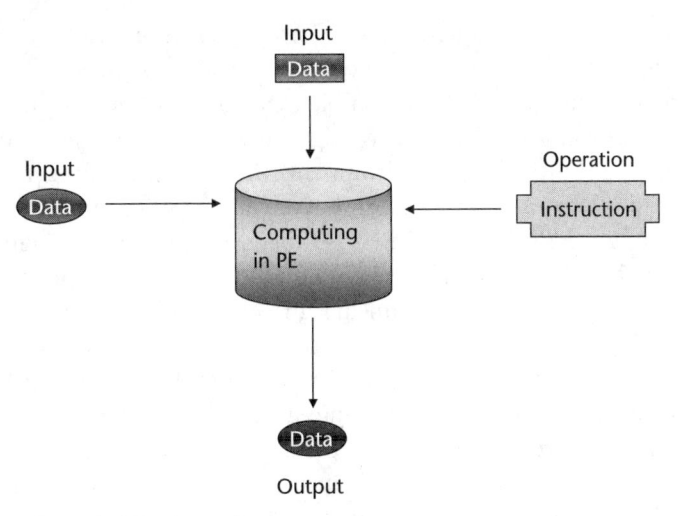

Figure 4.17 Operation of DNA computing.

sequence, and it equals the number of bp: base pairs). If 10^{15} motors were employed, the parallel communication speed in this nanobiosystem would become 2.5×10^{17} bits per second.

4.4 Computing by Biochemical Reactions in Microbes

In the previous section, we discussed the paradigms of molecular computing by nucleic acids—DNAs and RNAs. The biochemical operations on nucleic acids for solving NP problems are tailor-made technology. So far, there does not exist any complete system of molecular computing. Most existing experimental systems are for special purposes. At the same time, no one will deny that the semiconductor industry can easily produce a general-purpose computer from electrical circuits. In the methods discussed in Section 4.3, separated forms of individual DNA sequences/RNA sequences are mainly used. Using current biochemical technology to build a generic molecular computer isn't an easy job. For the goal of developing a complete functional molecular computer, a better choice appears to be using the microbe, a kind of naturally existing nanobiomachine in which the computing process can be designed under control. One of the characteristics of the microbe is that the computing process can be continuously maintained if we can control the signaling mechanism of microbial cells efficiently. In this section, we discuss the intercellular information processing mechanism of microbes in Section 4.4.1, the intracellular gene operation of ciliates in Section 4.4.2, and a control strategy for a cell culture array in Section 4.4.3.

4.4.1 Information Processing Mechanism of Microbes

4.4.1.1 Signal Processing Based on Microbial Cell Communication Mechanism

Microbes that contain bacteria are among the world's wonderful living beings, we cannot see them with the naked eye. *Vibro fischeri* is a marine prokaryote that has a symbiotic relationship with *Euprymna scolopes* and *Monocentris japonica*. *Euprymna scolopes* is a Hawaiian sepiolid squid. *Monocentris japonica* is a Japanese pinecone fish. The biological function of lighting is the physical basis of the controlled mechanism for signaling within cells and communication between cells in a multicellular environment. In 2000, an engineered cellular signaling system made from identical, unreliable cellular components of a microbe, where the Lux operon structure of genes is crucial to intercellular signaling, had been reported by Ron Weiss and Thomas F. Knight [76]. The components key to this biological implementation are biochemical signaling circuits in cells, where the cell-to-cell communication is carried out by controlling the interaction of gene and signaling proteins in microbes. The gene used to activate the cellular pathway is the input of the microbial information processing system. The protein generated from a microbial cell is the output. When these proteins interact with DNAs, control of genes will guide the reactions to generate new proteins. These generated proteins are the output. The control of the genes in cells is designed to activate the reactions toward the expected target of specific information mapping. The raw material is the microbial cells and the engineered control of the cellular signaling process makes the reactions predictable, and thus a certain degree of reliability can be expected for the derived biological circuits.

In the following, we summarize the main points of the biochemical information processing used by an engineered microbe system.

State-Level Description
The state refers to the gene sequence, or protein, in microbial cells. Transition: This involves the gene operations exerted on the plasmid in order for the microbe to undergo biochemical reactions in the organism. In this example, two microbial cells are used to demonstrate the information processing mechanism. One microbial cell is designated and controlled to send light, and it is called the *sender cell*. The other is designated to respond to the microbial sender cell, and it resends light after receiving the activation light signal from oustide, and it is called the *receiver cell*.

After digital information is input as the instruction/program code, the biochemical function of the microbe is activated. The biomachine-like mechanism of the microbe is used as moleware in the sense of hardware responding to the wet instruction/software.

Logical Programming
The primitives for logical programming are summarized as follows:

Gene-operation (Lux operator)
Metabolic pathway (VAI)
Send (light, sender)

Sense (light, receiver)
Gene-operator (LuxR)
Pathway (VAI, UV, GFP)
Resend (light)

As Figure 4.18 shows, within a sender cell, "gene-operation (Lux operator)" is activated by gene regulation in the plasmid. These DNA molecules will interact with the cellular proteins. Through the metabolic pathway in cells, VAI is generated and this causes the cell to emit light. In the receiver cell, when light passes through its membrane, the gene-operator (LuxR) is activated. Then, the pathway (VAI, UV, GFP) is activated to make the cell emit light as a response to its input-light. In a word, this example shows a quorum signaling process by engineered control of microbes. Recently, based on the cellular signaling mechanism of microbes, Ron Weiss reported a biochemical logical NOT gate and amplifier [99], laying an experimental foundation for microbial circuits applied to signal-level computing. Even though different forms of biomolecular circuits, such as the NOT gate, amplifier, and signal sender/receiver, have been developed, much work remains to be done toward a kind of general purpose biomolecular circuit based on element circuits.

The above example from Weiss and Knight is called amorphous computing [76], in which the information processing is realized by moleware at the microbe level.

4.4.2 Computing by Gene Operations in Ciliates

Computing in ciliates by Laura Landweber and Lali Kari [77], as well as by A. Ehrenfeught et al. [84], opened a new door in the field of DNA computing. Here, the naturally existing gene reproduction mechanisms have been formalized as rigorous computation models. Based on their investigations into formal languages and computability, excellent characteristics of molecular memory and information capacity in terms of theoretical computer science have been shown.

The ciliate, also called ancient bacteria, adopts a strange strategy for survival, which has evolved from nature. It selects asexual breeding when the nutrient environment is bad and sexual breeding when the nutrient environment is

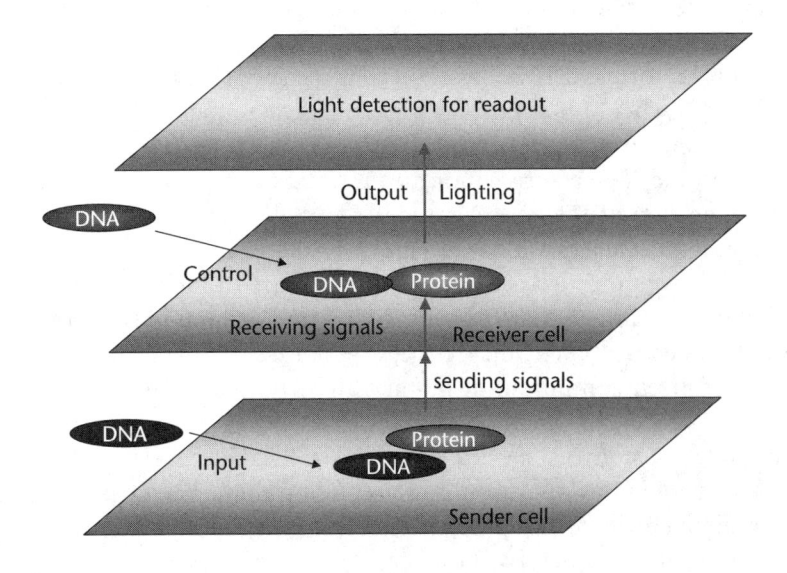

Figure 4.18 Illustration of communication among cells.

good. This resembles a kind of plant tree that uses a similar method for reproduction/generation. The asexual reproduction corresponds to one kind of gene operation. The sexual reproduction corresponds to the other kind of geneoperation. Corresponding to these biological processes, the corresponding biochemical phenomena for a gene information processing mechanism is explained by the observations that there exist two kinds of gene operations (that is, macro-gene operation and micro-gene operation) that use contiguously located and discretely located gene information in the ciliate chromosome, respectively.

These observations can be formalized into a computation model. In mathematics, the computing process is equivalent to the process of rewriting on strings. Through corresponding biochemical reactions, we can infer a process of rewriting on graphs. This is because the strings—gene strings—are interacted by corresponding biochemical reactions within a gene operation. The interactions of these strings occur in the biochemical reactions, and a corresponding graph can be generated where the node is the gene sequence and the edge is the reaction.

From the viewpoint of biochemistry, the gene materials for ciliate-based computing can be extracted from *Oxytricha nova* and *Oxytricha trifallax*. Gene unscrambling is one of the basic gene operations. There are two forms of gene representation: macronucleus and micronucleus forms. The micronucleus form represents the spatially contiguous gene information. The macronucleus form represents the spatially discrete gene information. In the micronucleus form of gene representation, genes are continuously encoded in either of the two gene strands of a chromosome. In the macronucleus form of gene representation, genes gather into several groups in either of the two gene strands of chromosomes. The mapping between two chromosomes is nonlinear.

Any gene sequence among chromosomes has the possibility of being combined. Many combinatorial forms of A, T, C, G can be generated by the method discussed above in theory. In Andrzej Ehrenfeucht et al.'s work [84], they used three kinds of gene operators that have different combinatorial forms of gene information from different locations in two-strand gene sequences. The two-strand gene sequences are bound into a chromosome. This chromosome is shaped with certain topological forms. It is understood that a general representation of gene operation is $\{X_iY_j\}^*$, where $i, j \in N$, X is a partial gene sequence located in one strand of a chromosome and Y is a partial gene sequence located in either of two strands of a chromosome. The asterisk indicates that this string can be repeated in a recursive way. One step of the operation is illustrated in Figure 4.19, and its mapping for generating a string is shown in Figure 4.20.

While we take into account the equivalence of the biochemical processes in ciliate and the mathematical process in theory, we should also take note of the difference between the biochemical process in nature—the gene operation in ciliate—and the formalized computing in mathematics. As mentioned in Section 4.1, the biochemical process has many outputs corresponding to specific inputs, but a computing model must give a specific output. As for ciliate-based molecular computing in vivo, control in the laboratory to select and stabilize the gene operations is requisite for building a ciliate-computer.

The space complexity of the method mentioned above is 2^n, and the time complexity is the steps of the rewriting. Because string rewriting only deals with the

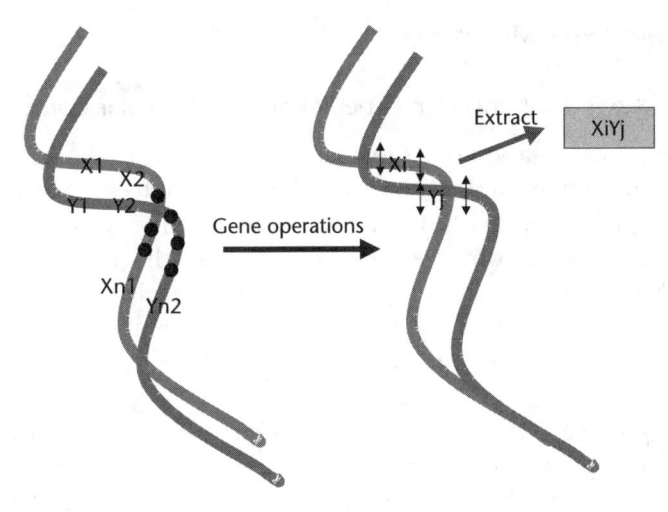

4.4 Computing by Biochemical Reactions in Microbes

Figure 4.19 Gene operations.

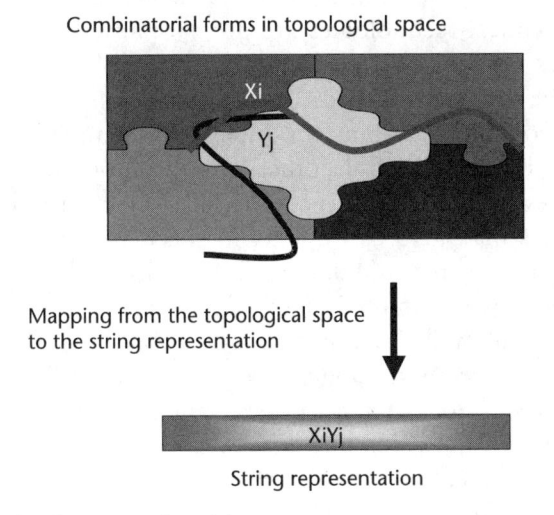

Figure 4.20 Mapping for generating string.

difference between the previous state and the current state, its efficiency has to be studied from concrete examples. In the case of applying ciliate-based molecular computing to solving HPP, the space complexity is the number of gene sequences representing the number of vertexes, and the time complexity is the time of operations on the gene sequences:

$$\text{time complexity} = n + m + \varepsilon \tag{4.3}$$

where n is the number of vertexes, m is the number of edges, and ε is a quantity that is much smaller than $n + m$. The logic operations include "micro-gene-operator()" and "macro-gene-operator()." Based on these two basic forms, combinatorial forms of the building blocks of genes can be derived.

4.4.3 Moleware Microarray

4.4.3.1 Considerations from Experiments on Molecular Computing

Possible implementation of a molecular computer is no doubt the biggest challenge for experimental technology. Based on biochemical technology, different choices of materials greatly influence the performance of an experimental system. Perhaps readers have already observed this characteristic through information representation schemes from the various paradigms of molecular computing. The realization of a molecular computing system in the laboratory requires at least the following three major improvements:

1. *Applying and modifying the existing automated devices in molecular biology.* According to the state-of-the-art biochemical technology, there are assistant-like devices such as the DNA chip, DNA microarray, protein micro-array, cell culture chip, multiple-path electrophoresis chip, and others [79–82]. These arrays and chips can be helpful for the readout of molecular computing systems. However, such widely used devices from biochemical technologies are not specially developed for molecular computers but for the needs of molecular cell biology. Consequently, they are unable to meet the needs of molecular computers. As we know, in molecular computing, inputs are often done manually, and some readouts are aided by devices while others are still done manually. Thus, automated devices would seem crucial. What has occurred in genomics and proteomics may give us impetus to speed up the application of automated machines. Automatic devices that act like robots have replaced the labor-intensive works in the human genome project by performing the functional protein analysis. In molecular computing, many tasks are expected to be done by automated machines, but the degree of automation nowadays is still very low. A tendency in molecular computing is to integrate test tubes with the chips into a system to upgrade this degree of automation (for example, the lab-on-a-chip [81] for DNA computing developed by Lim et al. in 2003 [83]).

2. *Introducing biological engineering technology.* Systematic equipment used for chemical engineering, such as microreactors, has been developed for use in industry for years. But just recently, microreactors and multiple pipe structures that can further upgrade the entire process of molecular computing have been introduced in DNA computers by McSkill et al. [100]. However, the current experimental systems of molecular computing are far from evolving into an integrated automated system. We need to consider the serious issue of how to apply such technology based on microreactors to molecular computing for more efficient molecular manipulation.

3. *Empirical schemes.* The biochemical experimental technology is often categorized into two biochemical forms—in vitro form and in vivo form. The in vivo form is more difficult than the in vitro form in the laboratory. Some reported DNA computing methods have been made in vivo, for example, ciliate-based computing by Landweber and Kari [77] as well as by Ehrenfeucht et al. [84]. In microbe-based computing systems, most of the

materials are extracted from cells and have to be purified. For example, the following empirical schemes based on test tubes are often used for molecular computing:

- Evolution in vitro;
- Aqueous computing by plasmid methods.

Evolution in vitro is a kind of biochemical method that is useful for biochemical operations in test tubes. Currently, it is being applied in molecular computing. The *E. coli* evolution in test tubes works well under the condition of engineered technology. This method uses bacteria, which are readily available material for directed evolution, or the so-called evolution in vitro. Using this method, the gene sequences can be operated to carry out computation. The cheap bacteria *E. coli* can reproduce a large number of copies. Even in one night, many generations of *E. coli* can be reproduced. The abundant knowledge of *E. coli* in textbooks helps us to make best use of this ever-available material. Using plasmid in test tubes is one of the cheapest and most convenient ways of DNA computing.

About a decade before Adleman's celebrated experiment on DNA computing, Tom Head had proposed an algorithm for applying DNA molecules to mathematical computation, which was named the H-system, or splicing system. The current method has been developed as a branch of DNA computing, especially from the viewpoint of theoretical computer science, by "evolving" from his early computation model. Recently, he and his coresearchers have been working on realizing gene operations by plasmid, which is called aqueous computing.

Aqueous computing, as its name implies, is water-like computing. The Latin origin of aqua is water, and water is so important for life and biochemistry that it cannot be taken for granted. As we have discussed in the previous section, the DNA recombinant technology needs equipment to enhance the handling ability of DNA sequences, but how wonderful it would be to directly write on the DNA. Normally, DNA sequences are designed to be used only once for specific computing tasks. If we want carry out the DNA computing process again for other computing tasks, we have to prepare a DNA sequence from scratch. In order to scale up the experimental schemes of plasmid-based operations for DNA computing in an aqueous environment, we need to study the cyclic gene structure to make best use of the greater information in a gene. Considering various types of media for molecular computing, more structural information becomes imperative. The complicated structure of moleware for computing requires us to control the system in order to achieve stability and autonomy.

4.4.3.2 Structural Control Inspired by Nanobiomachine

Structural information from biological cells has been emphasized in recent years by system biology. Structural biology not only tells us the spatial structure of molecules but also their functions. The physical features of the material such as interaction between molecules and chemistry characteristics of the enzymes, as well as reactions, are key factors for molecular computing. Because the sizes of cellular biomolecules range from 10 nm to 100 nm, we normally try to grasp the molecules in the sense of recognition, that is, we need to know which molecule is included in the result when the molecular computing is finished. In these cases, we do not care where these molecules are physically located at the nanometer-size level. A nanobiomachine works precisely at the nanometer scale, and its function is determined by the underlying structure, where the operators are designed according to the spatial constraints. Ion channels in a membrane are regarded as a kind of nanobiomachine in cells. The concept of a membrane has been proposed for computing by G. Pâun [78, 85]. A theoretical branch of molecular computing based on the concept of membrane and membrane-inspired information processing mechanisms has arrived as a welcome field. The mathematical concept of a multiset arrangement and the computation model by rewriting rules have been intensively and extensively studied in the field of theoretical computer science. Much more effort in theoretical membrane computing and membrane-based experimental technology still needs to be made if we want to realize an experimental membrane computer.

Toward the final target of realizing a molecular computer by microbial cells, the laboratory techniques are extremely important. Preparing and operating cell culture requires more techniques than manufacturing and operating on DNA molecules. The recognition of the desired molecules in a cell depends on the accuracy of molecules. In the cell, there are many proteins that interact, so it is necessary to control the states of these proteins. In the in vitro form of molecular computing using proteins in cells, each kind of protein can be handled in separate test tubes, and the combinatorial forms are designed according to certain expectations. In the in vivo form, the cell growth and activated states of the related proteins need to be carefully handled. The molecules that sustain the cell life and the molecules for operations of molecular computers have to be systematically taken into consideration and harmonically arranged.

If we had the experience of carrying out manual work and using partially automated devices in biochemical laboratories, we would realize the importance of developing new tools reflecting the key transition from detection of structural information to the structural design of a molecular computing system. We mean that in order to utilize the structures of biomolecules and the structures of cells, we have to design the structure of experimental systems in which engineered control schemes can be introduced to improve system performance. We discuss this topic through the example of engineered control of moleware microarrays [79, 80, 82] based on a quantitative measure.

We stress the control of moleware microarrays for three reasons:

1. The materials used in molecular computing are biomolecules extracted from cells in nature or prepared in an assay of cell culture in which the concentration has to be sustained at a certain degree of stability.
2. Under the natural existing state of cells, the proteins are strongly coupled in signals.
3. In order to obtain stable states of molecules in experiment, we have to control the related pathways based on quantitative technology by using the features of molecular interactions.

As Figure 4.21 shows, a microarray processes biochemical reactions in parallel. In order to keep the computing process stable, we have to design a controller for this microarray (see Figure 4.22). The control rule is given as follows:

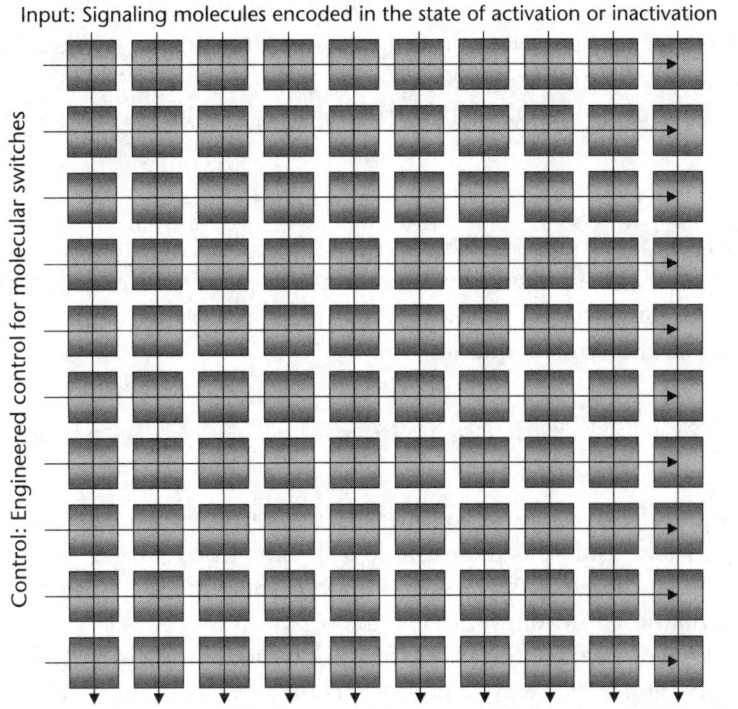

4.4 Computing by Biochemical Reactions in Microbes

Figure 4.21 Moleware microarray.

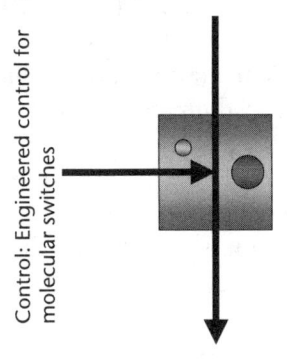

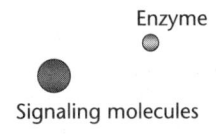

Figure 4.22 Control unit.

if variation of Y_j is smaller than empirical threshold $T(Y_j)$, then activation of Y_j is regulated by enzyme control.

Here, Xi (i=0,1,...,n-1, n∈ N), Yj (j=0,1,...,m-1, m∈ N) refers to the input of molecular sequences and the output of molecular sequences. The concentration is used as the measure for control.

The corresponding concentration of Y_j will be controlled by enzyme:

$$\Delta Y_j = (1 + \alpha) Y_j^* \tag{4.4}$$

where α is a random number in the range of (–1, 1) that excludes 0, which can be estimated by empirical data obtained from biochemical experiments.

The quantitative relation between Xi and Yj is represented by the following random differential equation:

$$X_i(t+1) = \beta X_i(t) + (\gamma_1 X_i + \gamma_2 Y_j) W_i(t) \tag{4.5}$$

where $W_i(t)$ is the random variable in a random process, β and γ_1 are empirical parameters, and γ_2 is the variable to be identified.

When these parameters are estimated in statistics, we can calculate the Y_j^* by the equation according to the value of Xi.

The cell is home to various types of biomolecules—DNA, RNA, and proteins—where cellular pathways adaptively control the complicated activities of a molecular network to functionally sustain the life of a cell. Based on what has been learned from biological cells in nature, materials extracted from cells have been adopted in the design of molecular computing. This methodology encompasses DNA computing, RNA computing, amorphous computing, ciliate computing, aqueous computing, and kinase computing (presented in Chapter 6). Not only do tailor-made empirical designs greatly affect the performance of a molecular computing system but also the implementation form of such a system varies according to the features of the molecular medium used. Fortunately, based on the identity of information storage and instruction in moleware, conceptual operators in terms of logic programming can be formulated to describe the computing model supported by a biochemically faithful mechanism of moleware. This may lead to explorations in a new theory of molecular computers, guiding the empirical designs of experimental nanobiosystems.

References

[1] An essay for "The Year of Physics," *Nature*, Vol. 433, January 20, 2005.
[2] "Dickerson's Formula: Biochemistry's Equivalent to Moore's Law," *Computational Medicine*, Vol. 18, No. 1, January–March 2002, http://www.npaci.edu/envision/v18.1/moore.html.
[3] Xiao, M., et al., "Electrical Detection of the Spin Resonance of a Single Electron in a Silicon Field-Effect Transistor," *Nature*, Vol. 430, July 2004, pp. 435–439.
[4] Elzerman, J. M., et al., "Single-Shot Read-Out of an Individual Electron Spin in a Quantum Dot," *Nature*, Vol. 430, July 2004, pp. 431–435.
[5] Kroutvar, M., et al., "Optically Programmable Electron Spin Memory Using Semiconductor Quantum Dots," *Nature*, Vol. 432, November 2004, pp. 81–84.

[6] Itatani, J., et al., "Tomographic Imaging of Molecular Orbitals," *Nature*, Vol. 432, December 2004, pp. 867–871.

[7] Stapelfeldt, H., "Electrons Frozen in Motion," *Nature*, Vol. 432, December 2004, pp. 809–810.

[8] Milliron, D. J., et al., "Colloidal Nanocrystal Heterostructures with Linear and Branched Topology," *Nature*, Vol. 430, July 2004, pp. 190–195.

[9] Hashimoto, A., et al., "Direct Evidence for Atomic Defects in Graphene Layers," *Nature*, Vol. 430, August 2004, pp. 870–873.

[10] Eisebitt, S., et al., "Lensless Imaging of Magnetic Nanostructures by X-Ray Spectro-Holography," *Nature*, Vol. 432, December 2004, pp. 885–888.

[11] "Imaging Techniques: Seeing Single Spins," *Nature*, Vol. 430, July 2004, pp. 300–301.

[12] Rohrer, H., "The Grand Challenges in a Pervasive Nanotechnology," *Proc. of Kyoto University SPM Workshop*, February 5, 2004.

[13] http://www-kawai.sanken.osaka-u.ac.jp/home_en.html.

[14] http://web.media.mit.edu/~nicholas/.

[15] Stix, G., "Little Big Science," *Scientific American*, Vol. 285, 2001, pp. 32–37.

[16] Gibbs, W. W., "Atomic Spin-Offs for the 21st Century," *Scientific American*, September 2004, pp. 57–65, in which the author says "The Einsteinian pachinko machine reliably separated the two viral genomes."

[17] Iijima, S., "Helical Microtubules of Graphitic Carbon," *Nature*, Vol. 354, November 1991, pp. 56–58.

[18] Hashimoto, A., et al., "Direct Evidence for Atomic Defects in Graphene Layers," *Nature*, Vol. 430, August 2004, pp. 870–873.

[19] Giot, L., et al., "A Protein Interaction Map of Drosophila Melanogaster," *Science*, Vol. 302, December 2003, pp. 1727–1736.

[20] Li, S., et al., "A Map of the Interactome Network of the Metazoan C. Elegans," *Science*, Vol. 303, January 2004, pp. 540–543.

[21] Han, J.-D. J., et al., "Evidence for Dynamically Organized Modularity in the Yeast Protein-Protein Interaction Network," *Nature*, Vol. 430, July 2004, pp. 88–93.

[22] Oh, P., et al., "Subtractive Proteomic Mapping of the Endothelial Surface in Lung and Solid Tumours for Tissue-Specific Therapy," *Nature*, Vol. 429, June 2004, pp. 629–635.

[23] Friedman, N., "Inferring Cellular Networks Using Probabilistic Graphical Models," *Science*, Vol. 303, February 2004, pp.799–805.

[24] Suharsono, U., et al., "The Heterotrimeric G Protein α Subunit Acts Upstream of the Small GTPase Rac in Disease Resistance of Rice," *PNAS*, October 2002, Vol. 99, No. 20, pp. 13307–13312.

[25] Wang, Y., and M. Lieberman, "Thermodynamic Behavior of Molecular-Scale Quantum-Dot Cellular Automata (QCA) Wires and Logic Devices," *IEEE Trans. on Nanotechnology*, Vol. 3, No. 3, September 2004, pp. 368–376.

[26] Andersen, J. S., et al., "Nucleolar Proteome Dynamics," *Nature*, Vol. 433, January 2005, p. 77.

[27] Fukuhara, A., et al., "Visfatin: A Protein Secreted by Visceral Fat That Mimics the Effects of Insulin," *Science*, 2005.

[28] Hug, C., and H. F. Lodish, "Visfatin: A New Adipokine," *Science*, 2005.

[29] Bhattacharya, K., and R. D. James, "The Material Is the Machine," *Science*, Vol. 307, January 2005, pp. 53–54.

[30] Leong, M., et al., "Silicon Device Scaling to the Sub-10-nm Regime," *Science*, Vol. 306, December 2004, pp. 2057–2060.

[31] Yan, H., "Nucleic Acid Nanotechnology," *Science*, Vol. 306, December 2004, pp. 2048–2049.

[32] "Supramolecular Chemistry & Self-Assembly," special issue, *Science*, Vol. 295, March 2002.

[33] "Bodybuilding: The Bionic Human," special issue, *Science*, Vol. 295, February 2002.

[34] Otomo, T., et al., "Structural Basis of Actin Filamnet Nucleation and Processive Capping by a Form in Homology 2 Domain," *Nature*, Vol. 433, February 2005, pp. 488–494.

[35] Burgess, S. A., et al., "Dynein Structure and Power Stroke," *Nature*, Vol. 421, No. 13, February 2003, pp. 715–718.

[36] Vallee, R. B., and P. Höök, "A Magnificent Machine," *Nature*, Vol. 421, No. 13, February 2003, pp. 701–702.

[37] http://www.res.titech.ac.jp/~seibutu/.

[38] Spitzer, N. C., and T. J. Sejnowski, "Biological Information Processing: Bits of Progress," *Science*, August 1997, Vol. 277, pp. 1060–1061.

[39] Baum, E. B., "A DNA Associative Memory Potentially Larger than the Brain," *Proceedings of a Dimacs Workshop*, April 4, 1995, Princeton University, edited by R. J. Lipton and E. B. Baum, Amer. Mathematical Society, pp. 23–28.

[40] Winfree, E., T. Eng, and G. Rozenberg, "String Tile Models for DNA Computing by Self-Assembly," in A. Condon, and G. Rozenberg (eds.) *DNA Computing—6th International Workshop on DNA-Based Computers*, DNA 2000, Leiden, the Netherlands, June 13–17, 2000, revised papers, Lecture Notes in Computer Science, Vol. 2054, Springer-Verlag, 2001.

[41] Adleman, L. M., "Molecular Computation of Solutions to Combinatorial Problems," *Science*, Vol. 266, November 1994, pp. 1021–1024.

[42] Ruggiero, C., et al., "The Nanobioworld," *IEEE Trans. on Nanobioscience*, Vol. 1 No. 1, March 2002, pp. 1–3.

[43] Tomita J., et al., "No Transcription-Translation Feedback in Circadian Rhythm of KaiC Phosphorylation," *Science*, January 2005, Vol. 307, pp. 251–254.

[44] Etienne-Manneville, S., and A. Hall, "Rho GTPases in Cell Biology," *Nature*, Vol. 420, 2002, pp. 629–635.

[45] Kaibuchi, K., S. Kuroda, and M. Amano, "Regulation of the Cytoskeleton and Cell Adhesion by the Rho Family GTPases in Mammalian Cells," *Annu. Rev. Biochem*, Vol. 68, 1999, pp. 459–486.

[46] Kawano, Y., et al., "Phosphorylation of Myosin-Binding Subunit (MBS) of Myosin Phosphatase by Rho-kinase In Vivo," *The Journal of Cell Biology*, Vol. 147, 1999, pp. 1023–1037.

[47] Hafen, E., "Kinase and Phosphatases—A Marriage Is Consummated," *Science*, Vol. 280, 1998, pp. 1212–1213.

[48] Helmreich, E. J. M., *The Biochemistry of Cell Signalling*, New York: Oxford University Press, 2001.

[49] Scott, J. D., and T. Pawson, "Cell Communication: The Inside Story," *Scientific American*, Vol. 282, 2000, pp. 54–61.

[50] Gershenfeld, N., *The Physics of Information Technology*, Cambridge, U.K.: Cambridge University Press, 2000.

[51] Blahut, R. E., *Principles and Practice of Information Theory*, Reading, MA: Addison-Wesley, 1987.

[52] Lipton, R. J., "DNA Solution of Hard Computational Problems," *Science*, Vol. 268, April 1995, pp. 542–545.

[53] Shannon, C., "A Mathematical Theory of Communication," *Bell System Technical Journal*, Vol. 27, July and October 1948, pp. 379–423 and 623–656.

[54] Zimmermann, K. H., "On Applying Molecular Computation to Binary Linear Codes," *IEEE Trans. on Information Theory*, Vol. 48, No. 2, February 2002, pp. 505–510.

[55] Mauri, G., and C. Ferretti, "Word Design for Molecular Computing: A Survey," J. Chen and J. Reif, (eds.) *DNA 9*, LNCS 2943, 2004, pp. 37–47, Berlin: Springer-Verlag.

[56] Liu, J. Q., and K. Shimohara, "A Biomolecular Computing Method Based on Rho Family GTPases," *IEEE Trans. on Nanobioscience*, Vol. 2, No. 2, June 2003, pp. 58–62.

[57] Liu, F., et al., "Direct Protein-Protein Coupling Enables Cross-Talk Between Dopamine D5 and γ-aminobutyric Acid A Receptors," *Nature*, Vol. 403, 2000, pp. 274–280.

[58] Tarricone, C., et al., "The Structural Basis of Arfaptin-Mediated Cross-Talk Between Rac and Arf Signalling Pathways," *Nature*, Vol. 411, 2001, pp. 215–219.

[59] Digicaylioglu, M., and S. A. Lipton, "Erythropoietin-Mediated Neuroprotection Involves Cross-Talk Between Jak2 and NF-κβ Signaling Cascades," *Nature*, Vol. 412, 2001, pp. 641–647.

[60] Katoh, H., and M. Negishi, "RhoG Activates Rac1 by Direct Interaction with the Dock180-Binding Protein Elmo," *Nature*, Vol. 424, 2003, pp. 461–464.

[61] Yoo, A. S., C. Bais, and I. Greenwald, "Crosstalk Between the EGFR and LIN-12/Notch Pathways in C. Elegans Vulval Development," *Science*, Vol. 303, 2004, pp. 663–666.

[62] Masip, L., et al., "An Engineered Pathway for the Formation of Protein Disulfide Bonds," *Science*, Vol. 303, 2004, pp. 1185–1189.

[63] Li, S., et al., "A Map of the Interactome Network of the Metazoan C. Elegans," *Science*, Vol. 303, January 2004, pp. 540–543.

[64] Han, J. D. J., et al., "Evidence for Dynamically Organized Modularity in the Yeast Protein-Protein Interaction Network," *Nature*, Vol. 430, July 2004, pp. 88–93.

[65] Mochizuki, N., et al., "Spatio-Temporal Images of Growth-Factor-Induced Activation of Ras and Rap1," *Nature*, Vol. 411, June 28, 2001, pp. 1065–1068.

[66] Adleman, L., "Computing with DNA," *Scientific American*, Vol. 279, August 1998, pp. 54–61.

[67] Adleman, L., "On Constructing a Molecular Computer, DNA Based Computers," R. Lipton and E. Baum, (eds.) *DIMACS: Series in Discrete Mathematics and Theoretical Computer Science*, American Mathematical Society, 1996, pp. 1–21.

[68] Braich, R. S., et al., "Solution of a 20-Variable 3-SAT Problem on a DNA Computer," *Science*, Vol. 0, No. 2002, January 1900.

[69] Faulhammer, D., et al., "Molecular Computation: RNA Solutions to Chess Problems," *PNAS*, Vol. 97, 2000, pp. 1385–1389.

[70] Liu, Q., et al., "DNA Computing on Surfaces," *Nature*, Vol. 403, January 2000, pp. 175–179.

[71] Chworos, A., et al., "Building Programmable Jigsaw Puzzles with DNA," *Science*, Vol. 306, December 2004, pp. 2068–2072.

[72] Liao, S., and N. C. Seeman, "Translation of DNA Signals into Polymer Assembly Instructions," *Science*, Vol. 306, December 2004, pp. 2072–2074.

[73] Yan, H., et al., "DNA-Templated Self-Assembly of Protein Arrays and Highly Conductive Nanowires," *Science*, Vol. 301, September 2004, pp. 1882–1884.

[74] *DNA and RNA Swap Roles*, Vol. 306, December 2004, p. 1997.

[75] Pease, P. J., et al., "Sequence-Directed DNA Translocation by Purified FtsK," *Science*, Vol. 307, January 2005, pp. 586–590.

[76] Weiss, R., and T. F. Kight, Jr., "Engineered Communications for Microbial Robotics," A. Condon (ed.), *DNA 2000*, LNCS 2054, 2001, pp. 1–16.

[77] Landweber, L. F., and L. Kari, "The Evolution of Cellular Computing: Nature's Solution to a Computational Problem," *BioSystem*, Vol. 52, 1999, pp. 3–13.

[78] The P Systems Web Page http://psystems.disco.unimib.it/.

[79] Gershon, D., "DNA Microarray Technology," *Nature*, Vol. 416, No. 6883, April 2002.

[80] Gershon, D., "Microarray Technology: An Array of Opportunities," *Nature*, Vol. 416, April 2002, pp. 885–891.

[81] http://www.lab-on-a-chip.com/home/index.aspx.

[82] http://www.microarray.org/sfgf/jsp/home.jsp.

[83] Lim, H. W., et al., "A Lab-on-a-Chip Module for Bead Separation in DNA-Based Concept Learning," in J. Chen and J. H. Reif, (eds.): *DNA Computing, 9th International Workshop on DNA Based Computers, DNA9*, Madison, WI, June 1–3, 2003, revised papers, Lecture Notes in Computer Science 2943, Springer 2004, pp. 1–10.

[84] Ehrenfeucht, A., et al., *Computation in Living Cells—Gene Assembly in Ciliates, Series: Natural Computing Series*, New York: Springer-Verlag, 2004.

[85] Paun, G., *Membrane Computing: An Introduction*, Berlin: Springer-Verlag, 2002.

[86] Winfree, E., et al., "Design and Self-Assembly of Two-Dimensional DNA Crystals," *Nature*, Vol. 394, 1998, pp. 539–544.

[87] Mao, C., et al., "Logic Computation Using Algorithmic Self-Assembly of DNA Triple-Crossover Molecules," *Nature*, Vol. 407, 2000, pp. 493–496.

[88] Benenson, Y., et al., "Programmable and Autonomous Computing Machine Made of Biomolecules," *Nature*, Vol. 414, 2001, pp. 430–434.

[89] Benenson, Y., et al., "An Autonomous Molecular Computer for Logical Control of Gene Expression," *Nature*, Vol. 429, 2004, pp. 423–429.

[90] Yurke, B., et al., "A DNA-Fuelled Molecular Machine Made of DNA," *Nature*, Vol. 406, 2000, pp. 605–608.

[91] Liu, J. Q., and K. Shimohara, "On Designing Error-Correctable Codes by Biomolecular Vomputation," *Proceedings of International Symposium on Information Theory 2005* (ISIT 2005), September 4–9, 2005, pp. 2384–2388.

[92] Liu, J. Q., and K. Shimohara, "Molecular Computation and Evolutionary Wetware: A Cutting-Edge Technology for Artificial Life and Nanobiotechnologies," *IEEE Trans. on Systems, Man and Cybernetics—Part C: Applications and Reviews*, accepted for publication.

[93] http://www.bio.nagoya-u.ac.jp/coe/2004pamphlet/01.html.

[94] http://biol.bio.nagoya-u.ac.jp/~blhome/04/overview.htm.

[95] Young, M. W., and S. A. Kay, "Timezones: A Comparative Genetics of Circadian Clocks," *Nature Reviews Genetics*, Vol. 2, 2001, pp. 702–715.

[96] http://www.weizmann.ac.il/mathusers/lbn/new_pages/new_pages/Cert_small.jpg.

[97] Ogihara, M., "Breadth First Search 3SAT Algorithms for DNA Computers," Technical Report TR-629, Department of Computer Science, University of Rochester, July 1996. http://www.cs.rochester.edu/u/ogihara/research/DNA/bfs.pdf.gz.

[98] Reif, J. H., "Successes and Challenges," *Science*, Vol. 296, 2002, pp. 478–479.

[99] http://www.ee.princeton.edu/people/Weiss.php.

[100] McCaskill, J. S., et al., "SteadyFlow Micro-Reactor Module for Piplined DNA Computations," A. Condon and G. Rozenberg, (eds.), *DNA 2000*, LNCS 2054, 2001, pp. 263–270.

CHAPTER 5
Theoretical Biomolecular Computing

We use electronic computers everyday. When we write a program and run the program on such a computer, we need not get deeply involved in the hardware in most cases. This attribute is called transparency. The transparency function of computers has even been extended to interactive and creative forms of computers for human interfaces. Unfortunately, the status of biomolecular computers is still far from reaching this level. The transparency of a computer is the capability of a computer system to execute program instructions automatically without the intervention of users, making it possible to write computing tasks in a conceptual form (e.g., the conceptual description in object-oriented programming) without directly specifying the hardware configuration. In order to obtain transparency in biomolecular computing, much remains to be done. As a fundamental step, we have to start from the primitive instance of any programming (i.e., designing an algorithm).

Algorithmic design is a traditional topic in theoretical computer science as well as in practical programming and software engineering. Considering the situation in molecular computing, it is not enough to simply resort to existing computer science theory. Rather, it is imperative to innovate new unconventional computing theories. For the computing process of molecular computing, we can see the material with the aid of nanotechnology, but the information created by the biochemical reaction corresponding to the algorithm needs to be understood with a certain semantic interpretation.

It might be helpful to view the process of molecular computing through the metaphor of a "Kabuki" drama (Kabuki is a traditional Japanese theatrical performance). The molecule would be the face of the actor with the makeup, and the information represented by the molecule would be the intended meaning of the facial expression. The actor's face can be easily seen, but understanding the meaning requires specific knowledge of Kabuki drama.

In Adleman's example of NP problem solving, the implementation of the DNA computing algorithm, which is equivalent to "protocol" in his work and in the performance of Kabuki drama, is done differently from that in electronic computers. Information in DNA computing is directly represented in the molecular complexes and the protocol realized by controlled biochemical reaction chains, while the information in an electronic computer is mapped from a software program to the semiconductor hardware.

In biomolecular computing, one of the most obvious features of informatics operators can be understood by the concept that "what the molecules encode is just what the molecules compute." The computing process in the information space is mapped into a signaling process of biochemical molecules in a physics space, that is, the functional space is designated by moleware in the form of a biochemical reaction that is comparable to the information representation given by the notation of information space. Furthermore, the input and output of computing need to be realized by engineered technology. The nontransparency characteristics of the current biomolecular computing system requires molecular-level design and testing. Accordingly, various models and algorithms have been proposed based on a biomolecular mechanism, and biomolecular structure has been inspired by the principles of informatics. Now, we would like to refocus our work on biomolecular computing from moleware operators to the study of molecular-level algorithms. The algorithmic design of biomolecular computing requires knowledge of theoretical computer science such as formal language, computation models, complexity analysis, and logic inference, but this knowledge must be based on the biochemical features of moleware structure and molecular signaling mainly in vitro, a kind of computation-oriented moleware informatics.

Although biomolecular computation needs to be realized by molecule materials, it also requires study of the informatics used in information processing mechanisms in order to achieve a rigorous design theory. Therefore, this chapter mainly discusses the following four aspects of molecular computing in terms of information engineering:

- Forms of formal language used to represent molecular sequences;
- Computation models as formal systems for theory study;
- Algorithms for NP problem solving;
- Logic description of molecular computing.

We will also introduce state-of-the-art theoretical biomolecular computation through methods, examples, and possible applications of algorithms [1–25].

5.1 Basic Concepts in Computer Science for Molecular Computing

Diverse methods using diverse materials for molecular computers were discussed in Chapter 4. As we discovered, the materials are directly related to the task of successfully building molecular computers. However, simply studying material at the material-science level does not lead to the realization of molecular computing. We need to remain focused on the inspiration from biological objects in order to derive the abstract computing process that is crucial for molecular computers. The molecules are spatially located in the molecule complexes. The corresponding symbols at different locations can be interpreted as different information forms. Different molecular structures also show different biochemical features in chemistry and biological functions in molecular biology, depending on their applications. Here, we use the molecular structure as the medium for carrying the information needed for computing. An important method for bridging material science and molecular computing is the mathematics tool of formal languages (model, grammar, and so forth). For example, the symbols of molecules such as DNA and RNA can be used to compose formal languages for informatics analysis. The molecule sequence looks like the symbols of music, and the derived biochemical reactions behave like sound effects depending on the rules we define.

As opposed to a conventional textbook on basic knowledge of computer science, here we present only the knowledge necessary for molecular computing: formal languages, automata, and algorithmic complexity. From the operators designed mainly according to the material features of molecules in the previous chapter, we try to derive abstract forms of molecular computing based on the knowledge of molecular biology in material science.

5.1.1 Formal Language

The following elements in formal language will be used for later discussion:

- Symbol—a symbol is normally represented by an alphabetic letter. The number of symbols is limited because the number of molecules used for molecular computation is normally limited, even in the case of significantly large numbers. In DNA computing, the symbols are A, T, C, and G.
- String—a sequence made by symbols. In DNA computing, the string appears, for example, as AATTCCGG.
- Operator—a process that can change the state of a symbol or a string (e.g., the hybridization in DNA computing).
- Formal language—a set of strings with certain constraints.
- Recognition—a process or a device in which the strings corresponding to a certain language are used as the input, and the signal "reject" or "accept" is given as the output according to the rules defined in advance for this process. If the output is "accept," it is inferred that this language can be recognized by the device.

For a more rigorous form of the recognition process, we need automata that will be presented in Section 5.1.2. In the basic knowledge of formal language, the commonly used main classes of languages in computer science theory are often given in a ranking called the Chomsky hierarchy.

The Chomsky hierarchy includes the following levels:

    4: recursively enumerative language: any computable strings
    3: context-sensitive language: aAb → aBb
    2: context-free language: A → B
    1: regular language: A → aA or A → a

where "A" and "B" refer to nonterminals, and "a" refers to a terminal. The terminal is a symbol that is accepted in the final state of the computing process.

Conversely, the nonterminal is a symbol or string that is not in the final state of the computing process. Corresponding to these language classes, there exist automaton classes, which will be discussed in the next subsection.

From the operation exerted on the molecular objects, the relationship between languages and related biochemical reaction processes can be studied, for example, Winfree et al.'s work, which is a helpful approach to gaining a clear understanding of the concept of DNA languages. From the formal language representation of molecules, we need to devise a way to use computer science terms to describe the objects in molecular computing.

Let's look at an example. Winfree et al. [1] used a two-dimensional structure of DNA tiles (for example, spatially located DNA complexes) to represent languages in which the generative capability of self-assembly comes from two DNA complexes of a two-dimensional structure. Here the tile refers to the structure where DNA sequences are connected into a tile-like shape as shown in Figure 5.1. The core of this work is the relationship between the molecules and the information. The molecules are used for operations. The information is represented by the molecules.

It has been reported that the tile-type of DNA complexes can generate sets of regular languages, context-free languages, and recursively enumerable languages. As an alternative to the context-sensitive language in molecular computing, a "noncontext-free" language generated by the DNA tile model has also been reported by Winfree et al. Although the domain of noncontext-free language is broader than context-sensitive language, it can probably be used to approximate the upper-bound of the set of context-sensitive language under certain conditions in molecular computing. It is speculated that context-sensitive language is generated by interacting double-stranded circular DNA complexes and splicing systems.

Formulating a generation mechanism, such as production rules, is one of the methods of describing languages. A more abstract way to do this is rewriting on string, tree, graph, or other data structures.

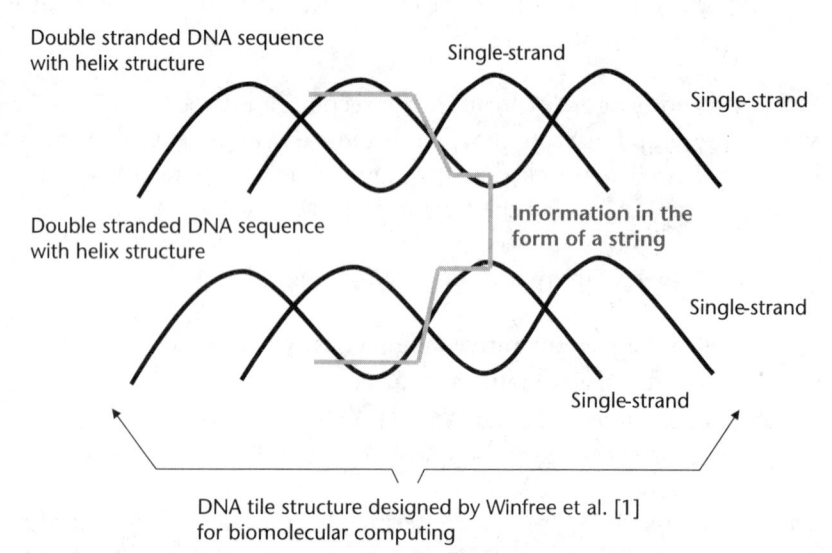

Figure 5.1 Tile structure of DNA complex in Winfree et al.'s DNA computing method.

5.1.2 Automata

As we have seen in Section 5.1.1, the symbolic sequence represented by formal language is useful for us to understand the semantic features of a molecular computer at the level of material structure. The molecular sequence is also used to represent the state of molecular computing. Intuitively, the transition between different states is the basis of molecular information processing. In contrast to the material-driven method of formal language description of the molecular complexes, the abstract description—and here we mean the form of automata—is the focus of this section. To study an automaton form of molecular computing as a formal system, even though it is abstract, is beneficial for developing a rigorous theoretical basis for algorithmic design.

To explain the concept of an automaton, we need five factors: the input, the state set, the alphabet set, the acceptable symbol set, and the rules of state transition. The alphabet set is used to define the symbols for representing the input. Basically, the state of the automaton is represented by symbols that are different from those representing the input. Briefly, computing here is a process of state transition. When the symbol that belongs to the acceptable set appears in the process of an automaton, the computing process will stop. The state updating is briefly described by the information processing units shown in Figure 5.2. This simplified form can be used to study the fundamentals of computing by moleware. The automaton is a kind of abstract computing device rather than a real computer.

An automaton consists of two parts including a tape and a head. As shown in Figure 5.3, the tape and the head are connected by the operation of "read/write." The tape is represented by a string:

$$\text{Tape} = A[0]A[1]...A[n] \tag{5.1}$$

where $A[i]$ takes the symbolic value from the alphabet, $i = 0, 1, ..., n$, $n \in N$. The index for the next reading operation of the head is increased when the current reading operation is finished. For example, at time t, when index = i, the head will

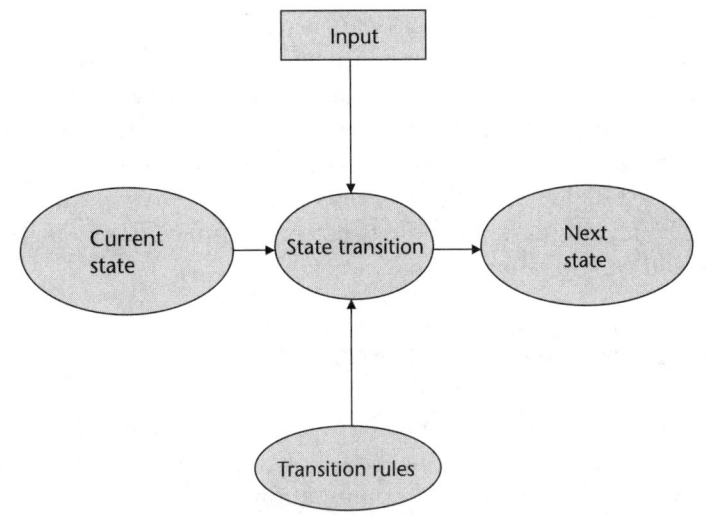

Figure 5.2 State updating mechanism of automaton.

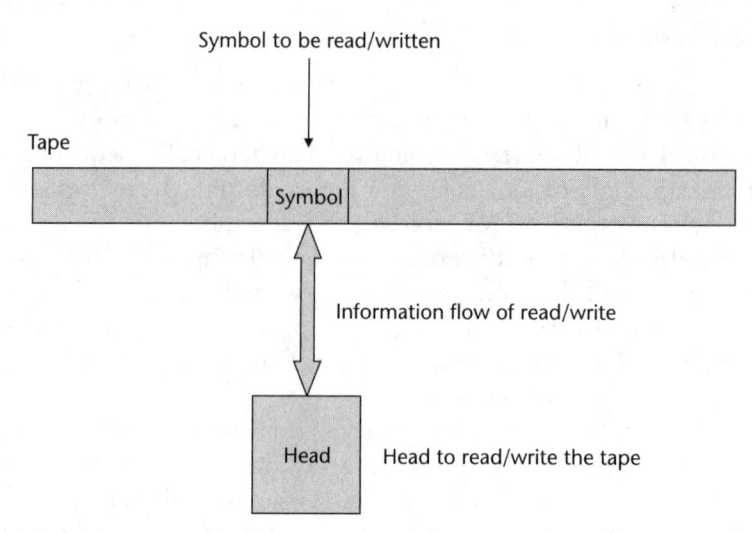

Figure 5.3 Schematics description of automaton.

read $A[i]$. Then, the index becomes $i+1$, which is used to denote the $t+1$ moment. The state of the automaton at moment t is denoted as $q(t)$. After the automaton reads the input $A[i]$, it changes its state to $q(t+1)$, according to the rules defined in advance, which takes the form

$$\text{IF } ((\text{input} = A[j]) \text{ and } (q[t] = b[k])) \text{ THEN } q[t+1] = b[k'] \qquad (5.2)$$

where $j, k, k' \in N$.

In a conceptual model of an automaton built by DNA molecules, the input, the state of the automaton, and the output are all DNA sequences. The related enzymes are used to make the automaton work continuously. In this sense, a DNA computer built by a DNA automaton is automatic in engineering.

Beneson et al. realized a DNA computer by using a DNA/RNA automaton. The input is the mRNA sequence. The output is a DNA sequence with a gene function that could be valuable for drug design. The transition rules of the automaton are used to compare various types of gene expression. The information obtained from the comparison operator is used to guide the control process of gene expression.

By using this automaton, an algorithm is designed by the logic forms of IF-THEN for bioinformatics in order to clarify the causal relationship between the gene and gene expression. This opens a new way of using molecular computers for automated design as well as for testing medical drugs. You may have noticed that the current automatic technology has been applied to automate experimental operations in genomics and proteomics.

5.2 Formalized Molecular Computing

After briefly introducing some basic background knowledge for studying theoretical molecular computing, in the following we analyze several representative formal systems in biomolecular computation.

5.2.1 H-System

The splicing system (H-system) was established by Tom Head in 1986. It is a formal system constructed with formal languages. The definition of the H-system is given as follows (see [10, p. 743]):

$$S = (A, I, B, C) \tag{5.3}$$

where

A is a finite alphabet, $A^* = A \cup \phi$
I is a finite set of initial strings in A^*
B and C are finite sets of triples (c, x, d) in which c, x, and d are in A^*

Here, we explain this model in terms of information engineering. The kernel mechanism of information processing is the focal point of the following discussion. In his classic paper (see [10, p. 752]), the "general combination" operator is defined to generate two new sequences "uxq" and "pxv" from two DNA sequences "uxv" and "pxq," where u, x, v, p, and g belong to A^*. This is a classic and basic operator in H-systems. Two DNA sequences denoted as "uxv" and "pxq" are objects to be operated on. They are cut at "x." The "u" in the first sequence and "q" in the second sequence are spliced into a new sequence "uxq." For other parts of the two sequences, the process is similar. The sequence "pxv" is made by splicing "p" and "v." Head successfully generated a regular language with the general combination operator that he proposed for the H-system.

The splicing process of genes and DNA sequences is realized by the operator. The biological splicing systems in nature refer to the biochemical mechanism of rearranging different pieces of gene sequence into building blocks. Biochemical technology allows people to do this kind of combinatorial operation on genes. The crossover of genes in the natural evolutionary process works in a similar way to exchanging two parts of the gene information. At the abstract level, this operator is used in biologically inspired algorithms such as the genetic algorithm (GA) and genetic programming (GP) in the general domain called evolutionary computation (EC) or evolutionary algorithms (EA). The difference between Tom Head's splicing system based on a biological splicing system and artificial EC is that the former mainly cuts and combines the strings (similar to crossover in EC) while the latter has operators that include, but are not limited to, crossover (e.g., reproduction, selection, even macro-level mechanisms for evolutionary processes over generations). This is because the H-system is an abstract model defined at the level of a formal language. The most important operation in his H-system model is recombination. The EC is basically a kind of algorithm that is oriented to engineering and applications.

Different structures of DNAs are used to define different operators in order to enhance the performance of the formal system, which can lead to reaching a different level of formal languages in the Chomsky hierarchy. Progress here includes Yokomori, Kobayashi, and Ferretti's work [4] on a circular splicing system and Sakakibara's work [5] on a tree-splicing system. Contrary to the Winfree et al. methods, the meaning of the operators in [5] is more generalized so as to encompass biologically inspired concepts. Also different from Adleman's paradigm of DNA computing, which started from experiments at first and was later applied to a specific NP problem, the method of H-systems came from an abstract representation model mainly based on the information description of the DNA mechanism. The comparison between the mathematical concepts and the biological objects can be made here within a broader framework of DNA molecules.

It is obvious that the H-systems of molecular computing were inspired by the molecular structure of DNA. The DNA sequence that takes a one-dimensional form may have provided a hint for Tom Head's first model of H-systems. This model is presented by formal language because there is an inherent relationship between the symbolic string and molecular sequences. Yokomori et al. [4] introduced a cyclic structure of DNAs into this system. When two DNA sequences are used as the objects for splicing operations, one can be a DNA sequence while the other can be a cyclic DNA sequence. After the cycle is split into two parts, a normal splicing operation is carried out. The sequences are then restored to a cycle. This is one of the basic operations. The splicing processes for the linear sequences and cyclic structure of DNAs are illustrated in Figure 5.4. Sakihabara et al. suggested use of the tree data structure to describe the biochemical secondary structure of RNA molecules. The splicing operator is thus exerted on trees. This is done by exchanging two subtrees in the way shown in Figure 5.5. In an abstract sense, the splicing operators on a string and on a tree can be compared to the crossover in GA and GP, respectively. As a general framework, there exist the following classes of the basic structures for DNA computing:

- One-dimensional DNA sequences;
- Circular DNA structure;
- Tree DNA structure;
- Hairpin DNA structure;
- Hybrid structure.

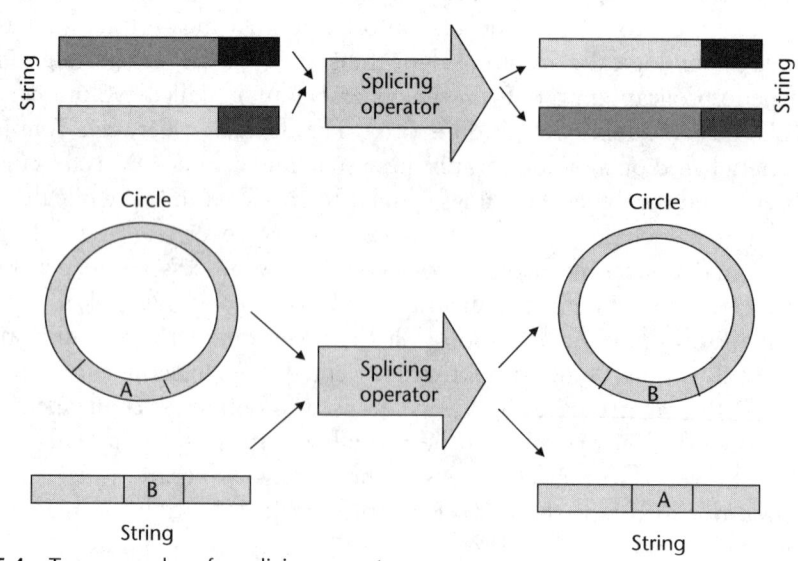

Figure 5.4 Two examples of a splicing operator.

5.2 Formalized Molecular Computing

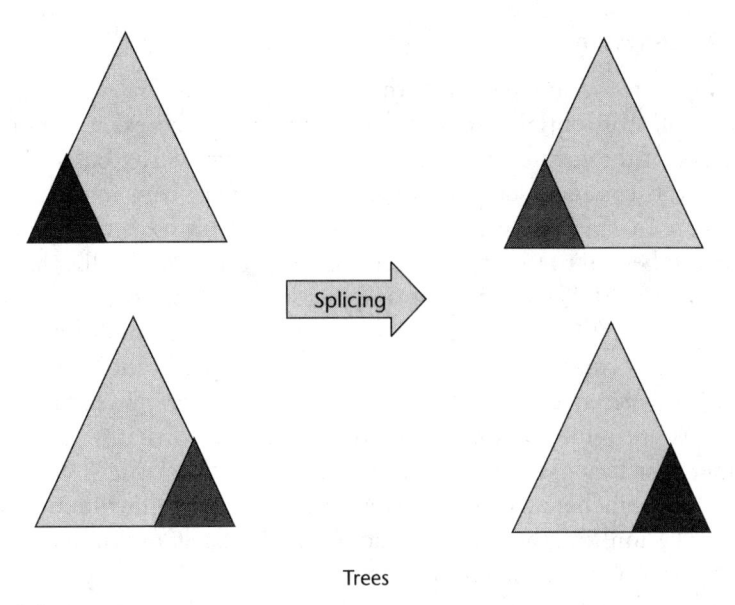

Figure 5.5 Splicing on trees.

For each of these classes, there are two forms of DNA sequences including single strand and double strand. From the shape of the structure and number of the strands, we can obtain the following major configurations.

- Two single strands of DNA sequences are spliced to form the classic H-system. After double-strand DNA sequences are separated into two single-strand DNA sequences, the splicing operation can be carried out. In the Adleman paradigm for DNA computing, two-strand and single-strand DNA sequences are used to carry out various types of molecular computing.
- Single-cycle DNA structure and single-strand DNA sequences are divided and recombined to form a classic circular H-system. This implies that two circular DNA structures can be combined.

The tree structure from RNA inspires us to explore data structures for formal systems of molecular computing. The string can be represented by data structures of list, stack, and queue in terms of programming. The data structure of the graph will be discussed in Chapter 6.

If these structures were spliced with each other, more patterns could be generated to provide even more powerful elements for parallel computing systems in the sense of multiple data flow and multiple instruction flow, which we define in different semantic modes.

It would be optimal to designate generalized interaction operators to realize interaction between the two circular structures of DNA sequences and the linear structure of DNA sequences in the Adleman paradigm of DNA computing.

5.2.2 P-System

As we discussed in Chapter 4, the membrane structure is an important mechanism in cells. It is essential to find a data structure to represent the biologically inspired concept of the membrane because data structure is the basis of programming. For this purpose, a multiset is one of the most proper forms to describe this kind of data structure in mathematics. Multiset refers to multiple sets that allow the same elements to belong to different sets at the same time. Normally, the intersection of two sets contains the elements that only belong to those two sets. However, the hierarchical information can be explicitly represented by a multiset structure.

The information processing structure in membrane computing can be understood by reference to the illustration in Figure 5.6. The circle represents the set in the sense of multiset. The arrow denotes the information flow around the membrane, which is assigned the direction of "in" or "out." Within the membranes, molecular biochemical reactions are used to carry out the information processing task. The multiple boundary structure is the basis of complex parallel computing behavior in membrane computing.

Based on the multiset structure and the membrane concept, differently derived models of membrane computing (P-system) can be easily understood. Now, let us recall the definition of a P-system. The membrane computing, or P-system, originated by Gheorghe Paun first appeared in [26], and then in [27]. In this section, we discuss the model of the P-system as a formal system in the sense of theoretical computer science. It is clear that membrane computing has become a major field of molecular computing, with potential in biological forms if sophisticated nanobiotechnology can be developed for its implementation in the future.

As indicated on the P-system Web site [9], "A P system is a computing model which abstracts from the way the alive cells process chemical compounds in their compartmental structure." The formal representation of a P-system can be

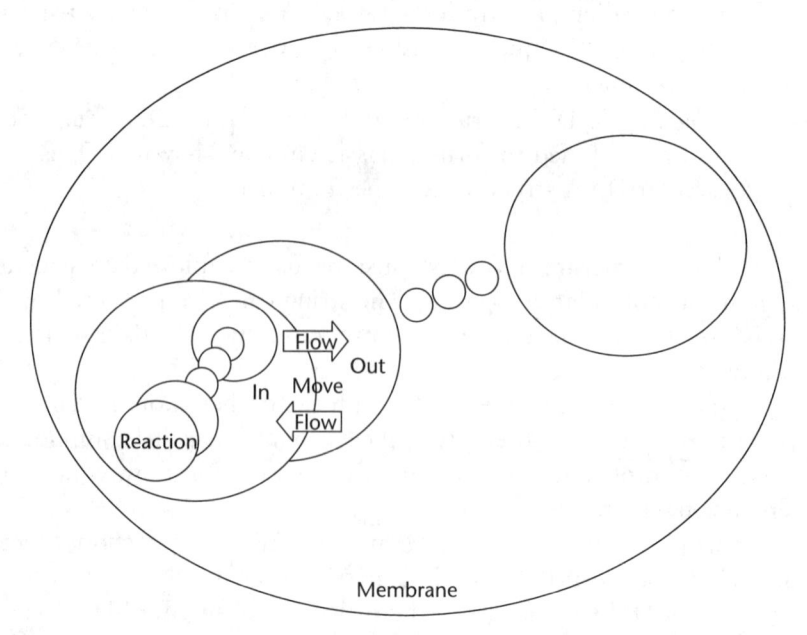

Figure 5.6 A multiset structure for membrane computing.

briefly summarized as the following form in terms of unconventional information processing:

<population-set, operator-set, constraint-set>

where the population set consists of individuals, the individual is the membrane structure in which the objects such as strings are included, and the population set is a whole membrane structure called skin.

The operators are classified into two kinds: operators on membranes and operators on the objects within the membranes. The former mainly includes dissolve, divide, create, and move. The latter mainly includes mutate, replicate, and split.

In membrane computing, it is possible to introduce different kinds of new operators, like the operators of gemmation and fusion designated by Besozzi et al. [23]. The objects in the membrane are also called axioms. The rules are designed to define the operators within the membranes for intramembrane computation and around the membrane boundary for intermembrane communication. The multiset is used to represent the concept of the membrane. The term "evolution" here refers to the fact that flow elements move in and out around these membranes. This process is often represented by rules.

To study the theoretical issue of P-systems, the number of test tubes is an important consideration. Evolution in the P-system is different from that in artificial life and evolutionary computation in the general sense, including GA, GP, ES, EP, and others.

The hierarchy of P-systems can also be simplified into a smaller number of sets at different levels. The simplification process is called "collapse." With the gemmation and fusion operations, the hierarchy of the membranes can be collapsed into seven membranes. Reducing the number of membranes is a key factor in designing the computing model of the P-system to be as small as possible. If the current known minimum number of membranes is expected to be reduced further, new powerful operators will have to be worked out in the future. We regard the process of membrane computing as a basic mechanism of possible future programming. Its flowchart is given in Figure 5.7, where the operators can be realized by a deterministic or nondeterministic process in either sequential or parallel mode.

The population is initialized at first. The membrane operators can create, delete, separate, and merge membranes. By using such operators, the elements in the related membranes are updated. When a new membrane is created, the elements to be contained in this membrane may come from other existing membranes or created as initial values. When a membrane is deleted, the elements in this membrane are deleted or moved to other membranes. When one membrane is separated into several membranes, the corresponding elements are also allocated to these different membranes. When several membranes are merged into one membrane, the corresponding elements are also put into one membrane. The object operators can operate the objects within membranes and are often defined by rules. The object is normally represented by the string. The updating operation on the string is basically equivalent to the rewriting operations in a string rewriting system. The conditions in different derivative models of the P-system have different effects on the rewriting process. The data structure of membranes is represented by a set. The

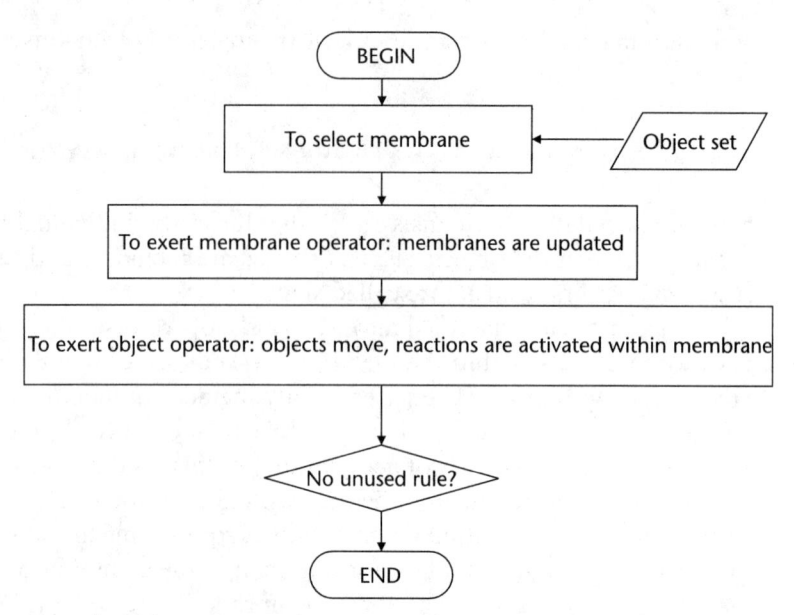

Figure 5.7 Flow chart of generic membrane computing.

hierarchical relationship is represented by a tree. The rule can be defined as a form of "IF condition THEN action" for programming. The rules designed in advance guide the information flow among the formal systems. The attributes include generating a set of formal languages in the Chomsky hierarchy. Thus, algorithms of P-systems can solve NP problem.

The cooperation mechanism of the P-system for parallel computing, which differs from traditional programming techniques, needs to be studied thoroughly in order to develop a programmable unconventional computing system, in which many autonomous membrane units that use the local rules are at work. If the membrane units are realized by the parallel distributed processing (PDP) mode in software, the communication represented in the form of rules will be the core of the parallel computing process. The central information processing issue is how to make collective behavior emerge from these autonomous units. From the viewpoint of applying a P-system to practical programming, we need to plot the communication among membranes and local processing within membranes. For example, to calculate $2 \times (3 + 4)$, a membrane computing system can be designed as:

Step 1a : a=2 in membrane 1.
Step 1b: b=3 adds c=4 in membrane 1.
Step 1a and step 1b are carried out in parallel.
Step 2: the result of step1a communicates with the result of step1b to get the result 14,
Step 3: the result 14 is the output from membranes 1 and 2.

The complexity of designed algorithms is determined by local and global communication processes.

R. Freund and F. Freund [11] demonstrated the P system through generating $\{a^n b^n \mid n \geq 0\}$. This is defined by two membranes. In membrane 0, the axiom set is {[+1][−2], [+1][−5], [+4][−2]}, and the recombination rule set is {([−4], [+4]), ([−5][+5])}. In membrane 1, the axiom set is {[+5]a[−1], [+2]b[−4]}, and the recombination rule set is {([−1],[+1]), ([−2],[+2])}.

A string that embeds $a^{n+1}b^{n+1}$ can be generated from the string that embeds $a^n b^n$ by using the rules and axioms. In a biologically inspired sense, this example can be explained as biochemical reactions of molecules represented by letter-pairs and match rules. The molecular operators on the strings are realized by match rules.

5.2.3 Rediscovering the Informatics Structure of the Biomolecular Computing System: An Informatics View of the Formal Processes of the Biomolecular Computing H-System

The computing process is dependent on the molecular mechanism in the biochemical medium. The components of this computing process include the spatial structure of the moleware operations and the temporal process of control processes of molecular signaling. Systematically, the whole system can be made with a kind of compromise between cost and performance. From the informatics viewpoint of the information process, let us analyze the biomolecular computing system by combinatorial theory. Graph theory and derived informational structure are representations chosen for our discussion here. The mathematics problems in graph theory have been used as the benchmark of biomolecular computing. This tendency started from Adleman's work on HPP by DNA computing. There are two extremes in graph theory: the HPP and the Euler graph problem. The HPP requires us to traverse all the nodes/vertexes and the Euler graph problem requires us to traverse all the links/edges. The two different parts of the graph are connected to the corresponding information network. TSP is a kind of weighted HPP, in which the minimum costs of weights are required for optimization.

As we previously discussed in Adleman's instance, the vertexes and edges are encoded by DNA sequences. The connected long edges are used to form the solution. The same encoding can be used for Euler graph problem solving in a complementary form. The denotation and detection are inverse processes relative to Adleman's HPP problem. The weighted HPP—TSP—can be designed by the DNA sequences based on the complementary features in biochemical biology. The weight representation and encoding schemes vary according to the detailed implementation. One way is to introduce a method of information representation for the weight, which is an additional information unit rather than edges and vertexes. This part of information for the weighted edge can be represented by the tile structure proposed by Winfree et al. [2]. The bridged part connecting two sequences representing edge and vertex is encoded for the weight information.

The information processing mechanism of biomolecular computing for problem solving is dependent on the biochemical signaling processes. The cost of biomolecular computing is determined by the space complexity and time complexity of algorithm realization forms. In a generalized description for this computing process, the nodes and links in networks are abstract representations for the computing process. The nodes refer to the molecular units such as DNA sequences or molecular complexes in certain forms in biochemistry, topology, or physics. The links refer to the biochemical reactions for the molecules. From the viewpoint of informatics, the abstract process of design algorithms is a kind of networking. The resultant network can be assigned to certain tasks designed in advance. To apply this idea to biomolecular computing, we need to study the methodology, theory, technology, and application of engineered informatics, which is related to industrial informatics. Here, engineered informatics refers to the studies of information processing systems aiming toward feasible engineered operators and designated control technology.

In using engineered informatics, the purpose of algorithm design is to determine the proper information flow and to design an "efficient" operator that can realize the computing task. The word "efficient" here means the amount of space and time complexity of the underlying algorithm under certain conditions. Combinatorial theory including graph theory is used to describe this informatics structure for networking. In this method, networking becomes a procedure that consists of the following tasks:

1. To design information processing unit. How is it possible to compute with molecules?
2. To design a parallel computing system. How is it possible to carry out parallel computing in a system?
3. To quantitatively study the system's complexity. How many molecules will be used and how long will the computing process last?
4. To test the designed algorithm by observing its performance and checking it against benchmarks.

The key to improving performance is the information process structure. In mapping the biomolecular computing process into a graph-represented network of information flow, we need to consider the number of molecules in a network made through molecular manufacturing to ensure that all of the molecules are prepared, or, in the case of using existing molecules, to ensure that the molecules are controlled. There are three major types of structures for biomolecular computing: the prefixed structure, the controlled variable structure, and the recursive nonexternal intervening structure. The first type can often be seen in DNA computing. The second is new but promising, and the last one is an ideal target. An entirely automated form of biomolecular computing would be possible if biochemical technology were developed further. The DNA automaton is a kind of exploration toward this ideal direction [24, 25]. From a single DNA or RNA molecule to the cell, each different level of nanobiomachines tells its own truth of the molecular biology from the viewpoint of engineered informatics, that is, the principle of self-assembly, self-organizing, and even the reconfigurable mechanism under the control of molecular engineering. At the stage of cells, signaling pathways also construct a signaling protein network. The idea of nanobiomachines can be realized through the signaling process of parallel biomolecular computing.

Among the key factors of programming biochemical computing systems, transparency is one of the major issues we have to address from the perspective of

5.2 Formalized Molecular Computing

algorithmic design. To smoothly transfer from the tailor-made molecular operator with ad hoc features to the mesoscopic-level synthesis of an algorithm is one of the most beneficial ways to practically design programs for biomolecular computing systems. The cost in space and time is expected to be reduced with the aid of novel algorithm paradigms provided by the rich features of molecular structure. The characteristics of molecules give us space to employ the interaction of the molecules through exploring the complex features of molecular biological systems. This is used to release the great power of moleware parallelism. The potential of molecular biological systems is due to their natural mechanisms existing billions of years before the birth of engineering science. However, we must learn more from nature in order to compute based on the concepts of natural computation. Of course, theoretical study on transparency would directly contribute to meeting the programming challenge in terms of making it easier for programmers.

The methodology used is "system of systems" for mesoscopic analysis of complex systems by molecular signaling control as well as synthesis of the middle layer of the program units that generate the molecular operators for specific algorithms. The technological schemes come from the integration of architecture reconfiguration, adaptive feedback control, and self-regulation. Soft computing [12] has been extended to include molecular (DNA) computing. The methodology of complex systems, system analysis and design, and biological molecular signal processing is combined here to better understand molecular signaling in the natural complex mechanism of nanobiosystems.

The system of systems refers to the systematical description of complex systems where many subsystems are interacted. This is an extension from the domain of system engineering for complex systems. In our work on biomolecular computing, the systems refer to the specific operator-level programs. The basic task is to design and test a program M that derives the molecular level computing operator for specific algorithms A1, A2...An ($n \in N$). The constraints of A1, A2...An are simply denoted as constraint (A1, A2...An). The molecular operators should be feasible and reflect the essence of biochemical reactions. The operators for computing in specific algorithms have two aspects of parametric information. One involves the parameters of computing in informatics, and the other involves the parameters of the biochemical reactions of the molecules. The problem solving in algorithms requires us to figure out the parameters of the problem and the derived parameters of the algorithms. The operator-level algorithms must handle these parameters as well as the parameters of the biochemical reactions for the realization of the computing processes, that is, mainly the coefficients of the biochemical reactions. The design, checking, control, and manipulation in biochemical experiments vary according to the molecular materials selected, the related techniques for manipulating these molecules, the equipment available in laboratories, and the special molecular structures required in the specific designed algorithms. The abstraction structure is mapped from the information structures determined by these factors. The mesoscopic structure is designed by a molecular-structure-independent architecture, which is a kind of networking process for operators. The derived networks are actually used to operate the changing part of graphs, which reflects the biochemical reactions. The systematic structure can be designed in an algorithmic way, that is, to automatically generate and test the operations of the computing processes.

The algorithmic units with parameters are represented in a network:

$$\text{operator-graph}^i (\text{parameter-set1, parameter-set2})$$

where parameter-set1 includes the parameters for molecular operators and related biochemical reactions, and parameter-set2 includes the parameters for determining the relation of the operator-level graphs (i.e., the quantitative measure of mesoscopic structure).

From {operator-graph^i (parameter-set1, parameter-set2)} (i = 0, 1...N), the network is constructed to describe the mesoscopic structure and used to infer the algorithms for the synthesis of M from A1, A2...An:

```
for (i = 0 to the number-of-operator-level-unit)
  { /* i */
  initialization;
  for (j = 0 to the number-of-the operator-unit)
     { /* j */
     the design of the operator structure;
     the design of the biochemical-reactions (molecules, coefficients);
     verification (the operator-unit);
     } /* j */
  construction of the entire mesoscopic structure (operator-graph);
  transparency measure calculation;
  } /* i */
set-the-global structure at the mesoscopic level;
model-check (global structure)
```

The logical verification of the global mesoscopic structure is then validated, and the feasibility of the local operator-unit is investigated. The term "validation" here refers to ensuring that the two processes—logical verification for the system of systems and the biochemical feasibility of the operation systems—are consistent and cooperate in a well-performed system. For example, we use two networks-network 1 and network 2—to represent the biochemical reactions of the operators in biomolecular computing. Network 1 for operation 1 is determined with parameters in the form of coefficient matrix A:

$$[a_{ij}]\ (i, j = 0, 1...9) \text{ whose size is } 10 \times 10$$

Network 2 for operation 2 is determined by the parameters in the form of coefficient matrix B:

$$[b_{ij}]\ (i, j = 0, 1...9) \text{ whose size is } 10 \times 10$$

The two networks reflect two operator schemes at the algorithmic level. In contrast to this, the global structure is represented by the graph/network in the form of matrix C:

$$[c_{ij}] \ (i,j = 0, 1...9) \text{ whose size is } 10 \times 10$$

The 10×10 size is selected here for convenience of discussion. The resultant network is represented in the logic form of

Graph (local parameters, global parameters)

which needs to be verified by the criterion:

logical verification (global graph) AND biochemical feasibility (local graph)

5.3 How to Design Algorithms for a Molecular Computer

The models of molecular computing are mainly inspired from the molecular mechanisms in nature. In applying them to applications, we need to adjust the computing mechanism design at the algorithmic level. Of course, the complexity analysis serves as a guideline for the entire synthesis process of molecular computing.

5.3.1 Observing Complexity from Benchmarks

Problem solving by molecular computing is one of the main issues in theoretical molecular computer science. Currently, one of the most important tasks in this field is improving the efficiency of the implementation. By comparing different techniques for feasibility and efficiency, some good algorithms are being adopted for the best usage. Solving NP problems (e.g., 3-SAT computation) is a touchstone for biomolecular computation. It is also beneficial to generic computing because it can promote exploration of potential operators and "parallelization" from the viewpoint of unconventional algorithm designs.

From the 3-SAT problem, which is one of the most important benchmarks for testing DNA computing algorithms, we can find a number of successful examples.

Basically, time and space are two quantitative measures used to evaluate the performance of algorithms: time complexity and space complexity. In parallel computing, space complexity is often traded for reducing time complexity.

Complexity is used to explain the difficulty of the problem to be solved by the program. There are several parameters related to program running. In order to obtain a unique measure to quantitatively describe the scale of a problem for computing, the phrase "the size of the problem" in computer science is defined as the most significant parameter of the problem, which greatly influences the cost of program running from the viewpoint of space or time. For example, in the SAT problem, the size of the problem means the measure of the number of clauses, the number of variables, and the number of literals. The complexity of problems or the difficulty of programs is defined as the upper-bound of the cost in time or space, like the function g(n) defined as the upper-bound of the cost of the problem. Here, complexity has two types including time complexity and space complexity.

The P class of the problems refers to those problems that can be solved in polynomial time. NP refers to a class of the problems that can be solved by a so-called "oracle" device in polynomial time. The full name is "nondeterministic polynomial time" class. However, this oracle device is a conceptual device whose implementation form depends on the specific process. In the case of the DNA computing process realized by Adleman, the oracle is a hybridization mechanism of DNA molecules. This mechanism is used to carry out a computing process by encoding HPP programs, and it is manufactured in test tubes. This is a set of the most difficult problems for a computer to solve. The P algorithm is defined as the one with polynomial time cost. However, in many cases, we have to handle an NP problem in which the problems in the NP classes are regarded as having the same level of difficulty. The NP-complete problem is the most difficult one in the NP class.

There are several good algorithms for SAT problem solving, in which different operators use different information representations, thus showing different computing performances. From the viewpoint of biomolecular computing algorithms, we are interested in how to make best use of the molecular mechanism to explore new information processing paradigms by algorithmic design. The kernel operators in the following algorithms selected for explanation are good examples for informatics.

1. *The Schöning algorithm for 3-SAT computing.* In the Schöning algorithm, the kernel operator finds the true value of the variable randomly. The 3/2 (1.33) value, the result from the statistics in SAT computing, is the base of the factor for the case of three variables in a clause. The time complexity is 1.33^n. Here, the operator can be briefly represented in the form of:

    Schöning-operator (SAT) = the operation that sets the
    assignment of the variable at random

2. *The evolutionary algorithm for SAT computing.* The operators of the evolutionary algorithm for SAT [21] mainly generate the candidates for the SAT solution and update the candidates according to the value of fitness defined by the difference between the candidate and clause constraint:

    EA-operator (SAT) = the operation that generates the candidates
    in the form of S1S2...Sn

    where Si (i=1,2,...,n) denotes the bit of the individual in the population of EA.
    The operator of EA is parallel and in a stochastic mode. The probability is an empirical set, and the match results guide the searching process. Generally speaking, the solutions are obtained if population size and generation number are sufficiently large. However, this is not absolutely guaranteed.

3. *The Suyama-Yoshida algorithm for 3-SAT computing.* The Suyama-Yoshida algorithm handles three variable clauses separately. At first, one variable is tried and checked, then the remaining two-variable part is searched as a 2-SAT. This strategy is extendable to large-scale problems if recursive operators are introduced for clause-partition recognition and used/unused variable detection.

4. *The Chen and Ramachandran algorithm for 3-SAT computing.* The DNA computing algorithm for k-SAT by Chen and Ramachandran considers the influence of the previous checked logical values on the variables checked at the current step. This is a kind of integration scheme of heuristic searching and random setting. The operators handle the molecular sequences by using several test tubes. Letting n be the variable number, 2n sequences are needed. Also, three sequences are needed for biochemical reactions. The main difference of this algorithm from the Lipton scheme is reflected in the variable index generation by random permutation and the force operation on the sequence for setting the logical values of the sequence representing the variable. This algorithm uses a rather large number of computing steps: The time complexity is k^2mn, which is the value in the worst case.

5. *The DNA computation method by Sergio Díaz, Juan Luis Esteban, and Mitsunori Ogihara [14] is a concurrent version of Schöning's algorithm [13].* This concurrent version of Schöning's algorithm [13] implemented by a DNA computation algorithm has realized a space complexity $O((2-2/k)^n)$ and a time complexity $O(k \times m \times n + n^3)$, where k is the literal number.

    The kernel operator of Schoening's algorithm is implemented by the DNA sequences. The variable representation is also made by the DNA sequence. The complementary form of DNAs is used to denote the true value or the false value of the variable. Multiple test tubes are used to operate the DNA sequences. This scheme is natural and useful in DNA computing. The core element of this algorithm is that the step number of the Schöning algorithm is decided by the parallel molecular operators. The space complexity—the number of DNA sequences that are traded for time complexity—implies the speed of DNA computing. The values of the sequences are randomly assigned by dynamic programming. The logical values of the variables in the sequences are randomly set by generating $F(k,n)$ candidates. The parallel operations for setting and checking are carried out at the DNA-sequence level.

6. *Surface-based DNA computing for SAT problem solving.* The method of surfaced-based DNA computing is developed by Qinghua Liu et al. [15] for the 3-SAT (4, 4) problem, whose time complexity is $O(m)$ and space complexity is $O(2^n)$ when it is scaled. It is well known that linear time complexity is feasible with most forms of DNA computing.

    With respect to space complexity, M. Ogihara and A. Ray pointed out that in surface-based DNA computing the DNA number used for 3-SAT problem solving is at least 10^{15} [16].

7. *Membrane computing for SAT problem solving.* We also need to mention the rigorous results obtained from using computation models of P-systems for solving 3-SAT with polynomial (especially linear) time.

Table 5.1 Complexity

Name of Algorithm	TIME	Space
SN	1.33^n	—
SR	M	2^n
SHD	$kmn+n^3$	$(2-2/k)^n$
CRD	k^2mn	$2^{(1-1/k)^n+\log\alpha}$ **
YSD	N+m	$2^{0.5n}$
LSK	M	m×n for 2^n

Notations
N: Schöning's algorithm [13]
SB: Surface-based DNA computing [15]
SND: The DNA computing implementation of Schöning's algorithm [14]
CRD: Chen-Ramachandran's algorithm [17]
YSD: Yoshida-Suyama's algorithm [18]
LSK: Kinase computing by Liu and Shimohara [19]
*: in the worst case
**: α is the parameter in the condition given in [17]

The comparative results of the representative molecular biocomputing algorithms are given in Table 5.1. These different forms of DNA computing algorithms provide us rich ideas for designing the informational operators and their reflected forms on molecular operations.

5.3.2 Obtaining Efficiency from Pathway Designs: Algorithmic Design Through Examples

Toward the design of a good molecular computing algorithm, we have different choices, such as modification of the scheme to speed up the biochemical reactions, increasing the containers to cover more molecules, or decreasing the number of manufactured molecules by using the existing biological pathway structure under protein enzyme control. For help in making these choices, efficiency is one of the most important ways to judge algorithmic performance in molecular computing. Here, efficiency refers to the quantity of computing resources used per algorithm, (i.e., the molecular number and the time consumed for biochemical reactions, whose abstract forms are space complexity and time complexity, respectively). In biomolecular computing, the normally adopted scheme is to use a huge number of molecules to carry out the computing process in parallel. The time complexity becomes linear in this scheme, and its merit is speed. The space complexity in biomolecular computing depends on the molecular structure. Reducing the number of manufactured molecules is one method to achieve a low-cost design scheme for biomolecular computing. There are several factors that should be considered for design algorithms. The materials vary greatly, and the corresponding biochemical reactions become different. In this subsection, we present two examples of biomolecular computing by signaling pathways in cells.

In the case of using cells for biomolecular computing, the crucial task is to determine the molecules to be controlled in computing, which is called control-space complexity. From the viewpoint of nanobiomachine-based technology for molecular computing, we introduce an algorithm from kinase computing (KC) to discuss the design issues on control-space complexity. In molecular computing processes, there are three kinds of molecules that are normally taken into consideration:

1. Naturally existing molecules in cells without any control;
2. Molecules in cells that need to be controlled;
3. Molecules that need to be prepared and put into the test tubes.

In the following example, we discuss the molecular number needed for control in molecular computing. The basic structure of the algorithm is the pathway structure illustrated in Figure 5.8 for biochemical application and Figure 5.9 for informatics application.

In the case of observing the features of the molecular computing process for solving NP problems, the molecular operators, algorithm design, and algorithmic complexity are key factors to be considered in theoretical study on molecular computing. In the sequential implementation of NP problem solving, the time complexity is NP, which makes a fast computing process impossible in most cases. In DNA computing, the pools of DNA sequences are used to represent all of the candidates of NP problems, and the DNA computing operators are carried out in parallel, so the time complexity becomes polynomial. However, the number of DNA sequences is still very large. Therefore, it is imperative to improve molecular manipulation technology in order to overcome this limitation.

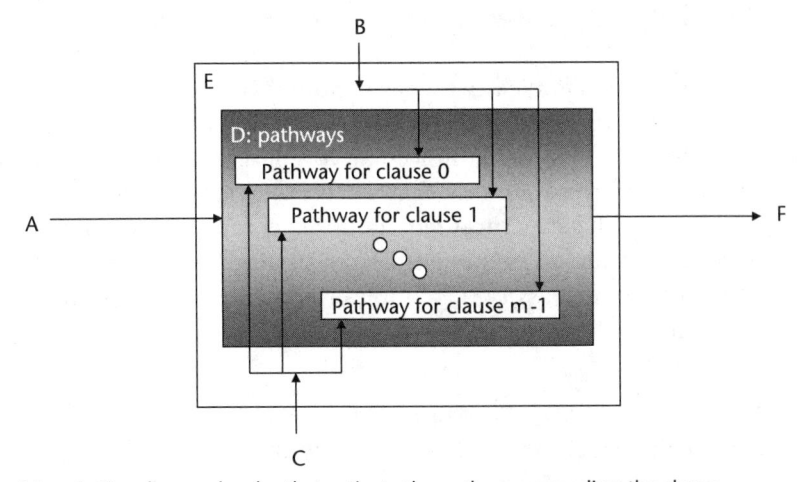

A input: Signaling molecules that activate the pathways encoding the clause
B control: Kinases or phosphatases that correspond to the constraints in clauses
C: SPK molecules that encode the variables
D: Pathways
E: Kinase computing process
F output: Molecules for solution

Figure 5.8 Overview of the computing process in mathematics scripts.

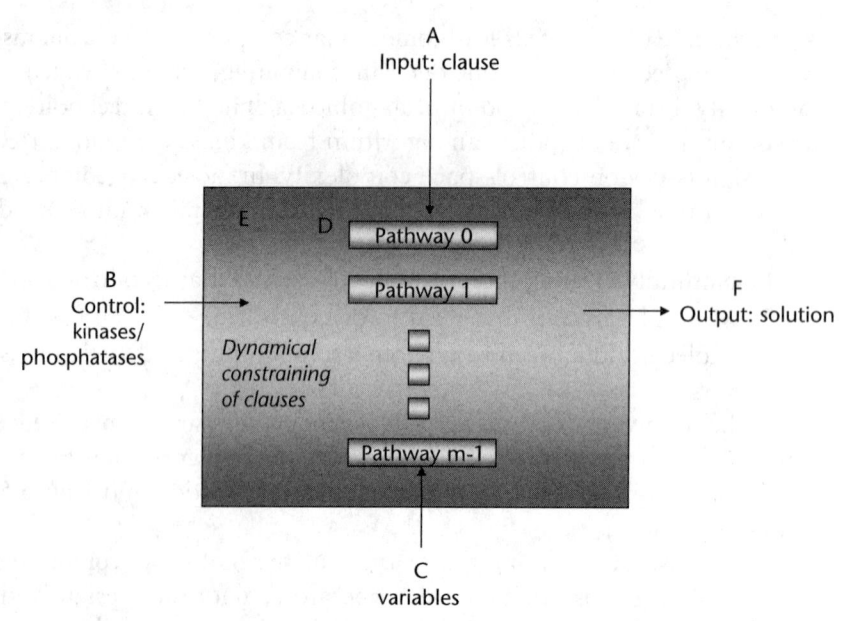

Figure 5.9 Biochemical conceptual description of the computing process.

For a k-SAT problem (k is the number of variables), the information structure includes:

Variable: total number of variables is n.
Clause: the conditions represented by the combinatorial forms of the variables. k is the number of variables in the clauses. If k = 3, k-SAT becomes 3-SAT. The number of clauses is m.
Candidates: combinatorial forms of the variables. All of the candidates are 2^n.
Satisfiability: the constraints that are reflected by the clauses.

Control space complexity means the number of signaling molecules to be controlled for pathway units in computing. These signaling molecules include kinases and phosphatases. A generic molecular computing approach is to use the parallel operators to find the solutions that satisfy the clauses. To design algorithms, we need to consider the structural factors of the computing mechanism. The different structures may bring different complexity features to the designated algorithms. In the following, we discuss two examples to demonstrate how to design algorithms for biomolecular computing.

5.3.2.1 Example 1: 3-SAT Problem Solving by Kinase Computing with Phosphorylation-Dephosphorylation Mechanism of Cells

As the basis of presenting the concept of algorithmic design, the molecular operators in the form of protocols are given and discussed with an instance of an SAT problem. Here, the term "protocols" refers to the molecular operators that correspond to the SAT computation algorithm. The protocol is designed to define the information processing based on the biochemical feasibility of the signaling pathway mechanism under engineered control and biological regulation of cells.

5.3 How to Design Algorithms for a Molecular Computer

Protocol:

1. The kinases/phosphatases are put into assays to activate signals of computing units in parallel. The kinases/phosphatases are designed to encode the clauses of the 3-SAT problem to be solved.
2. By the activation of the kinases/phosphatases for clauses given in 1, the pathways encoded for the clauses are regulated to generate the solution of each clause in individual assays. First, each sub-pathway encodes each variable constraint in the clause by molecule labeling in parallel. Then, the products of sub-pathways are collected as the outputs of the corresponding pathways in the related assays.
3. The common parts of outputs of these pathways are detected and collected based on the labeling states in 2.
4. The output in step 3 is read out as the final result of the solution to the problem.

In biochemical informatics, there are m groups of kinases/phosphatases to activate the corresponding pathways of encoding clauses:

K_{jl} = *kinase-X_i* if Ψ_{jl} is X_i.

or

Phosphatase-Xi if Ψ_{jl} is $\neg X_i$.

The variables of the SAT problem, Xi (i = 0, 1…n–1), are represented by the SPK molecules. They need to be activated for the biochemical reactions of pathways. The clause used to sieve the candidates for the solution is represented by the pathways of kinase/phosphatase. The form of the candidate molecular complex depends on how the molecular computation is realized in vitro and how the related molecules can be separated for detection. In the test tube form, these processes can be realized with the current lab technology when an antigen is controlled so that it is attached to the molecular complex for readout.

There are two basic ways to design the candidates: bottom-up and nonbottom-up. The nonbottom-up approach includes the mesoscopic method we mentioned in Chapter 4 and the top-down method. In the concrete sense of implementation by nanobiotechnologies, there are various choices in techniques and experimental skills. We select the mesoscopic method, which is effective for both concept presentation in information and kernel mechanism in biology. In terms of mesoscopic technology, the molecules are encoded as candidates. At first they are prepared separately. Then the controlled pathways are activated to produce the molecular complex to represent the designated combinatorial forms of the molecules. The molecular complexes of SPK and GTPase are the basic building blocks for information representation and code. From these biomolecular forms, we attempt to extract the informatics contents for algorithmic design.

We use an array to describe the candidates: candidate (i), i = 0, 1…n, n ∈ N. This is used for algorithmic study.

In the kinase computing process, we will discuss the various roles played by comparing mathematical (see Figure 5.8) and biological (see Figure 5.9) concepts.

In mathematics, input is the clause of the SAT problem. The clauses are denoted as the union set of $\{\Psi_j\}$ ($j = 0, 1, ..., m-1, m \in N$), where $\Psi_j = \Psi_{j1} \vee \Psi_{j2} \vee \Psi_{j3}$. Ψ_{jl} ($l = 1, 2, 3$) takes either positive form of the variable X_i or negative form of the variable $\neg X_i$.

Input: The kinase/phosphatase complexes for activating the pathways to encode clauses in the instance of the 3-SAT problem to be solved.

$$\{Kjl\}\ (j = 0,1,...,m-1;\ l = 1,2,3)$$

Output: The molecular complexes of SPKs in the form of $\{X'_0\ X'_1\ ...\ X'_{n-1}\}^{(v)}$ with the states of phosphorylation and dephosphorylation that encode the solution to the 3-SAT problem, where the SPK in the SPK complex encodes X'_i (0, 1, ..., $n-1$) and the state of phosphorylation or dephosphorylation corresponds to the state of X_i.

Control: The set of $\{K_{jl}\}$ activates the pathways that generate SPK complexes corresponding to the variables of $X_0, X_1, ..., X_{n-1}$ and, in turn, activates the individual SPK pathways encoded for the corresponding variables of the clauses.

Algorithm: A kinase-computing algorithm for 3-SAT problem solving consisting of the following steps:

Step 1: Generate activating signals for the pathways encoded for clauses $\{\Psi_j\}$ and generate n-bit SPK complexes.

Step 2: Generate the output of the subpathways for encoding the clause $\Psi_j = \Psi_{j1} \vee \Psi_{j2} \vee \Psi_{j3}$, where three so-called subpathways are used for encoding Ψ_{jl} ($l=1,2,3$) in 3-SAT problems, and then collect all of the outputs of the three subpathways for $[\Psi_j]$ as the output of the pathway of $[\Psi_j]$.

Step 3: Extract the common parts of all outputs of the pathway $[\Psi_j]$.

Step 4: Read out the SPKs in the form of a string:

$$\{X'_0\ X'_1\ ...\ X'_{n-1}\}^{(v)}$$

where $v = 0, 1...V_{sat}-1$. V_{sat} refers to the number of solutions with the states of phosphorylation and dephosphorylation of the SPK complex corresponding to the variables.

Regarding the biological aspects of the computing process, we briefly explain the following steps:

In step 1, the information representation is the basis for designing the related pathways. These kinases/phosphatases activate the related pathways encoding the clauses $\{\Psi_j\}$ in the 3-SAT problem. The molecular complexes of SPKs generated in step 2 are taken as the input to the pathways in step 3.

In step 2, the SPKs are regulated by the related pathways and are attached with label molecules denoted as *L*. These molecular complexes of SPKs bound with L molecules are detected and determined as the result of the output of the related pathway units in the molecular computation processes. The molecular complex of SPKs used to represent combinatorial variable forms [i.e., a molecular compound that encodes the combinatorial forms of the variable X_i ($i = 0, 1, ..., n-1$)] is denoted in the form of a string:

$$Q_0 Q_1 \cdots Q_{n-1}$$

where Q_i (0, 1...$n-1$) denotes the logical values of the related variables.

The cost of designing pathways for kinase computing mainly depends on activating the signaling molecules and related regulated pathways for immunofluorescence analysis, which is a practical and efficient technical tool in laboratories. In theory, it is inferred that the time complexity is $O(m)$ and the "regulation (control)-space" complexity is $O(m \times n)$ when the "kinase computing" algorithm is applied to solving the 3-SAT problem. We can reasonably expect scalability to be conditionally obtained because at least 2,000 kinases and about 1,000 phosphatases have been discovered in molecular biology, and these can be employed for word design and programming. One of the most important tasks in regulatory mechanisms of engineered phosphorylation-dephosphorylation processes is the decoupling of these processes and the control of directed signaling molecules that have a switch-like function in underlying pathways. Such a mechanism of kinase computing would be helpful in facing the challenges posed by biomolecular computation and nanobiotechnologies.

5.3.2.2 Example 2: 3-SAT Problem Solving by Kinase Computing with the Interaction Mechanism of Kinase/Phosphatase and GEF/GAP Pathways

Similar to the kinase/phosphatase pathways, GEF/GAP pathways also have a switch-like function. The basic idea of the algorithm design described above can be extended to introduce GEF/GAP pathways into the program-solving processes. The interaction of the kinase/phosphatase pathways and GEF/GAP pathways may contribute to decreasing the control complexity of related algorithms. As we will see and prove later, the complexity of the control space can be reduced to the logarithmic order by the interactions of two kinds of pathways. It is thus optimal to design KC algorithms for 3-SAT problem solving by the interaction of kinase/phosphatase pathways and GEF/GAP pathways. The constraints of clauses can be represented by the kinase/phosphatase pathways and GEF/GAP pathways. A single pathway encoded for a clause has a linear order of control-space complexity if two pathways are arranged in a two-layered cascade in which the information is mapped two times. Through the hierarchy of the pathways, the logarithmic order of the control-space complexity can be obtained in the top layer of the cascade.

Algorithm: A KC algorithm for 3-SAT problem solving with the interaction of kinase/phosphatase pathways and GEF/GAP pathways.

Input: The molecules corresponding to the clause pathways. The kinase/phosphatase combinatorial form for the clauses can be given in the following form:

$$\{\text{kinase/phosphatase-}W_{j1}(\Psi_{j1}, X_{i*}), \text{kinase/phosphatse-}$$
$$W_{j2}(\Psi_{j2}, X_{j*}), \text{kinase/phosphatase-}W_{j3}(\Psi_{j3}, X_{k*})\}$$

where

kinase/phosphatase-$W_{j1}(\Psi_{j1}, X_{i*})$ = kinase- X_{i*} if $\Psi_{j1} = X_{i*}$
kinase/phosphatase-$W_{j1}(\Psi_{j1}, X_{i*})$ = phosphatase- X_{i*} if $\Psi_{j1} = \neg X_{i*}$
kinase/phosphatase-$W_{j2}(\Psi_{j2}, X_{j*})$ = kinase- X_{j*} if $\Psi_{j2} = X_{j*}$
kinase/phosphatase-$W_{j2}(\Psi_{j2}, X_{j*})$ = phosphatase- X_{j*} if $\Psi_{j2} = \neg X_{j*}$
kinase/phosphatase-$W_{j3}(\Psi_{j3}, X_{k*})$ = kinase-X_{k*} if $\Psi_{j3} = X_{k*}$
kinase/phosphatase-$W_{j3}(\Psi_{j3}, X_{k*})$ = phosphatase- X_{k*} if $\Psi_{j3} = \neg X_{k*}$
and $i^*, j^*, k^* \in \{0, 1, ..., n-1\}$

Output: The molecular complexes of SPKs in the form of $\{X'_0 X'_1 ... X'_{n-1}\}^{(v)}$ with the states of phosphorylation and dephosphorylation that encode the solution to the 3-SAT problem, where X'_i (0, 1...n–1) in the SPK complex has the state of phosphorylation or dephosphorylation corresponding to X_i.

Control: The GEFs/GAPs $\{G'_{k'}, k' = 0, 1, ..., m'-1, m' \in N\}$ encoding the GTPase-words generate SPK complexes corresponding to the variables of $X_0, X_1...X_{n-1}$ and activate the individual SPK pathways encoded for the corresponding variables of the clauses.

The algorithm consists of the following steps:

Step 1: Generate the GTP-bound/GDP-bound words from the corresponding GEF/GAP pathways that have m' bits to activate the signals for the effector-set-based pathways: $S_0, S_1...S_{n-1}$.

Step 2: Generate activating signals for the pathways encoded for clauses $\{\Psi_j\}$ and generate n-bit SPK complexes for use in encoding n variables by m'-bit GTPase words.

Step 3: Generate the output of the subpathways for encoding the clause $\Psi_j = \Psi_{j1} \vee \Psi_{j2} \vee \Psi_{j3}$, where three so-called subpathways are used for encoding (Ψ_{jl} (l=1, l=2, l=3) in 3-SAT problems, and then collect all of the outputs of the three subpathways for [Ψ_j] as the output of the pathway of [Ψ_j].

Step 4: Extract the common parts of all outputs of the pathways [Ψ_j].

Step 5: Read out the SPKs in the form of a string:

$$\{X'_0 X'_1 ... X'_{n-1}\}^{(v)}$$

where v = 0, 1,...,Vsat. Vsat refers to the number of solutions having states of phosphorylation and dephosphorylation of the SPK complex corresponding to the variable.

5.3 How to Design Algorithms for a Molecular Computer

Here, we briefly explain these steps as follows:

In step 1, the key operation is the encoding of GEFs/GAPs for m'bit words, where m' refers to the number of GEFs/GAPs used in computing. Accordingly, a sufficiently large number m' is selected for encoding n symbols in the downstream pathways.

In step 2, the information representation is the basis for designing the related pathways. The kinases/phosphatases activate the related pathways encoding the clauses $\{\Psi_j\}$ in the 3-SAT problem. The molecular complexes of SPKs generated in step 2 are taken as the input to the pathways in step 3.

In step 3, the SPKs are regulated by the related pathways and attached with label molecules denoted as L. These molecular complexes of SPKs bound with L molecules are detected and determined as the result of the output of the related pathway units in the molecular computing processes. The molecular complex of SPKs used to represent combinatorial variable forms [i.e., a molecular compound that encodes the combinatorial forms of the variable X_i ($i = 0, 1, ..., n-1$)] is denoted in the form of a string:

$$Q_0 Q_1 \cdots Q_{n-1}$$

where Q_i (0, 1...$n-1$) denotes the logical values of the related variables.

In this step, much more attention is paid to the information flow in related pathways. As Figure 5.10 shows, an example is given for the case where the kinase/phosphatase pathways are activated by two specific activation-signals for the effector-set-based pathway S_{i*} and for kinase/phosphatase-$W_{j1}(\Psi_{j1}, X_{i*})$:

kinase/phosphatase-$W_{j1}(\Psi_{j1}, X_{i*})$ = kinase-X_{i*} if $\Psi_{j1} = X_{i*}$
kinase/phosphatase-$W_{j1}(\Psi_{j1}, X_{i*})$ = phosphatase-X_{i*} if $\Psi_{j1} = \neg X_{i*}$

The output (i.e., the variable in the n-bit constrained SPK complex corresponding to S_{i*}) is set as the phosphorylation state or dephosphorylation state

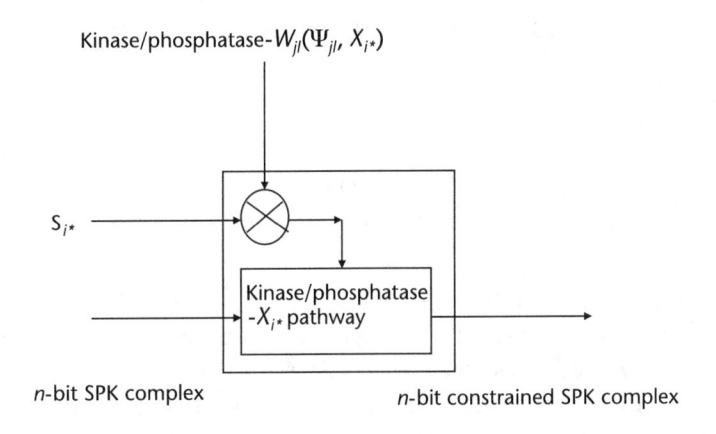

Here, constrained complex refers to the bit of X_i, which has been set according to the state required by the constraint in Ψ_{ji}.

Figure 5.10 Diagram of a subpathway encoded for the constraint of a variable in 3-SAT.

according to the constraint of Ψ_{j1} (the positive form or negative form of the related logical variable).

Complexity Analysis

Owing to the interaction of kinase/phosphatase pathways and GEF/GAP pathways, we can obtain the logarithmic order of the 3-SAT computation.

Complexity: The algorithm of 3-SAT problem solving by kinase computing given above has the control-space complexity of $O(log_2 n)$.

Proof: From the GEF/GAP pathways and kinase/phosphatase pathways of these processes, we can find that

$$\text{The control-space complexity of } O(m')$$
$$\text{exists for the GEFs/GAPs pathway part}$$

and

$$\text{the control-space complexity of } O(n)$$
$$\text{exists for the kinase/phosphatase pathway part.}$$

As a result, $O(log_2 n)$ can be inferred when transformation from GEF/GAP pathways to kinase/phosphatase pathways is realized for 3-SAT computation.

Proof for O(m'): Let m' be the number of GTPases. The GEFs and GAPs for activating the pathways given earlier are denoted as

$$\{G'_k\}$$

where $G'_k = G_k(g_k)$, $k = 0, 1 \ldots m'-1$. As above, m' refers to the number of GEFs/GAPs.

$G_k(g_k)$ = GEF, if g_k = GTP-bound state of Rho GTPase
$G_k(g_k)$ = GAP, if g_k = GDP-bound state of Rho GTPase

We define that

$\rho_k(g_k)$ = 1, if g_k = GTP-bound state of Rho GTPase
 = 0, if g_k = GDP-bound state of Rho GTPase

The binary words constructed by the set of $\{\rho_k(g_k)\}$ (k = 0, 1...m'-1) can be used to represent the related combinatorial switch-like forms of the activation/inactivation state of GTPases in the upstream pathways. Therefore, the number of words made by $\{G'_k\}$ is $2^{m'}$. By m' engineered-pathways, the number of related different proteins is still limited to the order of m', due to the key-hole mechanism of GEFs/GAPs for GTP-bound/GDP-bound GTPases.

Let there be n_1 for GEFs and n_2 for GAPs. The maximum of n_1 is 60 and the maximum of n_2 is 70. Consequently, these numbers are sufficient for encoding

GEFs/GAPs. The maximum number of words becomes 2^{60} because min (60, 70) = 60. This is used for encoding the information of combinatorial activation or inactivation forms of Rho GTPases. Now, let's consider the equation for a quantitative description of the GEF/GAP pathways that control the GTP-bound/GDP-bound states of GTPases:

$$d/dt\ (X) = A(t)\ X + B(t)\ U(t) \tag{5.4}$$

where

- X, a vector, refers to Rho family GTPases, whose total number is m'.
- $A(t)$, a matrix, refers to the cross-talks among the m' Rho family GTPases, whose size is $m' \times m'$.
- $U(t)$, a vector, refers to the control exerted on Rho family GTPases by GEFs/GAPs.
- $B(t)$, a matrix, is used to represent the cross-talk relations among the GTPases and GEFs/GAPs, where the interaction is physically available among the engineered pathways. Its rank is m'.

Here, controllability of the computing process refers to the ability to make the states of the m GTPases fall into ranges corresponding to all of the related combinatorial forms. This is achieved by utilizing the GEFs/GAPs through the pathways designed for the required clauses of the problems.

Let $A(t) = A$ and $B(t) = B$. The controllability of related pathways is dependent on the rank of

$$B{:}AB{:}\ldots{:}A^{m'-1}B \tag{5.5}$$

The rank of $B{:}AB{:}\ldots{:}A^{m'-1}B$ is m' because the maximum number of crosstalks generated by the GTPase pathway units encoded for m' GTPase is $m' \times m'$. Therefore, the signaling processes of GTPases regulated by GEF/GAP pathways are quantitatively faithful according to the necessary and sufficient condition of controllability that the related system (i.e., that for equation (5.1)) is controllable if and only if the rank of $B{:}AB{:}\ldots{:}A^{m'-1}B$ is m'. (For more details on related controllability theory for control systems, see *Discrete-Time Control Systems* by O. Katsuhiko, but note some of the notations used for controllability there are different from those used in this book [20].)

It is obvious that coupling only affects the GTP-bound/GDP-bound cross-talks among GTPase pathways. Consequently, we find that the control-space complexity of GEFs/GAPs is $O(m')$.

Proof for O(n): For the total number of variables n, the order of the kinases/phosphatases is 2^n. Therefore, the order of complexity is $O(n)$. Without loss of rigorousness, we define a virtual generator $\Theta\ (.)$ at the concept level:

$$\Theta\ (.) = \{S'_{i'1}, S'_{i'2}, S'_{i'3}\} \tag{5.6}$$

with respect to

$$\{\Psi_{j1} = X_{i'1} \text{ or } \neg X_{i'1}, \Psi_{j2} = X_{i'2} \text{ or } \neg X_{i'2}, \Psi_{j3} = X_{i'3} \text{ or } \neg X_{i'3}\} \quad (5.7)$$

where

IF $\Psi_{jl} = X_{i'l}$ THEN $S'_{i'l} = 1$
IF $\Psi_{jl} = \neg X_{i'l}$ THEN $S'_{i'l} = 0$
$(l = 1, 2, 3)$,

$i'1 \neq i'2 \neq i'3$, $i'l \in \{0, 1 \dots n-1\}$ ($l = 1, 2, 3$), which is defined as the set of indexes of the variables.

For kinase/phosphatase pathways that are used to encode the clauses of

$K'_{i'1} = K_{i'1}(S'_{i'1})$,
$K'_{i'2} = K_{i'2}(S'_{i'2})$,
$K'_{i'3} = K_{i'3}(S'_{i'3})$,

we have

$$K'_{i'l} \, (l = 1, 2, 3) = K_{i'l}(S_{i'l}) \quad (5.8)$$

that is,

$K'_{i'l}$ = kinase if $S_{i'l} = 1$,
$K'_{i'l}$ = phosphatase if $S_{i'l} = 0$

where $l = 1, 2, 3$.

In total, m pathways are used for encoding m clauses $\{\Psi_j\}$. Each of these consist of three subpathways corresponding to the related variable. We obtain $3m$ individual words in a combinatorial form of kinases/phosphatases. Here, we obtain two binary words: one has n bits for n variables, and the remaining variables have $3m$ bits for combinatorial forms of kinases/phosphatases according to the clauses. We discuss this topic in three cases:

1. $n = 3m$. If variables are not repeated in the subpathways, we get $O(n)$. If some variables repeatedly appear in the subpathways, the number of used variables is smaller than n. We can still get $O(n)$.
2. $n > 3m$. Even if all of the variables in subpathways are different, the number used is smaller than n. Thus, we have $O(n)$.
3. $n < 3m$. If different n variables have been used in subpathways, there exist $(3m-n)$ variables that are used repeatedly in subpathways. The control-space complexity is $O(n)$. This is true even if the number of different variables used in subpathways is less than n.

Now it becomes clear that the number of clauses, m, does not contribute to an increase in the control-space complexity any more than does the number of

The two networks reflect two operator schemes at the algorithmic level. In contrast to this, the global structure is represented by the graph/network in the form of matrix C:

$$[c_{ij}] \ (i,j = 0, 1...9) \text{ whose size is } 10 \times 10$$

The 10×10 size is selected here for convenience of discussion. The resultant network is represented in the logic form of

Graph (local parameters, global parameters)

which needs to be verified by the criterion:

logical verification (global graph) AND biochemical feasibility (local graph)

5.3 How to Design Algorithms for a Molecular Computer

The models of molecular computing are mainly inspired from the molecular mechanisms in nature. In applying them to applications, we need to adjust the computing mechanism design at the algorithmic level. Of course, the complexity analysis serves as a guideline for the entire synthesis process of molecular computing.

5.3.1 Observing Complexity from Benchmarks

Problem solving by molecular computing is one of the main issues in theoretical molecular computer science. Currently, one of the most important tasks in this field is improving the efficiency of the implementation. By comparing different techniques for feasibility and efficiency, some good algorithms are being adopted for the best usage. Solving NP problems (e.g., 3-SAT computation) is a touchstone for biomolecular computation. It is also beneficial to generic computing because it can promote exploration of potential operators and "parallelization" from the viewpoint of unconventional algorithm designs.

From the 3-SAT problem, which is one of the most important benchmarks for testing DNA computing algorithms, we can find a number of successful examples.

Basically, time and space are two quantitative measures used to evaluate the performance of algorithms: time complexity and space complexity. In parallel computing, space complexity is often traded for reducing time complexity.

Complexity is used to explain the difficulty of the problem to be solved by the program. There are several parameters related to program running. In order to obtain a unique measure to quantitatively describe the scale of a problem for computing, the phrase "the size of the problem" in computer science is defined as the most significant parameter of the problem, which greatly influences the cost of program running from the viewpoint of space or time. For example, in the SAT problem, the size of the problem means the measure of the number of clauses, the number of variables, and the number of literals. The complexity of problems or the difficulty of programs is defined as the upper-bound of the cost in time or space, like the function g(n) defined as the upper-bound of the cost of the problem. Here, complexity has two types including time complexity and space complexity.

The P class of the problems refers to those problems that can be solved in polynomial time. NP refers to a class of the problems that can be solved by a so-called "oracle" device in polynomial time. The full name is "nondeterministic polynomial time" class. However, this oracle device is a conceptual device whose implementation form depends on the specific process. In the case of the DNA computing process realized by Adleman, the oracle is a hybridization mechanism of DNA molecules. This mechanism is used to carry out a computing process by encoding HPP programs, and it is manufactured in test tubes. This is a set of the most difficult problems for a computer to solve. The P algorithm is defined as the one with polynomial time cost. However, in many cases, we have to handle an NP problem in which the problems in the NP classes are regarded as having the same level of difficulty. The NP-complete problem is the most difficult one in the NP class.

There are several good algorithms for SAT problem solving, in which different operators use different information representations, thus showing different computing performances. From the viewpoint of biomolecular computing algorithms, we are interested in how to make best use of the molecular mechanism to explore new information processing paradigms by algorithmic design. The kernel operators in the following algorithms selected for explanation are good examples for informatics.

1. *The Schöning algorithm for 3-SAT computing.* In the Schöning algorithm, the kernel operator finds the true value of the variable randomly. The 3/2 (1.33) value, the result from the statistics in SAT computing, is the base of the factor for the case of three variables in a clause. The time complexity is 1.33^n. Here, the operator can be briefly represented in the form of:

    Schöning-operator (SAT) = the operation that sets the assignment of the variable at random

2. *The evolutionary algorithm for SAT computing.* The operators of the evolutionary algorithm for SAT [21] mainly generate the candidates for the SAT solution and update the candidates according to the value of fitness defined by the difference between the candidate and clause constraint:

    EA-operator (SAT) = the operation that generates the candidates in the form of S1S2...Sn

    where S_i (i=1,2,...,n) denotes the bit of the individual in the population of EA.
    The operator of EA is parallel and in a stochastic mode. The probability is an empirical set, and the match results guide the searching process. Generally speaking, the solutions are obtained if population size and generation number are sufficiently large. However, this is not absolutely guaranteed.

variables, n. In the cases of kinases and phosphatases, 2,000 kinases and 1,000 phosphatases are available, respectively. Multicell arrangement by assays in vitro can also provide a huge number of related signaling proteins for representing variables. Thus, the derived information capacity is large enough for encoding a large number of variables. The feasibility of the scheme for connecting the kinase/phosphatase pathways and GEF/GAP pathways of the computing process can be assured by the fact that the encoding schemes between these two parts are biologically faithful. According to the information flows presented above, we can see that the mapping scheme between the GEFs/GAPs and kinases/phosphatases can be designed when the number of m' is sufficiently large to represent the words used for activating kinase/phosphatase pathways with n bits. Therefore, it is obvious that the final result of $O(log_2 n)$ can be inferred from the encoding schemes between the two parts, since the effector-set-based pathway units only provide n signaling proteins for their downstream pathways.

Q.E.D. (End of Proof)
Through the cross-talked two-layer signaling structure of regulated pathways of GEFs/GAPs and kinases/phosphatases, a logarithmic order of control-space complexity has been achieved in molecular computing based on Rho family GTPase pathways. This result lays the foundation for further moleware implementation of the GTPase-based kinase computer in low-cost biochemical experiments conducted in vitro. One application of this kind of molecular computing is expected to be medical drug design by nanobiotechnologies. Here, the process of molecular computing, in the form of programmable experiments in test tubes, can be applied to logically study the relationship among the GTPase-affected signaling proteins. These proteins are expected to be the keys to understanding the complex signaling mechanism and cellular functions of GTPases as well as their effects on curing neurodiseases.

5.4 Touchstone for Nanobio-Oracle: Moleware Logic

Biomolecular computing is rooted in the nature-inspired idea of the biomolecular information processing mechanism, so the molecular operators are well studied based on the biochemical characteristics in laboratories. Considering the development of a future biomolecular computer toward easy accessibility for the user, programmability is an inevitable issue. To realize this ability requires a method that can verify the programs of biomolecular computing in logic. This means that we have to check the logical attributes of the biomolecular programs based on biochemical reliability.

5.4.1 Consistency of Computing Operators and Feasible Experimental Supports: Verification of Logic Process

In order to make it possible to program a biomolecular computing system easily, in a manner like object-oriented programming, moleware program verification is an important basis to connect the mathematical models of theoretical molecular computing with empirical experimental implementation of molecular computational operators. A computer test is one of the crucial steps in connecting design and implementation in molecular computing systems. Even if automated biochemical experiment technology were available in laboratories for biological purposes, we would still find that the biochemical feasibility of biomolecular computing operators mainly depends on moleware experiments. For the purpose of experimental implementation of biomolecular computing at the level of molecular operators, the logical verification of biomolecular computing programs is a basic step. In the case of a molecular computer, the unique feature is that the configuration of the computing process integrates the information storage with the information operation. This is in sharp contrast to the traditional electronic computer. Currently, the molecules representing information are updated to correspond to the computing sequences. The program codes are "written" on the molecules that are run by chemical reactions. The information flow is produced by moleware in parallel:

$$\text{information}(t) \to \text{information } (t+1) \to \text{information } (t+2) \to ..., \to ...$$

where t refers to the time. A new task emerging from the discussion in the form of information flow is to study the logic process of the molecular program itself based on operators, data structure, and program structure in unconventional computing. In semiconductor computers, these are realized in the mode of software engineering. With logic programming, the formalization of the computing process can be studied by logic systems. The kernel task for logic verification of biomolecular computing is to find and verify the consistency of logical information flow of the primitive instructions in concept and the corresponding feasible biochemical material flow in implementation. For example, information flow represented as an abstract form of an input-output relationship can be discussed using the predicate form that is commonly used in logic programming. If the basic operation process of molecular computing is defined with the form of

hybridization (DNA-sequence-1, DNA-sequence-2)

which combines two DNA sequences into one DNA sequence, the corresponding logic form of the computing by biochemical reactions of two double-stranded DNA sequences for molecular computing becomes

computing (DNA-sequence-1, DNA-sequence-2)

where "computing" is defined as a predicate corresponding to the above hybridization operator. This yields a new DNA sequence as the result of computing the two DNA sequences. For transition from the molecular operator form to formal biomolecular computing processes, the information processing mechanism represented by rules is the kernel. Each rule is used to describe the computing process. Each single computing process that corresponds to a single information flow can be obtained from experiments. The information is represented in the form

5.4 Touchstone for Nanobio-Oracle: Moleware Logic

$$\text{input} \rightarrow \text{output} \quad \text{(pathway number)}$$

Information flow should be designed to match the computing process (e.g., string rewriting process. In terms of string rewriting, the left side of the rule is rewritten into the right side of the rule). The computing device that realizes the information flow is not explicitly represented here, since the informatics representation of the I/O structure is sufficient to help us understand and study the MIMD architecture of the parallel computing systems, where logic is important for program verification of molecular computation. The logic form is inspired by software engineering and is used to describe the molecular computing process with biological faithfulness. The sample experiment corresponds to a single information process. The synthesis in logic verification opens the way for automated design in simulators. The theoretical topic for design and testing requires us to study a kind of automatic technology of program validation. Actually, we were inspired by "model check," which is an important automatic verification technology for bridging computer design and test.

5.4.2 Formalized Method for Moleware Logic

We touched on the logic mechanism previously in NP problem solving. However, the computing process itself is not designed in logic. From the opposite viewpoint, we now discuss the logic forms of the computing process. The correctness of the logic mechanism of molecular computing is the foundation of the entire process of the molecular computer.

Moleware logic is defined as a logic form of molecular computing. It is useful to test and verify the molecular computing process with software engineering technology and electronic computer assistance in current IT technology for biomolecular computer building. This is also helpful to broaden the applications of molecular computing. The application of DNA molecular automata for logical wet programs applied in medical investigations is one of the motivations for this idea. The diverse methods for molecular computers have been discussed in Chapter 4. The materials are directly related to the task of building molecular computers at the material level. The logic method should be universal and systematically similar to the software engineering for a semiconductor computer.

With respect to the feasibility of the designed operation, simulation can benefit the test work for implementation. The reliability study from the viewpoint of implementation is important for construction of biomolecular computing systems. Therefore, before designing experimental schemes for implementation of molecular computers, we should work on models for logical description having consistency of the feasibility of biochemical reaction with the logically proved operations in a generalized form:

$$\text{biofunction (molecular computing operator)}$$

which takes true value or false value corresponding to the availability of this molecular computing process according to the biochemical feasibility in the case of biochemical molecular implementation.

For one information-processing operator, several biochemical operators are used to implement it. Here we discuss a logical method to design the rules for computing. The logical inference process works according to the logical values defined by the features of biochemical reactions. By inferring logical values, the causal chain of the biochemical reactions can be constructed to compare the efficiency of different programs with economical solutions.

As mentioned, the concept of model check inspired our approach to the verification work. Model check refers to the process of finding the configuration of the system, which is the parameters of the logic form of the designed programs that are often represented as the logical formula in the state space corresponding to the model. In the following instance, the idea of "model check" for verification of molecular programs is applied to the verification of molecular computing.

5.4.2.1 Example 1: A Model for Abstract Cellular Pathways Based on String Rewriting Representation

The formalization of the molecular computing process can be done by the logic expression used for inferring the molecular operation scheme. The inference process will be conducted by the primitive form of the program codes that directly describe the molecular operators. Here, we consider an example in the form of string rewriting for molecular computation.

Let V be the alphabet set, the set of terminals

$$V_T = \{a, b, c, d, e, f, g, h, u, v, p, q\} \tag{5.9}$$

and the string set of

$$\{S1, S2, ..., S14\}$$

representing the strings made by V.

We define the following string rules for the computing system by pathways:

1. $a \to S_1$ (pathway 1)
2. $S_1 \to S_2 b S_3$ (pathway 1)
3. $S_2 \to c$ (pathway 1)
4. $S_3 \to d$ (pathway 1)
5. $c \to S_4$ (pathway 2)
6. $d \to S_5$ (pathway 3)
7. $S_4 e \to S_4 p e$ (pathway 2)
8. $S_4 p e f \to S_4 e f$ (pathway 2)
9. $S_4 e f q \to S_4$ (pathway 2)
10. $u S_5 g \to u S_5 p g$ (pathway 3)
11. $u S_5 p g h \to u S_5 g h$ (pathway 3)
12. $u S_5 g h q \to u S_5$ (pathway 3)
13. $S_4 \to u$ (pathway 2)

14. $vuS_5 \to S_5$ (pathway 3)
15. c (pathway 1) $\to c$ (pathway 2) (pathway 4)
16. d (pathway 1) $\to d$ (pathway 2) (pathway 4)
17. a (pathway 4) $\to a$ (pathway 1) (pathway 4)

The biochemical meaning of the pathway structure for computing is discussed in Chapter 8 in terms of molecular bioinformatics. Obeying the convention of theoretical computer science, we describe here the string rewriting form for pathway-based computing. String rewriting is an abstract way to describe the information flow. Pathway 4 is connected to pathways 1, 2, and 3 and influences these pathways in terms of the intracell communication processes by rules (15)–(17). The string rewriting processes within one pathway unit for the corresponding biochemical reactions are described by rules (1)–(14). If we replace (7) by the form

$$S_4 e \to S_4 ep$$

it will yield the features in context-sensitive languages (CSLs), which is possible by a pathway of multiple molecules. From the viewpoint of the structure of information flow, it is obvious that graph rewriting can generate interacted string systems under certain conditions. The graph rewriting form may lead to the construction of a signaling network for a more complex mechanism of computing systems. If the rules are assigned with logic values, the set of rules can be used to infer the information processing systems by the logic sequences. Inspired by the model check methodology, the computation model constructed by the above string rewriting rules, where the interaction mechanism is embedded and the rules are discussed by pathway units, can be oriented to the biochemical reaction forms. We discuss four pathway units in this instance.

Pathway unit 1

1. $a \to S_1$
2. $S_1 \to S_2 b S_3$
3. $S_2 \to c$
4. $S_3 \to d$

Pathway unit 2

5. $c \to S_4$
7. $S_4 e \to S_4 pe$
8. $S_4 pef \to S_4 e f$
9. $S_4 efq \to S_4$

Pathway unit 3

6. $d \to S_5$
10. $uS_5 g \to uS_5 pg$

11. $uS_5pgh \to uS_5gh$
12. $uS_5ghq \to uS_5$
13. $S_4 \to u$
14. $vuS_5 \to S_5$

Pathway unit 4

15. c (pathway unit 1) $\to c$ (pathway unit 2)
16. d (pathway unit 1) $\to d$ (pathway unit 2)
17. a (pathway unit 4) $\to a$ (pathway unit 1)

The corresponding biochemical forms of the pathways represented by the information flow of input-output are presented for quantitative study of the relationships among the signaling pathways. The biochemical reaction form of the corresponding string rewriting rules is used for verification of the reactions as well.

The molecular complex for string S_6 is the input to this pathway unit and denotes the variable X_1 in the clause of an SAT problem. In pathway unit 1, the subpathways are represented as follows:

$$a = \text{pathway}(S_6) \tag{5.10}$$
$$S_1 = \text{pathway}(a) \tag{5.11}$$
$$(S_2, b, S_3) = \text{pathway}(S_1) \tag{5.12}$$
$$c = \text{pathway}(S_2) \tag{5.13}$$
$$d = \text{pathway}(S_3) \tag{5.14}$$

c and d are output of the pathway unit 1 and input to the pathway units 2 and 3, respectively.

In pathway unit 2, c is the input.

$$S_4 = \text{pathway}(c) \tag{5.15}$$

S_4 is the output of pathway unit 2.
p is produced by the pathway

$$p = \text{pathway}(M_1) \tag{5.16}$$

where M_1 is the signaling protein molecule in the cell, under the condition that S_4 and e act as protein enzymes.

$$(S_4, e, f) = \text{detection}(S_4, e, f, p) \tag{5.17}$$

As output of the above pathway, protein enzymes: S_4, e, and f are in the test tube/array.

$$S_4 = \text{detection}(S_4, e, f, q) \tag{5.18}$$

The result is given by the detection of the output of the above pathway and q in test tubes/assays

$$S_4 = \text{detection}(S_4, e, f) \tag{5.19}$$

This produces the output of pathway unit 2.

Pathway unit 3:

$$p = \text{pathway}(M_2) \tag{5.20}$$

where u, S_5, and g are protein enzymes and M_2 is a protein molecule in a cell for signaling. The new molecule p is produced.

$$(u, S_5, g, h) = \text{detection}(u, S_5, p, g, h) \tag{5.21}$$

The detection is from (u, S_5, p, g) with respect to the above pathway and h.

This process gives rise to the contextual rewriting function of $h \leftarrow p$ under the context of prefix u, S_5 and suffix g.

The two processes are equivalent to the conceptual process:

$$h = \text{operation}(p) \tag{5.22}$$

where u, S_5 g, act as protein enzymes.

$$(u, S_5) = \text{detection}(u, S_5, g, h, q) \tag{5.23}$$

g, h, and q are consumed by the detection function on molecules.

$$u = \text{pathway}(S_4) \tag{5.24}$$

$$S_5 = \text{detection of}(v, u, S_5) \tag{5.25}$$

From the containers of this pathway output and S_5, v, the output of this pathway unit is given as S_5.

In order to verify the logical operator by biochemical feasibility, the parameters/coefficients of the biochemical reactions that sustain the pathway units for computing are the basis of selecting the qualitative and quantitative measures. We need to analyze the pathway coefficients and logic reasonability/rationality of the interacted complex system (SoS). The coefficients are calculated for judgment of the logic correctness of the interacted pathways in cells. The criterion for verification of the program characteristics/features is given as follows:

$$X_1 \text{ AND } (X_2 \text{ V } X_3)$$

S_6 is used to represent X_1. X_1 is denoted as "a" and kept sustainable in the pathway units. The other variables are set as the other pathway units. The "AND" logic is reflected in the structure of interaction of the pathway units. S_4 is made to denote X_1 and S_5 is for X_2.

Consequently, the question is how to verify the physically feasible operators of pathway units to carry out this computation in logic.

The process of finding the parameters in the state space can be constructed by the SAT-based or SAT-inspired search mechanism. This could be a logic inference process or made by a biochemical process of moleware as well. It will be possible to integrate the computing process with the program verification of a biomolecular computing system into one body in the near future, owing to the computing/signaling mechanism discussed earlier.

In this chapter, we focused on algorithmic complexity and design schemes of biomolecular computing. From comparison of different complexities derived from different algorithms of biomolecular computing, we can conclude that the operators constructed by different signaling mechanisms play a crucial role in complexity performance of the derived algorithm. The biomolecules greatly differ in both physical structure and biochemical behavior. The resultant biochemical information flow generated by the complex molecular systems for computing corresponds to different specifications of biochemical reactions and configuration of moleware architecture. Inspired by the naturally existing mechanism of the nanobiomachine, the parallelism of moleware architecture and corresponding high-intensity information storage makes efficient biomolecular computing possible. The theoretical studies in mathematics and experimental technology provide a sound foundation for design schemes, which shows that information science and genomics/proteomics can work together in a computing process constructed by molecular bioinformatics structures.

The biochemical features of biomolecules bring us nature-inspired design schemes for biomolecular computing. The results for the designed algorithms and corresponding computing processes are interesting because we can observe new phenomena in computing science and thus open new information technology as well. The examples we discussed in this chapter mainly are the NP problems, which have great significance and broad applications.

Informatics and biochemistry have been used in tandem for information engineering, in which the information processing of biomolecular computing is constructed based on the biochemical reactions of biomolecules. This effort itself is a great step forward in unconventional computing paradigms. After the formalization of the biomolecular computing processes are made, a systematic model of biomolecular computing can be established, and this model will be discussed in the next chapter.

In summary, the nanolevel of biomolecular systems is rich in operable space for our designs of computing devices. Nano-architecture provides a kind of media that is "free of gravity," which works just like a molecular network for very large-scale parallel computing. In the case of applying cells as the building blocks of computing devices, the signaling pathways provide the basis for self-assembling the computing units and self-organizing the control units of the underlying biomolecular computing systems.

References

[1] Winfree, E., et al., "Design and Self-Assembly of Two-Dimensional DNA Crystals," *Nature*, Vol. 394, August 1998, pp. 539–544.

[2] Winfree, E., T. Eng, and G. Rozenberg, "String Tile Models for DNA Computing by Self-Assembly," *DNA6*, LNCS 2054, 2000, pp. 63–88.

[3] Lewis, H. R., *Elements of the Theory of Computation*, Upper Saddle River, NJ: Prentice-Hall, Inc., 1981.

[4] Yokomori, T., S. Kobayashi, and C. Ferretti, "On the Power of Circular Splicing Systems and DNA Computability," *Proc. of IEEE Intern. Conference on Evolutionary Computation*, Indianapolis, April 1997, pp. 219–224.

[5] Sakakibara, Y., and C. Ferretti, "Splicing on Tree-Like Structures," *Theoretical Computer Science*, Vol. 210, 1999, pp. 227–243.

[6] Sage, A. P., Conflict and Risk Management in Complex System of Systems Issues, Systems, Man and Cybernetics, 2003. *IEEE International Conference*, Vol. 4, October 2003, pp. 3296–3301.

[7] Yon, Z., "Improved DNA Algorithm of Chinese Postman Problem, 2004," *WCICA 2004, Fifth World Congress on Intelligent Control and Automation*, 2004, Vol. 3, June 2004, pp. 2313–2317.

[8] Ibrahim, Z., et al., "Molecular Computation Approach to Compete Dijkstra's Algorithm," *5th Asian Control Conference*, Vol. 1, July 2004, pp. 635–642.

[9] P-system Web site, http://psystems.disco.unimib.it/main.html.

[10] Head, T., "Formal Language Theory and DNA: An Analysis of the Generative Capacity of Specific Recombinant Behaviors," *Bulletin of Mathematical Biology*, 1987, Vol. 49, pp. 737–759.

[11] Freund, R., and F. Freund, "Molecular Computing with Generalized Homogeneous P-Systems," *DNA6*, LNCS 2054, pp. 130–144.

[12] http://www.cs.berkeley.edu/~zadeh/.

[13] Schöning, U., "A Probabilistic Algorithm for k-SAT and Constraint Satisfaction Problems," *Proc. of the 40th Symposium on Foundations of Computer Science*, 1999, pp. 410–414.

[14] Díaz, S., J. L. Esteban, and M. Ogihara, "A DNA-Based Random Walk Method for Solving k-SAT," *DNA6, Lecture Notes in Computer Science*, Vol. 2054, A. Condon, and G. Rozenberg, (eds.), Berlin: Springer-Verlag, 2001, pp. 209–219.

[15] Liu, Q., et al., "DNA Computing on Surfaces," *Nature*, Vol. 403, 2000, pp. 175–179.

[16] Ogihara M., and A. Ray, "DNA Computing on a Chip," *Nature*, Vol. 403, 2000, pp. 143–144.

[17] Chen, K., and V. Ramachandran, "A Space-Efficient Randomized DNA Algorithm for k-SAT," *DNA6, Lecture Notes in Computer Science*, Vol. 2054, A. Condon and G. Rozenberg, (eds.), Berlin: Springer-Verlag, 2001, pp. 199–208.

[18] Yoshida, H., and A. Suyama, "Solution to 3-SAT by Breadth First Search," *DNA Based Computers V*, E. Winfree and D. K. Gifford, (eds.), DIMACS Series in Discrete Mathematics and Theoretical Computer Science, Vol. 54, AMS, 2000, pp. 9–22.

[19] Liu, J. -Q., and K. Shimohara, "A Biomolecular Computation Method Based on Rho Family GTPases," *IEEE Trans. on Nanobioscience*, Vol. 2, No. 2, June 2003, pp. 58–62.

[20] Katsuhiko, O., *Discrete-Time Control Systems*, 2nd ed., Upper Saddle River, NJ: Prentice-Hall, 1995.

[21] Gottlieb, J., E. Marchiori, and C. Rossi, "Evolutionary Algorithms for the Satisfiability Problem," Vol. 10, No. 1, Spring 2002, pp. 35–50.

[22] Pedrycz, W., G. Succi, and O. Shai, "Genetic-Fuzzy Approach to the Boolean Satisfiability Problem," *IEEE Trans. on Evolutionary Computation*, Vol. 6, No. 5, October 2002, pp. 519–525.

[23] Besozzi, D., et al., "Gemmating P Systems: Collapsing Hierarchies," *Theoretical Computer Science*, Vol. 296, No. 2, 2003, pp. 253–267.

[24] Benenson, Y., et al., "Programmable and Autonomous Computing Machine Made of Biomolecules," *Nature*, Vol. 414, 2001, pp. 430–434.

[25] Benenson, Y., et al., "An Autonomous Molecular Computer for Logical Control of Gene Expression," *Nature*, Vol. 429, 2004, pp. 423–429.

[26] Paun, G., *Computing with Membranes*, TUCS Research Report, No. 208, November 1998.

[27] Paun, G. "Computing with Membranes," *Journal of Computer and System Science*, Vol. 61, No. 1, 2000, pp. 108–143.

CHAPTER 6
Cellular Biomolecular Computing Based on Signaling Pathways: Kinase Computing

Signaling pathways form one of the most important mechanisms in the cellular system [1–5]. They have a complex biological function in which the Rho GTPase, also called Rho family GTPase, plays an important role (e.g., the morphology and movement and behavior of cells [1, 5]) and the protein enzymes, such as kinases and phosphatases, greatly influence cellular life [2] through "uncontrolled cell growth and tumorigenesis." To study the biological functions of cells, we have to systematically understand the underlying mechanism of cells. Systems biology is one of the promising ways toward achieving this goal. Explorations of the engineering issues related to biological cells have been launched in the field of cell engineering. In these explorations, a critical task is to apply system science and engineering to the synthesis and analysis of cellular information-processing systems in order to obtain the desired cellular function. The advances made in molecular biology permit observation of a single cell [3], even though the cell is still very complex for engineering processes by nanobiotechnology. The methodology of "system of systems" is appropriate for this case. From the viewpoint of molecular biology, there are many subsystems within cellular systems. From the viewpoint of parallel computing, an MIMD computer is a suitable metaphor for cellular information processing. The unconventional nature of biomolecular computing is reflected in the materials used for building computers—molecules/moleware—and the methodology for design and testing of molecular computers. The related studies and technology are also often regarded as unconventional. To achieve the mapping from the molecular signaling process to the computing process, we need to derive new knowledge and innovative methods from the existing technology (in multiple disciplines [1–24]). In this sense, we consider this new research field unconventional. The mainstream information sciences, those that might be considered conventional, can provide a great amount of useful knowledge for now and the future.

This chapter presents kinase computing (see [25]), developed from biological inspiration into unconventional computing paradigms. The key element of kinase computing is the signaling pathway mechanism in cells. The great potential of parallel information processing can be realized by exploring the naturally existing capacity of cellular communication processing.

6.1 Cellular Pathway: Another Ubiquitous Society in Another Universe

Nearly everyone acknowledges the impact and practical advantages of ubiquitous computing for telecommunication in our IT society. Here, we will travel to another ubiquitous society called cellular society, which is so tiny that we have to use an unfamiliar molecular language to communicate with our moleware friends. The inhabitants in cellular society are various types of molecules, and the common units constructing the cellular community are biochemicals. The cell communication among these molecules is ubiquitous, but wet ubiquitous rather than electrical ubiquitous. Recently, H.A. Hundley et al. discovered that ribosome-associated Mpp11 has a ubiquitous function in regulating molecular messages in metazoan cells [4]. An enormous number of molecules work tirelessly for the survival of the cell. The Latin origin of the word "ubiquitous" means "anywhere and anytime." The cell signaling processes by molecules also ubiquitously work for cell communication, and it is even possible that these can be used for molecular computation. The uniqueness of cellular computing is reflected in the unique mechanism of biochemical moleware signaling in the cell itself.

In cellular society, everyday traffic is the transportation of biochemicals. Cell signaling is a process realized by the biochemical reactions that happen in cells according to the molecular-structural features and activation status of enzymes at a given time. These biochemical flows form a graph or a network. The derived communication updates the biochemicals in cells through the production of new chemicals and the removal of existing chemicals. This provides us a new way of information processing in which the methodology and technology differ from those of computing in telecommunications networks. Communication here refers to molecular communication as a process of transferring and transporting molecular messages through biochemical form in cellular systems. Cell communication in this section includes intercell communication and intracell communication. The former refers to the signaling process of transmitting chemicals between cells, which in turn can cause biochemical functional reactions in cells. The latter refers to the biochemical reactions used for signal transduction within cells, such as cascades within cells from the membrane receptors to the proteins' interaction with the gene in the nuclei.

6.1.1 Ubiquitous Cell Communication for Parallel Information Processing

Cell communication at the signal transduction level is conducted in three ways in cells: through endocrine, paracrine, and synaptic modes. In intercell communication, molecules are sent from certain cells and received by so-called target cells, in which the signaling cascades are activated by the second messengers involved in metabolic and/or signaling pathways through the receptors. The receptors around membranes include two major types: membrane-bound receptors and intracellular receptors (there are also four minor types). The signaling mechanism of cell communication is illustrated in Figure 6.1 from the viewpoint of information processing. The cross-talk among the cellular pathways leads to a richness of nonlinear complex behavior. The cellular pathways even influence the growth and development of cells. Apoptosis

6.1 Cellular Pathway: Another Ubiquitous Society in Another Universe

(programmed cell death) and cancer are two important products of the natural signaling mechanism of cells. The cell community works like a society where energy, information, and materials are organized in a collective mode.

Communication in information terms can be traced back to Shannon's information theory and even earlier. Engineered-level understanding of communication can be traced back to Wiener's cybernetics and maybe even earlier. The possible relation between communication and computation can be explained by the naturally existing cellular signaling mechanism in which each molecule is regarded as an autonomous information-processing unit. The related pathways in cells act as rules that determine systematic behaviors. Local behavior of these molecules can be observed in purified materials in test tubes. Aggregation of these molecules in living cells eventually leads to globally emerging behavior under the definition of nonlinear dynamics. From its appearance, we might consider this system a kind of biochemical "mobile wireless ad hoc network" built by moleware. By looking at the biological pathway network through the concepts of its electronic ad hoc counterpart, we can easily understand the behavior of the cellular pathways and networking. In an ad hoc telecommunication network, a lot of work needs to be done to improve scalability, which is an important performance criterion that guides the system design and testing. In contrast to this mechanism, the signaling network in cells can be easily scaled up in an autonomous and robust way based on its self-organizing mechanism, a phenomenon from which we can learn something.

As shown in Figure 6.1, the cellular communication process can be observed from A to D. Molecule A is regarded as an information source and molecule D as the destination. A chain of biochemical reactions carries out the message delivery in the order A → B → C → D. The channel is the cellular pathways in cells.

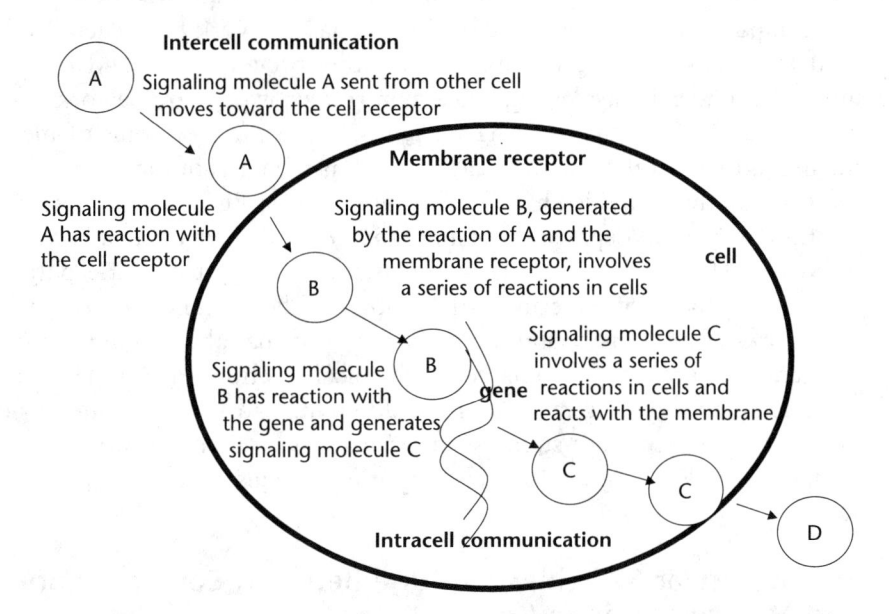

Figure 6.1 Schematic description of basic cellular communication.

There are several commonalities between a mobile wireless ad hoc network and a parallel cellular information-processing network: mobility of the nodes, ubiquity of the local information process in the neighborhood, and high complexity due to interaction networks. The difference between them is the protocol used: the former is designed in advance and monitored according to its requirements, while the latter has its own biochemical protocol that can be carried out autonomously for biochemical reactions without any human/artificial interaction. In electrical ad hoc networks, engineers focus on the information flow among networks and improvement of network services (for example, information transportation and related protocols [6]). In the sense of optimization in nature, the adaptation ability of the pathway network can adjust the "configuration" of the metabolic and signaling pathways in terms of biomass, coupling, and regulation. The entire pathway network works as a content-addressing memory that is accessible by cellular function in an ad hoc mode.

Cell communication by pathways can provide a parallel distributed processing (PDP) mechanism for information processing. The biochemical features of the molecular interaction mechanism of the cellular pathways are keys to this PDP, providing a kind of bio-ubiquitous computing. One of the inspirations from this PDP mechanism is the structure and related method in cellular information processing. Actually, different structures in different parts of cells can be used to conduct molecular computing. In Landweber and Kari's method [20], the gene mechanism of evolution in a bacterium of ciliates is used to derive universal computation. In Ehrenfeucht et al.'s formal model [21], micronuclear genes in ciliate are modeled as a rewriting system. Weiss and Knight [22] focus on intercell communication for amorphous computing in the form of logic circuits. S. Basu et al.'s work on a synthetic multicellular system deserves attention [23]. The membrane is the object in P-systems, or "membrane computing," originated by Paun in theoretical computer science (see [26, 27]). In Magnasco's work [24], chemical kinetics is used to design a Turing universal computing process. Cells and related cellular molecules under kinase/phosphatase controls provide a framework of designing nanobiomachines for computing purposes. From the viewpoint of methodology, the ubiquitous parallel hierarchical network is the basis of the entire cell system in information theory, which handles the pathways for PDP (for example, the process of phosphorylation/dephosphorylation and GTP/GDP exchange). For the moment, the molecular switch in cells is the most realistic way to build a pathway-based computer capable of experimental feasibility. The biofunction-oriented architecture works as an MIMD under the condition of spatial configuration of wetware and temporal orders of the related biochemical reactions in cells. The symbols corresponding to signaling molecules are different, and multiple kinds of chemicals flow in multiple types of streams simultaneously; consequently, we have a variety of options in designing molecular computing systems.

6.2 The Molecular Switch as a Bridge Between Cell Communication and Molecular Computing

In cell communication, signaling molecules move around and carry messages in chemical forms. The locations of these molecules do not rigidly constrain their actions, as is the case with their electronic counterparts. The biological ubiquity mainly affects the logic relations among the molecules. Our basic idea is to start with communication to design a computation process using cellular pathways.

In molecular biology, the term "pathway" refers to the functional biochemical reactions in living cells. Concretely, a pathway is represented as a directed graph G. The input A is the reactants that start the biochemical reactions corresponding to the conceptual graph. The output C is the result of this biochemical reaction, which may be an indivisible or an integrated pathway. Other molecules involved in this reaction are the intermediate parts of the pathways.

In biology, the signaling pathway is mainly studied for its biological functions in cell communication. The transportation of a message from molecule to molecule produces the rich biochemical features of cells. If these molecules in the cells are treated as symbols that correspond to different biochemical functions, we can schematically describe the pathway system of cell communication in terms of Shannon's information theory.

From cell communication, we expect to extract a new kind of unconventional computing paradigm. We define information as follows: X and Y are inputs to the pathway, and Z is the output of the pathway. Consequently, we want to obtain

$$Z = X + Y$$

The molecules for X, Y, and Z are designated and set as corresponding signals. For example, if the states of X and Y are in binary 1 and binary 0, respectively, we need to design and control the pathway to give the state of Z as binary 1. In order to denote binary information by molecules, we need to introduce the binary information representation method into biomolecular computing. The molecular switch is one of the best molecular mechanisms for doing the work of information representation.

6.2.1 Binary Information Representation by Molecular Switch

Cellular molecular switches provide the basis of the possible architecture of kinase computing. We consider two classes of molecular switches in cells, which are phosphorylation/dephosphorylation switches and GTPase switches.

The GTPase switch is a molecular switch in which the GTP-bound/GDP-bound state of GTPases is regulated by GEF/GAP; this is also called the GEF/GAP switch.

The SPK switch is the molecular switch for SPK molecules in cells in which the phosphorylation/dephosphorylation state is regulated by kinases/phosphatases; this is also called the kinase/phosphatase switch.

The GEF/GAP pathway refers to the signaling pathways of cells that regulate the GTP-bound/GDP-bound states of GTPase by GEF/GAP.

The kinase/phosphatase pathway refers to the signaling pathways of cells that regulate the phosphorylation/dephosphorylation states of the molecules by kinases/phosphatases.

In this notation, we use the state molecules to name the molecular switches and the regulator molecules to name the pathways. The switch and pathway are two aspects of the cellular signaling process according to their functions in informatics and biology.

In physical biochemistry, GTP-bound and GDP-bound states of GTPases are determined by GEFs and GAPs, respectively. Phosphorylation and dephosphorylation of signaling proteins are determined by kinases and phosphatases, respectively. During the process of GTP hydrolysis, the GTP-bound GTPase is hydrolyzed into a GDP-bound state. The signaling pathway mechanisms can provide two states of signaling molecules under certain conditions of cellular regulation. Accordingly, the molecular switches form the basis of designing a molecular computing system by cellular pathways in cells.

We use the following abbreviations of terms in molecular cell biology for later discussion.

- GEFs: guanine nucleotide exchange factors;
- GAPs: GTPase-activating proteins;
- GTP: guanosine triphosphate;
- GDP: guanosine diphosphate;
- GTPase: the proteins in cells with two states—GTP-bound state and GDP-bound state;
- GDI: guanine nucleotide exchange inhibitors;
- Kinase: a kind of protein in cells that can phosphorylate the signaling molecules in cells;
- Phosphatase: a kind of protein in cells that can detach any phosphate from phosphorylated signaling molecules;
- SPKs: signaling-protein molecules in cells regulated by kinases/phosphatases and their corresponding pathways; these proteins mainly include serine, threonine and tyrosine;
- p: phosphate.

The molecular switch is reversible for binary states of signaling molecules in cells. The GTPase switch and kinase switch are the two major types of molecular switches. In the computing process of kinase computing, these two switches are used to code the binary information. They can be controlled through the signal transduction around the membrane receptor. In one of the possible design schemes, the kinase switch is used for data processing and the GTPase switch is used for instruction. Among the relations between the two switches, some kinase switch pathways regulated by GTPase are already known (for example, the Rho-MBS-MLC pathway reported in [11, 12]).

Here, we present the basic framework for information representation based on these two types of molecular switches. The information representation and operation of the GTPase switch and SPK switch is discussed by reference to the following formal description.

Axiom 6.1
Under the condition that GEF/GAP pathways of cells in vivo are activated, we have

$$G_l = 1 \text{ if } G_{gef}(l) \neq \phi \text{ and } G_{gap}(l) = \phi \qquad (6.1)$$

6.2 The Molecular Switch as a Bridge Between Cell Communication and Molecular Computing

and

$$G_l = 0 \text{ if Ggef}(l) = \phi \text{ and Ggap}(l) \neq \phi \tag{6.2}$$

where G_l is the lth GTPase in the GTPase set. Ggef (l) and Ggap (l) are the lth GEF and GAP sets, respectively.

Axiom 6.2
Under the condition that GEF/GAP pathways of cells in vitro are activated, we have

$$G_l = 1 \text{ if GEF}(l) \neq \phi \text{ and GAP}(l) = \phi \tag{6.3}$$

and

$$G_l = 0 \text{ if GEF}(l) = \phi \text{ and GAP}(l) \neq \phi \tag{6.4}$$

where G_l is the lth GTPase in the GTPase set. GEF (l) and GAP (l) are the lth elements in the GEF and GAP sets, respectively. The corresponding relationship between the GTPase and GEF/GAP can be defined empirically.

In GEF/GAP pathways, 1 is used to represent the GTP-bound state, while 0 is used to represent the GDP-bound state. We assume that the quantity of GEF or GAP is sufficient to catalyze the GTPase activation or GTPase inactivation, respectively.

Axiom 6.3
Under the condition that kinase/phosphatase pathways of cells in vivo are activated, we have

$$SPK_l = 1 \text{ if kinase}(l) \neq \phi \text{ and phosphatase}(l) = \phi \tag{6.5}$$

and

$$SPK_l = 0 \text{ if kinase}(l) = \phi \text{ and phosphatase}(l) \neq \phi \tag{6.6}$$

where SPK_l is the lth SPK in the SPK set. Kinase (l) and phosphatase (l) are the lth kinase and phosphatase sets, respectively.

Axiom 6.4
Under the condition that kinase/phosphatase pathways of cells in vitro are activated, we have

$$SPK_l = 1 \text{ if kinase}(l) \neq \phi \text{ and phosphatase}(l) = \phi \tag{6.7}$$

and

$$SPK_l = 0 \text{ if kinase (l)} = \phi \text{ and phosphatase (l)} \neq \phi \qquad (6.8)$$

where SPK_l is the lth SPK in the SPK set. Kinase (l) and Phosphatase (l) are the lth elements in the kinase and phosphatase sets, respectively. The corresponding relationship between the SPKs and kinases/phosphatases can be defined empirically. In kinase/phosphatase pathways, 1 is used to represent the state of phosphorylation; 0 is used to represent the state of dephosphorylation. We assume that the quantity of kinase or phosphatase is sufficient to catalyze the activation of phosphorylation or dephosphorylation, respectively.

For the convenience of discussion, we have already unified the terms of the two types of molecular switches as the GTPase switch and the kinase switch. Based on the binary representation mechanism, a simple form of automata is constructed to show how to carry out computation in terms of the GTPase pathway structure in cells.

6.2.2 Computing Formalized as an Automaton

With the function of signal transduction in cellular systems, the state of the molecules in cells can be changed by enzymes that are normally proteins. The transition from state to state can be described by the automaton form of computing. The state of the automaton is represented by the state of molecules.

From the viewpoint of molecular biology, the existence of extra-cellular signaling molecules is the general condition needed for the GTPase and SPK switches to be activated through the membrane receptors. This is also the fundamental principle in biological cells. In the sense of engineered pathways, in vitro biochemical technology can provide different concrete control schemes for regulating the signaling molecules with this kind of molecular switch.

To construct computing devices based on molecular switches, there are necessary information conditions. Here we discuss one instance of the GTPase switch.

Condition 6.1
There exists a set of extracellular signaling molecules $\{Q(l)\}$ ($l = 0, 1...L-1, L \in N$) that can be used to activate the GEF/GAP pathways through the cellular membrane. The signaling cascades from these molecules to GEFs/GAPs are in the following order:

Extracellular signaling molecules $\rightarrow$ membrane receptors $\rightarrow$ molecules that transfer messages $\rightarrow$ GEFs/GAPs

There are several types of realized forms from this idea. We can use any form because different experimental forms of the same computing process are equivalent for the information processing function. As an example, we first discuss automaton design by using a GTPase switch. Later we will generalize this into a generic form of biomolecular computing.

The automaton can update GTPase states of the molecular complexes by regulating the molecular switch of GEFs/GAPs, where the signals $\{Q(l)\}$ can be employed directly or indirectly. We select in vitro cells as the materials used under laboratory conditions. Here, the GTPase switches are adopted for computing. The schemes of designing automata by using kinase/phosphatase pathways are similar to those of designing automata by using GEFs/GAPs.

We define the following notations: Let A be an alphabet. G_w is the set of all states $\{w(2^L)\}$ encoded by $\{G_{gef}(l) \cup G_{gap}(l)\}$ based on axiom 1 under in vitro conditions in a laboratory. $s0$ is the start state in G_w. G_{wq} is the set of final states that are acceptable when a halt occurs. $G_{wq} \subseteq G_w$.

We can conceptually design a deterministic finite state automaton as

$$W = (A, G_w, s_0, G_{wq}, \delta) \tag{6.9}$$

where δ is defined as the following function:

$$\delta: G_w \times A \rightarrow Gw \tag{6.10}$$

The input to the automaton is defined as a single molecule in $\{U(l)\}$ and denoted as q in A. The current state of the automaton is denoted as $G_w(t)$, and the next state of the automaton is denoted as $G_w(t+1)$. According to the automaton structure, any possible implementation of the state transition from the current state of GEF/GAP to the next state of GEF/GAP is controlled by the input in A. Here, we assume that a single molecule $U(l)$ exists in the above process. This fits the definition of automaton in its basic form. For the concrete design of $\{U(l)\}$, we will discuss this topic's technical considerations later within the framework of an engineering system operated by controlling the related signaling.

Although the single-automaton form of computing given here is simple, it can be extended to a multiple-automata system under a certain condition Θ. In addition to the assumption of a single molecule from $\{U(l)\}$, other possible situations include manual operation of the state transition based on detection of GTPase as well as automated state transition by controlling engineered pathways. In theory, multiple automata for this process can be formalized in an equivalent way to a single automaton, based on the abstract representation given above.

By adopting automaton W, the configuration $C_0, C_1 \ldots C_n$ can be defined as

$$(q, U_w) \tag{6.11}$$

where U_w refers to the sequence that is constructed by the molecules from $\{u(l)\}$ and is activated in succeeding steps. Designed in concept and in experiments, such sequences can be generated by controlling the related pathways. Thus, certain sequences for computation can be formulated by the sequence

$$\{C_i\} \ (i = 0, 1 \ldots n) \tag{6.12}$$

Here, the configuration process is presented only by formalization. We suppose that the configuration $C_0, C_1 \ldots C_n$ can lead to a computing process of a deterministic finite state automaton under the condition that the corresponding pathways are controlled through the above process.

We can define the order of U(l) corresponding to the configuration given above. The states of the GTPase complex are the result of the activation of the GEFs/GAPs, which is already defined in the automaton. They can be read out by FRET. These states are used to automatically or manually activate the designated biochemical reactions for activating GEF/GAP:

$$U(l) \text{ activates } Gw \qquad (6.13)$$

where Gw refers to the state represented by the combinatorial words of GEF(l)/GAP(l).

The mapping from U(l') to Gw is a one-to-many mapping. The biochemical reactions of this mapping can be realized in different physical forms. One of the direct ways is to activate the designated "metaphor molecules" based on the detection of the readout of GTPase complexes and control of the designed GEF/GAP pathways. Under the condition of in vitro cells, these "metaphor" molecules can be controlled to activate the molecular switches by the reactions of extracellular signaling pathways and the membrane receptor.

Another type of molecular switch for an automaton—the kinase switch—is presented in the following example. We have two options under the above situation. Either type of molecule for the molecular switches can be selected as the state of the automaton. In the previous example, the motivation came from the fact that the entire signaling process used can be controlled by an engineering approach for the whole system. As opposed to the previous style, we apply a different design scheme to an automaton using the Rho-MBS-MLC pathway in the following example. This scheme comes from empirical knowledge, and it is nearly a tailor-made one. Although we are not sure this configurable structure is also scalable, this simplistic pathway architecture can make best use of the biochemical feature.

6.2.3 Example: Designing an Automaton for Kinase Switches Guided by GTPase

Consider two signaling proteins, MBS and MLC, under the regulation of Rho-GTPase. Because the MBS pathway has an MLC phosphorylation function on the MLC pathway, we have the phosphorylation state of MBS, dephosphorylation state of MBS, and dephosphorylation state of MLC when the two pathways are coupled by this function. Let the phosphorylation state be denoted as 1 and the dephosphorylation state be denoted as 0. We define the state sets of automata as $S_0=(0, 0)$ and $S_1=(1,0)$, where the binary state refers to the phosphorylation/dephosphorylation states of MBS and MLC. For the virtual tape, we denote the kinase of MBS as a, the phosphatase of MBS as a', the kinase of MLC as b, and the phosphatase of MLC as b'. At the initial moment of the computing process for the automaton, this virtual tape becomes the string $\{a\ b\ b'\ b\ b'\ a'\}$. The configuration series are given as follows:

$(S_0, a\ b\ b'\ b\ b'a')$
$(S_1, b\ b'\ b\ b'a')$
$(S_1, b'\ b\ b'a')$
$(S_1, b\ b'\ a')$
$(S_1, b'\ a')$
(S_1, a')
(S_0, Z)

where Z refers to the stop symbol.

Transition rules are given in the form of a state transition table (see Table 6.1).

6.2.4 Information Structure for Automaton-Based Computing

The automaton structure directly describes the state transition and its condition. The pathway units are used to designate the signaling process for realization of the automaton. The word "realization" here refers to the information structure of the computing model, which can tell us how the model works logically. It is a kind of information description. The automata form shows the preliminary computing behavior of the molecular computing process. The state transition is made according to the input.

We use the abbreviation MSP-automaton (molecular switch based pathway automaton) to denote the automaton constructed by the signaling pathways having the function of molecular switches in cells under the condition of engineered regulation. To make the state transition of the MSP-automaton work smoothly, we need to design a signaling process. Let X be a binary word to represent the state of the molecular complex used to describe the data (for example, the SPK complex or GTPase complex). We call a word for this kind of molecular complex an S-word and the molecules for representing the binary state S-molecules. The signaling molecules that regulate the binary state of S-molecules in molecular switches are called regulator molecules, or R-molecules. The fundamental structure of the MSP-automaton is consistent with the nature of the signaling pathway mechanism. The biochemical reactions are carried out so that the molecules are updated dynamically. The structure of the molecular computing process is reconfigurable because the progression of pathway signaling can be realized through different pathways using the causal chains of biochemical reactions. This provides ample information capacity for the structural design of a molecular computing process by signaling pathways.

Table 6.1 Transition Rule

Rule Number	Current State	Input	Next State
0	S_0	a'	S_1
1	S_0	b'	S_0
2	S_0	b'	S_0
3	S_1	b'	S_1
4	S_1	b'	S_1
5	S_1	a'	S_0

The binary states of the molecular complex represented in an S-word can be updated by controlling the molecular switch. This is because the molecular switch is binary and there is a one-to-one relationship between the R-molecule and the S-molecule under the condition of an in vitro cell in the current biochemical technology. Consequently, any binary values of S-words can be generated by the MSP-automaton. These words include 0. The state of an automaton is thus reflected in the R-molecules, or R-words.

Using the MSP-automaton, we can define a conceptual computing device as a computation model that uses the signaling pathway units. This can be designed by an MSP-automaton and a control process. The control process starts from the regulation of S-molecules and proceeds to the regulation of R-molecules. The operators—detection, readout, and control—are designed for the MSP-automaton and are used for information processing through a feedback mode, as shown in Figure 6.2.

Here, the related notations are given as follows:

n_x: the input of the pathway unit constructed by R-molecules;
S_x: the state of R-molecules;
$V(l)$: the signal for detection in the control process.

The computing units require detection of S-molecules for readout $\Delta(n_x)$ and a manual or automated process of controlling R-molecules $\Gamma(.)$ to generate the input to the MSP-automaton $Q(l)$.

The state of the automaton is represented by the molecular complex. This molecular complex is regulated by the molecular switches under the condition of in vitro and in vivo cells, where engineering technology is necessary for molecular manufacturing. For example, let's look at $V(l)$: The output can be controlled by this molecular switch to get any binary value.

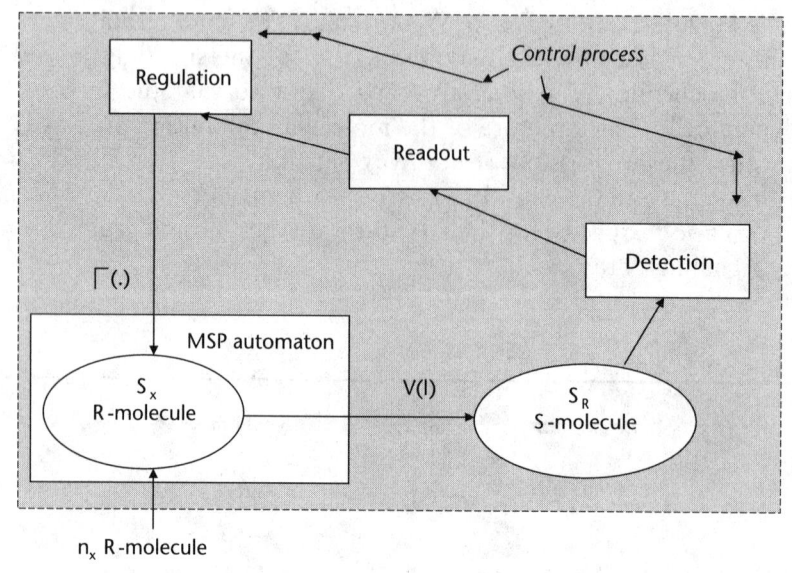

Figure 6.2 Pathway unit of computation model.

6.2.5 A Computing Model Based on Pathway Units with Turing Computability

As we have shown above, the structure of cellular signaling can be used to control the MSP-automaton for the computing process of a finite state automaton. Within living cells, the signaling proteins aggregate and assemble adaptively by the internal nanobioengines constructed by signaling pathways for cell communication. By controlling the signaling pathways, the biochemical information can be processed logically and robustly. The pathway units constructed based on molecular switches provide us different engineering choices for designing models by using different configurations and different reconfigurable operators, which are the main topics we are studying. Beyond the specifications of computing systems, computability is the key factor that determines the performance of the corresponding computing model. Among the computing models, the Turing machine is the most powerful one. A computation model with high computability in theoretical computer science would be a powerful one in the sense of its information processing ability. To equip a computation model with Turing computability is naturally our goal in the model design of biomolecular computing systems.

There are several ways to prove that a computation model has the equivalent computability of a Turing machine. The μ recursive function is one of the models that approach the Turing machine in terms of computability. The definition of the μ recursive function and the μ recursive function's relationship with the computability of Turing machines can be found in [7]. The basic notations of the μ recursive function are adopted according to [7, pp. 232–233]. Here, our methodology is to prove that the underlying computation model is equal to the μ recursive function, and thus by deduction that it possesses Turing-equivalent computability. To satisfy the requirements for a μ-recursive function, the structural components of the underlying computation model based on MSP-automata-centered pathway units will be studied.

Generally speaking, the conditions of a μ-recursive function can be classified into two classes: the conditions of a primitive recursive function and the conditions of enhancing the primitive recursive function to a μ-recursive function by introducing the unbounded minimalization function. The stages of forming a primitive recursive function include (1) the initial functions, (2) composition, and (3) primitive recursion. The initial functions also include (1.1) the 0-place zero function, (1.2) the ith k-place projection function, and (1.3) the successor function.

Our discussion is organized according to the stages indexed above. We present our proof as follows.

From the biochemical faithfulness of the proposed idea, we employ the abstract form of a biomolecular computing process based on the structure of pathway units given in the previous subsection for the theoretical studies on the derived computation model. From this information-processing mechanism, we try to find the relationship between the biomolecular computing model and the μ-recursive function in terms of computability.

The stages seem to be distinct, but their interactions are actually closely related to the entire computing process. We start from the simplest one—the 0-place zero function.

Point (1): Working on initial function.

Point (1.1): Designing the 0-place zero function.

The 0-place zero function is defined as a mapping from 0 to 0, such that

$$\zeta(0) = 0$$

This function means a map from 0 to 0.

Without giving a detailed explanation, the pathway unit in this section refers to the mathematics concept based on the MSP-automaton discussed above. The signaling process of this kind of automata needs the condition of in vitro cells to support the one-to-one mapping between the enzymes and signaling molecules, that is, S-molecules and R-molecules. The states of S-molecules and R-molecules are regulated by the corresponding mechanism of molecular switches.

In a physics sense, the state of one molecule in the molecular complex is only activated by the corresponding enzyme under the above condition. In principle, there is no limitation on the combinatorial states of molecular complexes. Therefore, we can obtain a one-to-one mapping denoted as F_{oto}.

From F_{oto}, the following form can be inferred:

$$\zeta_{pathway}(0) = 0$$

where $\zeta_{pathway}$ refers to the pathway-unit form of computing corresponding to the function of F_{oto}.

The operation of setting the state of the S-molecule to 0 by activating the R-molecule yields the result of the 0-place zero function. The monotonic characteristics of the function F_{oto} are a kind of by-product of the basic configuration.

The mathematics foundation for this conclusion can be found in code theory by algebraic tools. The designated information by pathway unit satisfies the constraint of word and code in mathematical code theory. Since the abstract form of biomolecular computing can be selected as the code form, we can study the computing principle of biomolecular computing through the data structure of biomolecular programming, which is independent of the physics constraint of molecular compounds in biochemical reactions. After the information represented by the molecular complex having topological features is transformed into the data structure form, it is convenient for us to use the sequence as a one-dimensional form of the data structure. The sequence is described by the form of words in mathematics because each state of the sequence is deterministic at each moment of the automaton transition; this is also represented by the code form in mathematics because the sequence satisfies the "uniqueness" feature of the word (for definition, see [8]). The uniqueness of the word means that there is no other word with identical information.

Point (1.2): Designing the *i*th k-place projection function.

The *i*th k-place projection function

6.2 The Molecular Switch as a Bridge Between Cell Communication and Molecular Computing

$$\pi^k_i(n_1...n_k) = n_i \tag{6.14}$$

is defined as a mapping from N_k to N.

The codes corresponding to the values of $n_1...n_k$ are different owing to their uniqueness feature. The function needs the operation of selecting a signal from multiple signals by the appropriate pathway unit. Based on the biochemical features of the signaling mode of the related pathway unit, the pathway concept $\pi^{(k->i)}_{pathways}$ is defined as the process

$$\text{pathway } \pi^{(k->i)}_{pathways} (\text{reactant}(n_1)... \text{reactant}(n_k)) \rightarrow \text{reactant}(n_i), \tag{6.15}$$

which realizes the projection function in terms of the selection operator of the MSP automaton-based model. The output states of k pathways are generated by the pathway unit that is controlled according to the reference signal i. This selection process can be described by the function

$$\text{control }() = \Gamma(\text{ (regulation }(Rx), \text{ reference signal for control in } Rx) \tag{6.16}$$

It is natural to extend this single-input/single-output (SISO) into a multiple-input/multiple-output (MIMO). The MIMO is described by the function

$$\begin{aligned}&\text{multiple-input-control }() \\ &= \Gamma(\text{ (regulation}(Rx), \text{ multiple-readout, reference signal for control in } Rx)\end{aligned} \tag{6.17}$$

Here, the "multiple-readout" signal needs several devices. The reference signal is used to guide the control process manually or automatically. When the reference signal is n_i, the project function will produce the selected signal by controlling the underlying pathway unit.

The operator for selecting the multiple-routes decides the information flow from multiple codes to a single code. Since the code is unique in value, the difference in the states needing to be updated is 0. This means that we need not change the state of the molecular complex in a physics sense. The total number of steps in the projection operation, in a mathematics sense, is 1. When the reference is determined as n_i, the control process of this function becomes

$$\begin{aligned}&\text{multiple-input-control }() = \Gamma(\text{ (regulation}(Rx), \text{ multiple-input }(n_i...n_k), \\ &\text{reference-signal for control in } Rx \text{ specified for } n_i)\end{aligned} \tag{6.18}$$

under the control constraint according to the target information that is expected to generate the projection function.

Based on manual and automated control of conditions, the following two control processes are formulated:

$$\text{multiple-input-control}\ () = \Gamma\ (\text{regulation}(Rx), \text{multiple-readout}, \text{reference signal for control in Rx (condition)}) \tag{6.19}$$

$$\text{multiple-input-control}\ () = \Gamma\ (\text{regulation}(Rx), \text{multiple-input}\ (n_i...n_k), \text{reference signal for control in Rx specified for } n_i\ (\text{condition})) \tag{6.20}$$

The design of a pathway unit is possible based on the in vitro condition for pathway units, which is the basic assumption for all of the discussion in this section.

Point (1.3): Successor function sigma.

The successor function sigma

$$\text{sigma}\ (n) = n + 1 \tag{6.21}$$

is an operation of increasing by one. This function can be realized for the regulation of molecular switches. The binary number is represented by the signaling molecules in the form of code. To construct the "counter" by increment, we only need to operate on the binary values of the codes by the state transition of the related automaton. Consequently, the pathway unit

$$\text{sigma}_{\text{pathway}}\ (n) = n + 1 \tag{6.22}$$

is feasible in informatics.
Now we return to

Point 1: Initial function.

In concept, the initial function is actually a set. Therefore, the existence of the initial function is inferred by the operations of pathway selection and interactions based on the availability of the pathways in points 1.1–1.3.

Point 2: Designing composition function.

$$f(n) = g(h_1(n), h_2(n)...h_l(n)) \tag{6.23}$$

The realization of the composition function requires a continuous computing process of the MSP automaton for the molecular computing process. Combinatorial forms of the signaling molecules are available for generating this function:

6.2 The Molecular Switch as a Bridge Between Cell Communication and Molecular Computing

$$f(n) = g(h_1(n), h_2(n)...h_l(n)) \qquad (6.24)$$

The minimum configuration of the composition operator includes a state transition of automata with control processes, a selection process, and a control process. The construction of f(n) is shown in Figure 6.3. As the figure shows, we assume 1 pathway each for the MSP automata corresponding to $h_1(n)$, $h_2(n)...h_l(n)$. In order to realize the function, we set the number of the pathway units to 1. The selection operation of pathway signals is feasible owing to the ability to construct multiple pathway units and the controllability of the information flow from multiple pathway signals to a specific pathway signal through the pathway units. The construction process of the above computing process only requires a few steps, within a linear order of time complexity; this process can be directly controlled by the related enzymes under the condition of in vitro cells.

Point 3: Primitive recursion.

Let ñ refer to 1, 2...n. Here, we construct two parts of the primitive recursion function:

$$f(ñ, 0) = g(ñ) \qquad (6.25)$$

and

$$f(ñ, m+1) = h(ñ, m, f(ñ, m)). \qquad (6.26)$$

This is called primitive recursion.

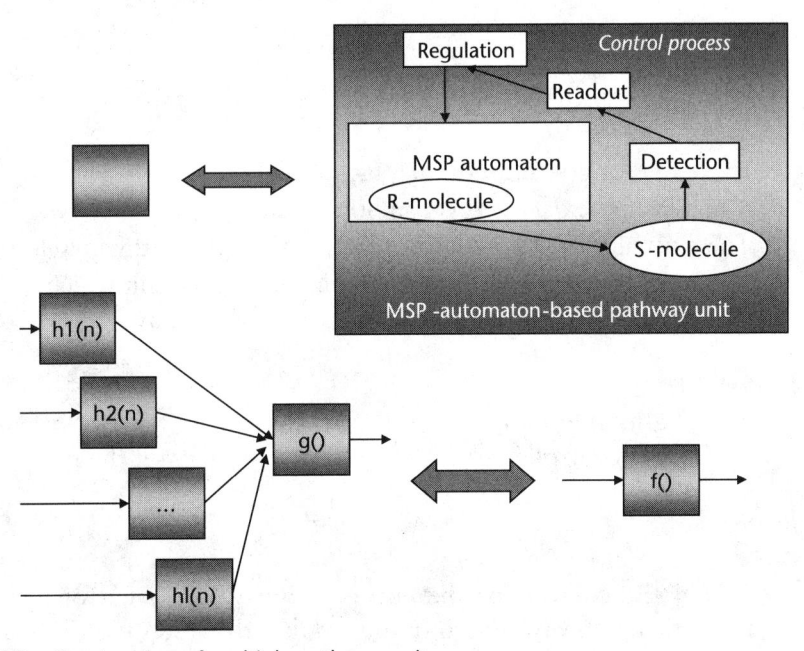

Figure 6.3 Construction of multiple pathway units.

The existence of the first formula is obvious owing to the fact that an input 0 to a pathway can be assigned with a protocol that sets the reactant, and this has no effect on the pathway. In other words, we introduce a reactant that is independent of the enzymes of the underlying pathway. In a biochemical sense, this function is equivalent to the pathway form as

$$f(\bar{n}, 0) = \text{the pathway } (\bar{n}, \text{*input-reactant}) \qquad (6.27)$$

where *input-reactant does not activate the enzyme of the underlying pathway.

Furthermore, $f(\bar{n}, 0)$ is designed to be equal to $g(\bar{n})$.

For the second formula, we construct $f(n,m+1)$ by using the MSP automata based on $f(n, m)$.

Let $f(n,m) = P$ and $f(n,m+1) = Q$. First, the input 1, 2...n and m represented by molecules activate the automata. Then, the pathway selection rule is used to select the output of the automata that is identical to P. The output of the automata is the readout result of the molecular complex regulated by the controlled pathways. Therefore, $f(n, m+1)$ exists. The difference of P and Q is assumed to be 1 bit without loss of generality. Let us define an automaton to implement a function $w(n,m)$, which can be constructed to generate U.

When either P or U equals Q, we use the pathway selection rule, and Q is obtained as the result.

The existence of the primitive recursive function can be inferred according to the definition in [18, p. 233].

Designing an Unbounded Minimalization Function
According to the definition in [18, p. 249],

$$f(\bar{n}) = \mu\, m\, [g(\bar{n}, m) = 0] \qquad (6.28)$$

which is interpreted as:

   if m exits for $g(\bar{n}, m) = 0$, then $f(\bar{n}) = m$;
   if m does not exist for this condition, $f(\bar{n}) = 0$.

The existence of g() and f() is introduced by the pathway selection rule.

The output is dependent on the designed protocols in which g(n, m) is used as a signal for detection and activation of the pathway.

Protocol 1:
    IF $g(n,m) = 0$,
  THEN the controlled pathway for automata will set the f() as m (6.29)

Protocol 2:
    IF $g(n,m) \neq 0$,
  THEN the control for the entire pathway is exerted on the automata to set this pathway as zero-output so that the result f() will be zero
(6.30)

Here, the zero value is defined and designed among the binary forms of the states of the molecular complex.

In a word, the conditions for g(n,m) are used to configure the control process in an engineering way. The derived computation model under these conditions is actually an abstract machine constructed by pathway units based on the MSP automaton, which is summarized as a computing device that consists of the operators and the process of the MSP automaton. The fact that one from multiple signaling molecules can be selected and activated in the multiple molecular switch pathways supports the composition function. Here, we find that the pathway (.) can be designed in an indivisible function form and used to construct other functions used in our model by pathway interactions. The interaction (.) and selection (.) are the basic operations at the pathway level.

6.3 From Automaton to Rewriting: Toward General Parallel Computing

An automaton can show the basic function of computing in terms of state updating. Based on the automaton structure, we can acquire a deep understanding of the essence of the pathway signaling process as a kind of parallel information processing. To describe the pathway signaling process, a graph is one of the most appropriate forms. When the states of the pathways are updated in the signaling process, the computing process is carried out in parallel. The concept of graph rewriting suits the features of the signaling mechanism in molecular computing. In this section, the word graph refers to a hypergraph when we discuss graph rewriting. The difference between a hypergraph and a conventional graph is that the link between any two vertices that are not adjacent is defined as a hyperedge in the hypergraph, while only the link between two adjacent vertices is defined as an edge in a conventional graph. The prefix "hyper" in hyperedge means that the edge is not a normal edge in a graph. A hypergraph consists of vertices and hyperedges.

6.3.1 Formalization

Signaling molecules and related pathways form the biological foundation for studies on molecular computing using pathways. The abstract forms of the objects in biomolecular computing become the vertices and edges in hypergraphs, where related pathways and their interactions can be formalized in logic. It is natural to use a graph as the representation form of biochemical reactions. The vertices in a graph represent the molecules that are involved in the related biochemical reactions. The edges represent the biochemical reaction processes for those molecules. The weighted edges can be assigned with the coefficients of the biochemical reactions. If we want to trace the molecules in a chain of biochemical reactions, the hyperedges are appropriate forms. The graph-based informatics of pathways is closely related to visualization, which can help us to understand the complex network constructed by the pathways in cells.

Now we begin to discuss the hypergraph representation for the formalized form of biomolecular computing.

Let A be an alphabet set;

The vertex set V_H and hyperedge set E_H are defined based on the symbols in A.

H: a hypergraph, which consists of vertex set V_H and hyperedge set E_H, for example,

$$H = < V_H, E_H > \quad (6.31)$$

where $V_H \cap E_H = \emptyset$.

The basic factors of a graph rewriting system are defined based on graph representation and graph operators. The structure here is oriented to the task of information processing.

Studying pathway structure in biomolecular computing requires an understanding of pathway behavior through measuring the input and output of pathways. The two parts of a graph—input-output relation and internal connection—show different facets of pathways. After being transformed into the two parts, a graph is represented by the structures of two partial graphs. The difference between the two partial graphs and the bigraphs of Roger Milner [19] is that the two graphs are simple and mainly used to describe the features of pathways while topographs and monographs and bigraphs are defined by rigorous mathematics with strong constraints [19]. The vertices that represent the input and output of pathways and the hyperedge connecting them are formed into a graph called the input/output graph (IOG for short). The vertices other than the input and output are the internal structure of the pathway. These vertices and the edges connecting them are formed into a graph called the internal structure graph (ISG for short). The pathway takes the form of two graphs:

$$\text{pathway} = (\text{IOG}, \text{ISG}) \quad (6.32)$$

For the convenience of discussion, we call this graph consisting of two partial graphs a "pathway graph" (PG for short). It is simpler than the bigraph in Milner's work because the constraint is not the one that he defines for bigraphs. The pathways denoted by the pathway graph form can reflect the information flow by networking the biochemical reactions among the molecules involved.

Next, the hypergraph H is represented by a two-partial-graph structure representation. A predicate can be used to summarize the logic form of the pathway graph:

pah (IOG(pathway-input($x_1, x_2,...,x_{L1}$))
pathway-output ($y_1, y_2,...,y_{L2}$)
IOconnection (in-set($x_1, x_2,...,x_{L1}$)
out-set ($y_1, y_2,...,y_{L2}$)

$$\text{ISG (node-set}(z_1, z_2,...,z_{L3}) \\ \text{intraconnection (node-set}(z_1, z_2,...,z_{L3})) \tag{6.33}$$

where

$x_1, x_2,...,x_{L1}$ refers to the input/vertices of the pathways, $L1 \times N$;
$y_1, y_2,...,y_{L2}$ refers to the output/vertices of the pathways, $L2 \times N$;
$z_1, z_2,...,z_{L3}$ refers to the internal vertices of the pathways, $L3 \times N$;
pathway-input() refers to the predicate for the input of the pathway;
pathway-output() refers to the predicate for the output of the pathway;
IOconnection() refers to the predicate for the I/O relation;
node-set() refers to the predicate for the node set of the pathway;
intraconnection() refers to the predicate for the internal connection of the pathway.

The information processing function of the pathway is represented by two predicates:

$$\text{pathway-input}(x_1, x_2,...,x_{L1})$$

and

$$\text{pathway-output}(y_1, y_2,...,y_{L2}) \tag{6.34}$$

The input and output of the pathways are crucial factors for constructing the pathway units for information processing.

$$\textit{pah}\,(IOG(x_1, x_2,...,x_{L1}, y_1, y_2,...,y_{L2}), ISG(z_1, z_2... z_{L2})) \tag{6.35}$$

where

$x_1, x_2,...,x_{L1}, y_1, y_2,...,y_{L2} \in V_H,$
$z_1, z_2,...,z_{L3} \in E_H$
$L_1, L_2 = 0, 1...$

This representation is used to explicitly describe the two parts of the whole pathway in order to separately study the input/output relation and the internal structure.

In order to describe the operations of pathways such as the temporal process of biochemical reactions, interactions of pathways, and controlled engineered pathways, it is necessary to introduce graph-rewriting systems. For more on the mathematical systems for rewriting, readers may refer to [18]. While we can benefit from introducing the terms of graph rewriting in terms of high-level operation for computing, we must also acknowledge that the formal system of graph rewriting has certain constraints (e.g., topology or logic). Actually, the rewriting system for hypergraphs can be interpreted as the interaction of string rewriting systems, if one prefers sequence operations.

The rewriting done on hypergraphs by pathways is denoted as GRP, that is,

$$G \rightarrow G', \text{ w.r.t. } \zeta \tag{6.36}$$

where G, G' refer to hypergraphs and ζ refers to a set of constraints. In the case of topological graphs, the constraint is the topological relation of the related vertices in the graph. The graph-rewriting description for pathway-based biomolecular computing is convenient because the chemical reactions corresponding to the signaling pathways are temporal processes in which the objects of reactants are always changing. The reactants such as kinases and phosphatases are directly operated to change the structural characteristics of the chemical compounds through the activation operation of the related pathways, that is, phosphorylation/dephosphorylation. It is clear that the studies of theoretical computer science and experimental chemistry have different aims. Topological features are key elements in mathematics, while the conformational features are the elements in biochemistry.

In the graph-rewriting operation using the logic forms of

$$pah\ (IOG(x_1, x_2...x_{L1}, y_1, y_2,...,y_{L2},), ISG(z_1, z_2,...,z_{L3}))$$

$$\rightarrow$$

$$pah\ (IOG(x'_1, x'_2,...,x'_{L1}, y'_1, y'_2,...,y'_{L3}), ISG(z'_1, z'_2,...,z'_{L3})) \tag{6.37}$$

the vertices and edges have been changed by the rewriting operators, that is, the rewriting operations of

$$\begin{aligned}
x_1, x_2,...,x_{L1} &\rightarrow x'_1, x'_2,...,x'_{L1}, \\
y_1, y_2,...,y_{L2} &\rightarrow y'_1, y'_2,...,y'_{L2}, \\
z_1, z_2,...,z_{L3} &\rightarrow z'_1, z'_2,...,z'_{L3}
\end{aligned} \tag{6.38}$$

The related operations are listed as follows:

- ER (edge replacement): the operation of addition or deletion of edges;
- VR (vertex replacement): the operation of addition or deletion of vertices;
- HR (hypergraph replacement): the operation of addition or deletion of parts (hyperedge and/or hypervertex) of hypergraphs;
- HER (hyperedge replacement): the operation of addition or deletion of hyperedges of hypergraphs;
- HVR (hypervertex replacement): the operation of addition or deletion of hypervertices of hypergraphs.

Rewriting on these graphs, networks, and pathways can be realized by adding/deleting edges/vertices. The elements in the derived sets of underlying graphs form certain equivalence classes that can be used to quantitatively define the corresponding equivalence relation.

The rewriting operators are defined in two major forms:

Graph transformation: $G(t) \to G(t+1)$ for pathways $\hspace{2cm}$ (6.39)

Interaction: $G_1, G_2,\ldots,G_L \to G_i$ for pathways $\hspace{2cm}$ (6.40)

The graph-rewriting operators are oriented to the structure of the information processing. This makes the work of designing predicates easier than a concentration-level calculation.

6.3.2 Transition from Hypergraphs to Bigraphs

The transition from hypergraphs to bigraphs involves the following:

1. Activation *activatePATH()*: this takes the value of {T, F}.

$$\begin{aligned} activatePATH\,(pah(.)) \\ = True \quad &\text{if} \quad pah \text{ is activated} \\ = False \quad &\text{if} \quad pah \text{ is not activated} \end{aligned} \hspace{1cm} (6.41)$$

2. Interaction of two pathways *interact (pah1, pah2)*: this takes the value of {T, F}.

$$interact\,(pah1, pah2) \hspace{2cm} (6.42)$$

The value of this function is used to describe the interaction mechanism of pathways:

$$\begin{aligned} interact\,(pah1, pah2) \\ = True \quad &\text{if pathways } pah1 \text{ and } pah2 \text{ are interacted} \\ = False \quad &\text{if pathways } pah1 \text{ and } pah2 \text{ are not interacted} \end{aligned} \hspace{0.5cm} (6.43)$$

3. Graph transformation *trans(.)* for rewriting: this takes the value of {T, F}.

$$trans(pahi\text{->}pahj)$$

$$\begin{aligned} = True \quad &\text{if the graph rewriting is available} \\ = False \quad &\text{otherwise} \end{aligned} \hspace{2cm} (6.44)$$

The logic form of graph rewriting operations as presented above is used to verify the computing process derived from the abstract model, for example, the model presented before has the μ-recursive function. Using the mapping: $υ_{l1} \to υ_{l2}$ in G_H, we can define the input and output of pathways *pah(.)* in $R_p(A)$ in a rigorous logic form.

Based on the earlier discussion, the formalized form of biomolecular computing processes requires rigorous analysis of the logic operations inspired by the biological pathways of cells from the viewpoint of theoretical computer science. This is because the sequence of empirically designed moleware operators needs to be designated as a continuous process in which the current state of the computing system activates the next state of the computing system. Working out the state transition process and corresponding state updating operator for computing is the central task of automaton theory in computer science. For the purpose of analyzing the interaction of pathways of cells, the algebraic operators of biomolecular computing based on these pathways are appropriate forms for describing the interaction mechanism. In a word, the molecular computing processes that are constructed by the signaling pathways are formalized in terms of a special kind of hypergraph rewriting for massive parallel computation, where the set of pathways is also defined as a kind of conceptual object.

To connect the mechanism of signaling pathways to the graph-rewriting system, a pathway unit is designed by automata that are oriented to the state transition in the parallel mode. The derived pathway units, where rewriting operators on pathways are introduced, are used to logically implement the automaton, that is, to functionalize the automata in the form we design and expect in advance. The pathways of the molecular switch provide the signaling mechanism for the automaton. In the sense of parallel computing systems, the "pathway" concept in rewriting is defined as the pathway object represented by a graph/hypergraph, and it is used to describe the structure of the control process and the automaton (in single or multiple mode). The pathway in rewriting refers to the pathway object that is regulated by the related control process, where the pathway objects form a temporal series of molecular signals through the controllable engineered process to activate the R-molecules. The rewriting system based on pathway units still uses the control signaling for the control process in manual, semiautomated, or automated mode and for the state of R-molecules with functions in MSP automata models. In the graph-rewriting system, the feature of parallelism requires multiple operators that are similar to the simple operation process. The rewriting model on the pathway signaling process describes the state update of the MSP automaton in parallel. Multiple inputs of a rewriting system are the molecules used to activate the R-molecules.

Since we have introduced the predicate form for the pathway in two partial-graph structures, we can summarize the following predicates:

Input of the pathway denoted as pah() = the molecules that activate the R-molecules (6.45)

Output of the pathway denoted as pah() = the S-molecules (6.46)

Pathway (G) → pathway (G') (6.47)

where the rewriting process is operable by pathway control, which has the procedure function form of a pathway unit when the corresponding programs are made to demonstrate computing behavior.

The concurrent mechanism of graph rewriting for biomolecular computing can be studied through the technology of logic programming. This means that the transparency of the biomolecular computing from the signaling-level description of the biochemical reactions to the computational-operator-level description of the computing mechanism can be achieved in terms of high-level description of the underlying predicate form. This makes it convenient for us to study the simulation for a computer-aided designing system and a development tool based on software engineering technology for verification and testing of the moleware programs. It is natural that the programming language be linked to the formal language and logic.

6.3.3 McNaughton Language, Confluent Rewriting, and Controlling with the Structural Characteristics of MSP-Automaton

Since the MSP-automaton is directly used for describing the pathway units of molecular switches, the construction of pathway units can be converted from a binary representation at the signaling level to a symbolic representation at the logical level. After the binary-switch-based biochemical information processing is extended to symbolic form in the automaton, we can select the predicate form operators in logic to check the correctness of the inference of molecular programs. Through discussion on the application of biomolecular computing algorithms for SAT problem solving, we can understand how the logical mechanism is constructed by the moleware signaling mechanism. Although the logical operators based on the above signaling mechanism vary in different forms, the basic macro-structure operators can be defined in a logical form with a biologically faithful structure for computing. In the sense of a Gödel number, it can be easily understood.

The rewriting operators on graphs are defined by the simultaneous updating operations of multiple objects for different pathways. These are located in different neighborhoods, and the overlapping areas of objects are different. Now, the activation process is operated by the current access of the related objects. This feature comes from the biochemical reactions of the pathways with the interactions. For the current operators that lead to parallel computing behavior, we can use three objects for discussion, say pathway i, pathway j, and pathway k, which give results in McNaughton language. By using the rewriting operators, the pathway i gives "a" and "b," assuming that "a" is in pathway j and "b" is in pathway k. Then, "a" and "b" can go to pathway l, where "i" is reproduced.

We use the abbreviation MNL to refer to the McNaughton family of languages [9].

Proof for the above feature of McNaughton language can be made by reference to the pathway structure and pathway operators:

Let
W be a class of string-rewriting systems,
F be a family of languages,
L be a language,

$L \subseteq \Sigma^*$,
where Σ, Γ is a finite alphabet,
  * refers to the situation where an empty symbol is included in the set Σ,
$\Gamma \supset \Sigma$.
  FSRS: a finite string-rewriting system
$$FSRS \in W \text{ on } \Gamma,$$
the symbols in Σ: terminal,
the symbols in $\Gamma \setminus \Sigma$: non-terminal.
$$\text{string } t1, t2 \in (\Gamma \setminus \Sigma)^* \cap IRR(FSRS)$$

IRR(FSRS): the set of FSRS that does not give out any rewritable results.
FSRS refers to finite string rewriting system. Its state is finite when the rewriting process is carried out for the objects that are rewritten.

Y: a letter that satisfies the condition

$$Y \in \Gamma \setminus \Sigma \cap IRR(FSRS)$$

The process for MNL is defined by the following mechanism: All of the elements (denoted as w) belong to the set of Σ^*. They are the terminal symbols (called terminals for short), and the empty symbol is also included here.

The objects considered in this discussion have the following two routes:

(a) W belongs to L
(b) $t1 \, w \, t2 \to^*_R Y$, that is, the string t1 w t2 belongs to the set of M_Y (M_Y: the set that can reduce the string Y)   (6.48)

The set of all elements of M_Y forms a relation set on set Y, which consists of string. * refers to the empty symbol included in the right-hand side of rewriting. R is a relation that is defined according to related set of strings in the rewriting systems.

We define the derived operators as follows:

The operator "->" is defined by the rule:

$$u \, l \, v \to u \, r \, v$$
$$\text{if } l \, (\text{``->''rewriting into}) \, r \qquad (6.49)$$

where u and v belongs to the set of Σ^* and l and r are two strings in the string rewriting systems.

From u', if v' is reduced, u' can be reduced and is called reducible, where u' and v' belong to Σ^*. If v' does not exist, u' cannot be reduced.

Let L be a language set specified by the four-tuple (R, t1, t2, Y). According to the equivalence between the two points (a) and (b), we adopt the definition of

McNaughton languages at the conceptual level of information science by the following two rules:

1. Strings $t1$ and $t2$ are the nonterminals that cannot be reduced, that is, they cannot be rewritten further, and include empty symbol(s).
2. String Y is the nonterminal that cannot be reduced, that is, it cannot be written further, and does not include any empty symbol.

(6.50)

where w represents any string that includes an empty symbol.

By the rule of

$$t1 \ w \ t2 \ -> \ Y \qquad (6.51)$$

we can get a string w, a set including all of the strings denoted as w that satisfy the above condition that L become a set of MNL.

The generating process of McNaughton language by the string rewriting system can be studied by the composition of different strings. Here, there are two concepts that need to be considered: rewritable strings and complete rewriting systems. Rewriting in the understanding of the information process is reflected in the operator or the process in which the symbols of a string are updated. Rewritable refers to a string that can be rewritten further. The complete rewriting system for string rewriting that is confluent is used for parallel computing.

Basically, the process of generating language L can be designed by the construction of different units that can generate three strings. This is because the definition of MNL requires three parts of a string. These three strings are used to construct one string for MNL; two of them are not terminals and cannot be "rewritten" further in the sense of a rewriting system. Also, no empty symbol is used for rewriting. The two strings used as two ends of a string are composites of another longer string. The remaining one is used as the middle composite of the longer string. It can be any string that includes an empty string consisting of empty symbols. The resulting string is not terminal and cannot be written further. This sandwich-type structure of a new string made by splicing/combining three string parts is different from the ones we discussed before (e.g., the operators in a splicing system). The set of {w} is formed, and this consists of the elements of any string w that is generated and constrained by the sandwich-like rule and related operator mechanism. This generating mechanism fits the MSP automaton and related computing process. The two ends of the generated strings are "fixed," meaning not changed or updated. In the sense of rewriting they cannot be rewritten further. The middle part is "flexible" because it can be rewritten further. The "fixed" parts in the sense of rewriting can be built by specific operators of a pathway automaton whose outputs are deterministically controlled. In designing the "flexible" part, skill in the design of automata is helpful.

The MSP-automaton mainly includes two kinds of molecules: state represented by R-molecules and output given by S-molecules. The three parts of the control process for generating the expected string are modeled as a controller:

$$\text{regulation signal} = \text{controller (S-molecule)} \quad (6.52)$$

The output of the automata constructed by the pathways consists of a graph that is composed from the strings in a molecular complex. The automata constructed by the pathways can be used to infer confluent information flow.

Such a confluent process is a phenomenon in string rewriting systems. There exist symbols "a, b, c, d" that belong to an alphabet and are different symbols. "a" generates "b and c" by rewriting, and then "b and c" generate "d" by rewriting. The basic process of pathway automata (MSP-automaton) is feasible in the mode of multiple designated pathway units. The limited steps of using other pathway automata also are available for generating different results of different R-molecules. The same symbols from two pathway automata will be rewritten in later steps under different controls of pathway units. Consequently, the two strings will be different.

Owing to the reversible features of the molecular switches employed for pathway units, the remaining part of the entire expected process can be inferred by the generating process of "b" and "c" from "a." A confluent result can be inferred by the automata in a graph configuration for generating "a," "b," and "c." Thus, the confluent string-generating mechanism can be constructed by the automata we designated above, that is, the confluent process, which we call PA for short. The representation for this automaton is

$$\text{PA (A, S0, state set, acceptable state set, transition rule)} \quad (6.53)$$

where A is the set of alphabet, S0 is the start state.

Using the operators defined above, a parallel molecular computing device can be designed logically to produce the signals in the MIMD mode. The input is represented as the set {start signals, activating-computing signal, activating-control signals, activating-recognition signals}. This can also be understood as instruction at the beginning of the computing process, which is a molecular program. We unify the term for molecular operators as "molecular primitive." We design a logic form for the pathway units in the automaton-based confluent process. The pathway units for this process work in the following logic rules:

- *Input*: the molecular primitive for starting the operation procedure of information processing;
- *Output*: the molecular primitive for ending the operation procedure of information processing;
- *Control*: the continuation of the molecular primitive for starting the operation procedure of information processing.

The control is the key factor for the entire process. In contrast to the physical operators of moleware and rigorous mathematics operator for computing, the control has more engineering meaning, and its concrete task is to work out the control schemes so that they make the pathway signaling what we need for computing. Here, an instance of the above process shows that the underlying information flow of the sandwich-like structure can generate the confluent routes of symbolic strings.

6.3.4 Designing a Rewriting Process by Pathway Units Based on MSP-Automata

We have discussed the automaton oriented to pathways, the graph description of pathways, and a pathway-based signaling control mechanism. Now we discuss the construction process of rewriting for MNL. The predicate form is a direct scheme for the description of the inference process.

Let the input be

$$S0ScSrSf$$

where

$S0$: the signal for starting the computing process;
Sc: the signal for computing;
Sr: the signal for pathway (recursive computing process);
Sf: the signal for feedback.

The structural information processing mechanism of pathway units is used to control and activate the molecular signals for computing:

$S0$ -> to activate all pathway units;

Sc -> to activate the pathway unit a, whose output value is constrained by the bound-minimum mechanism of the μ-recursive function.

"->" refers to the information processing that corresponds to the string rewriting operator. The left side of "->" is the state before rewriting, the right is the result after the rewriting operator. To produce the signal flow of the sandwich-type string, three pathway units denoted as a, b, and c are constructed. The automaton structure within unit a is mainly used to carry out the computation. The primitive recursive functions are used as basic operators for mapping the pathways.

Sr -> the pathway unit b

The Sr signal is represented by the symbols of the rewriting systems that are used as the recognition signal regardless of the communication mode among pathways that are synchronous or asynchronous. The communication function is made by the engineered pathways and related control. The confluent process can be designed and inferred here. The confluent symbol at the end of the confluent rewriting process is an important message that helps us to identify the "frame" of "message blocks" in the sense of a telecommunication-like concept.

Sf -> the pathway unit c

The feedback for control uses the MNL function for the detection and activation, where the units a, b, and c are constructed for the three string parts in the sandwich-like string structure. This involves graph rewriting and interactions of the string rewriting processes. The proof for the MNL can be inferred in an inverse way. The computing process is constructed for parallel computing. Its scalability comes from the homogenous pathway units and heterogeneous control under the condition of in vitro cells, where the pathway regulation is exerted in a bioengineering way. The computing process can provide parallelism, which is important in computer science. The rules are given in logic forms. This computing process leads to the design of a compiler of molecular computing.

The test on the designed pathways is parallel inference by multiple logic rules with interactions for pathways, in which each of the operators in the form of IF x THEN y is used for constructing MNL.

6.3.5 A Compiler: Translating Moleware Language into Programmer-Friendly Informatics Operators

The construction process for the automaton and formal languages that lead to a parallel information process needs development of the compiling technology for molecular computers.

A compiler for biomolecular computing by pathway units is helpful for releasing molecular computing's power of parallelism in biology, mathematics, and logic toward unifying the programming process. The studies on the automaton and formal language of pathway units are the basis of programming, reprogramming the pathway units, and reconfiguring the architecture of pathway-based computing devices. The common part of a compiler for this molecular computing system can be made by software using an electronic computer for translating the high-level programming language into the molecular level operations (i.e., primitives). The major difference between the compiler for molecular computing and that for electronic computers is the operators. We have to consider the technical features of the molecular operations. Two pathway units i and j are selected as an example to discuss this issue. The basic logical operation in logical programming can be described by the pathway units. The technology of programming languages such as LISP is mature, and the related knowledge can help us to understand logical programming. The basic operation corresponding to logic programming language is

IF x and y THEN
    (parallel) y1: y1←y1-value;
    (parallel) y2: y2←y2-value;
    communication (y1,y2): to exchange the value.

The action is designed as parallel operations. The processes for calculating the logic value of x and y are parallel operations.

After we assign the value of T/F to the logical variables x and y, we can use pathway units to construct the AND function.

The conditions and actions can be represented by the data structure of a list. According to the list, the control primitives on pathway units can be activated. The condition judgment is made to select the molecular operation. The reprogram operation is exerted on the molecular structure. By using the MSP-pathway automaton and the μ operators, the pathway units are used to describe all of the above processes. It is reasonable to expect scalability for multiple pathway units (say, n pathway units) through a reconfigurable architecture of the information processing system. The pathway-based computation model is a computing process that is designed and constructed by the three pathway units labeled a, b, and c. The pathway units are constructed by the automata and their control processes, and they are designated as:

pathway unit a
pathway unit b
pathway unit c

The details of these three pathway units are explained as follows.

The input and output of the three pathway units are in the form of information processing modes:

pathway unit a (input-a, output-b, connection-matrix-a)
pathway unit b (input-b, output-b, connection-matrix-b)
pathway unit c (input-c, output-c, connection-matrix-c)

where input, output, and connection matrix are defined to describe the information processing mechanism of the MSP-automaton model.

As shown in Figure 6.2, the corresponding automata and control processes are given as follows:

pathway unit a (input-a, output-b, connection-matrix-a) by pathway
   automaton a and control process a
pathway unit b (input-b, output-b, connection-matrix-b) by pathway
   automaton b and control process b
pathway unit c (input-c, output-c, connection-matrix-c) by pathway
   automaton c and control process c

The pathway units a and c are designed and controlled to produce the output of terminals. The terminal generation can be obtained by the configuration of the pathway automata a and c. In the pathway unit b, the automaton b is made to generate the nonterminal. All three pathway units generate letters. These letters consist of strings generated by the pathway units. The strings are different and represent different molecular complexes. The output letter and string are used as information to be communicated to other pathway units. The pathway unit b is used to generate letter x. Equivalently, two identical pathway units b are built for generating different output letters y and z, under the control designed in advance.

Because of the recursive mechanism of the signaling pathways by molecular switches, the reversible processes of the information processing mechanism discussed above are feasible by inverse design. Structural confirmation of the characteristics of the computation model for confluent processes forms the basis of the logic inference mechanism.

6.3.6 Systematically Understanding the Interaction Structure in Pathway Computing

Biomolecular computing by signaling pathways in cells employs the signaling mechanism of the cellular pathways in terms of biochemical characteristics. A complex network of signaling is constructed when the single pathways built by the molecular switches are interacted. This is naturally a good method for parallel computing. The complexity in the definition of complex systems and system engineering implies that we need to determine the description of structural interaction operations by programmable design and controllable schemes, for example, observing this computing process from the viewpoint of pervasive computing for adaptability and autonomy. One of the most helpful methodologies is pervasive computing. In information processing, terms such as confluent strings, confluentness (confluence) and congruence in systems, and confluent for adaptation (by MNL) are helpful for discussing the interaction of pathways. The pathway automata and controllable process are beneficial to the autonomy of pathway-based computing. Congruence for heterogeneous systems reflects the asynchronous feature and thus suits the nature of biomolecular computing. One of the measures for pervasiveness in pathway-based computing is having a structure that is updated by evolvable/adaptive evolutionary computation and that can be represented in a metric form.

6.3.7 Generalized Form for Computing

Features for nano (hard, soft, wet) forms (i.e., the features of molecular computing for compiler design) are different from those of the electronic computers we are already familiar with. Two important issues for the compiling technology are universal representation (e.g., the formal description of the information in rewriting systems) and ALU operators. Note that all of the operations are carried out on a set of pathways formed as a special kind of hypergraph under certain topological conditions. The formal system that we propose here reflects the idea of modeling a kind of molecular computing that can transfer its systematic state from one set of pathways to another set of pathways, where "pathway rewriting" is conceptualized by operators defined by rewriting rules.

Let L be a category, and we define a construct

$$< Ob(L), Mor(X,Y), M >$$

where

6.3 From Automation to Rewriting: Toward General Parallel Computing

$Ob(L)$ denotes the set of pathway objects
$Mor(X,Y)$ is the morphism of X into Y ($X, Y \in Ob(L)$)
M is the law of composition designed by rule Q here.

This is the core of the formal system. The model can be formalized by the following construct:

$$W_{cells} = \; < V, T, D, U, E, Y, Z, PTs, Q >$$

where

V is the alphabet set;
T is the terminal set ($T \subset V$);
D is the set of $\{0, 1\}$;
U is the set of vertices ($U \subset V$);
E is the set of edges;
Y is the set of hypergraphs $<HE, U>$

in which
HE is the set of hyperedges in Y,
    and the corresponding HR (hyperedge replacement)
    and VR (vertex replacement) are defined based on HE;
Z is the set of local concentrations (discrete values) $V \supset Z$;
PTs is the set of pathways, that is, the directed graphs that have inputs and outputs in U and contain hyperedges in HE with interactions that fall into HE.

The pathways can be categorized into a class of special directed hypergraphs. Each one is defined as

[pathway] ::= [pathway] $\cup$ [single-hyperedge-with-two-vertices]

under the operation of Q.
Q is the set of operators for operations on hypergraphs in HE from V and E, that is, we have the set of

$$Q = \{Q1, Q2, Q3, Q4\}$$

for the objects in PTs. For the interactions in the set of all pathways, the operational processes carried out by the operator set of Q are formalized by the following four rules in terms of rewriting on hypergraphs:

1. Rule $Q1$ (rule of interaction) for the interactions of pathways as defined in PTs;
2. Rule $Q2$ (rule of feedback-making) for the addition of feedback in the pathways;

3. Rule Q3 for the addition of new pathways;
4. Rule Q4 for pathway deletion.

From this model, we can derive the following forms of rewriting on "pathways":

$$PTs(Gh) \rightarrow PTs(Gh')$$

the objects of rewriting are limited to pathways obtained by Q, and

$$Gh'' \rightarrow Gh'''$$

the objects of rewriting are limited to hypergraphs by Q, where Gh, Gh', Gh'', and Gh''' refer to hypergraphs in Y and rewriting is carried out by executing the Q operators.

In summary, logic forms are convenient in the analysis of corresponding molecular computing processes and can also be generalized in describing the characteristics of operations designed for molecular computing. If we could develop a computer-aided tool for the design and testing of biomolecular computing systems, the logic operator would be the central factor in the underlying moleware programming, which is based on the theoretical models that include graph-rewriting systems.

6.4 Blueprint of a Kinase Computer

6.4.1 Quantitative Description for Biochemical Features

From the viewpoint of the biophysics of the biochemical reactions described above, the quantitative features of the proposed process in terms of biochemical faithfulness of the related pathway units should be studied so that the molecular switches can be designed and controlled as stable ones. The Michaelis-Menten equation is the form used to describe the biochemical reaction, and it's often used in cellular biology for quantitative analysis of the biochemical processes:

$$E + S \underset{K_2}{\overset{K_1}{\rightleftarrows}} ES \overset{K_3}{\rightarrow} E + P$$

E: enzyme;
S: substrate;
ES: compound of enzyme and substrate;
P: product.

The molecular computing performance is dependent on the reaction speed of the biochemical process. The formula

$$V = d[S]/dt = Vmax[S]/Km+[S]$$

6.4 Blueprint of a Kinase Computer

is the main equation to quantitatively describe the behavior of the related molecular switches. V is the velocity of the substrate consumption.

$$K_m = K_1/K_2$$

For the GTPase switch, we have the formula for the activation state of GTPase:

$$\begin{aligned}d[\text{GTPase-GTP}]/dt &= K3GEF[GEF][\text{GTPase-GDP}]/([\text{GTPase-GDP}]+KmGEF) \\&- K3GAP[GAP][\text{GTPase-GTP}]/([\text{GTPase-GTP}]+KmGAP)\end{aligned}$$

where

GTPase-GTP: GTP-bound GTPase;
GTPase-GDP: GDP-bound GTPase;
GEF: acts as enzyme;
GAP: acts as enzyme.

For the reverse direction of the molecular switch, we have

$$\begin{aligned}d[\text{GTPase-GDP}]/dt &= K3GAP[GAP][\text{GTPase-GTP}]/([\text{GTPase-GTP}]+KmGAP) \\&- K3GEF[GEF][\text{GTPase-GDP}]/([\text{GTPase-GDP}]+KmGEF)\end{aligned}$$

We assume that the concentration is denoted by density in an arbitrary unit and time is set as minute (min) without explicit explanation [12].

The pathway for the corresponding signals in the molecular switch can be simulated by the above equations. If we design a molecular computing system with multiple test tubes, the issue of pathway coupling need not be considered because it does not significantly influence the performance of the molecular computing. But if we design an in vivo form of molecular computing by cell culture, we should study the coupling mechanism and the related regulation schemes for computing. In the case where one signaling protein is regulated by many protein enzymes (i.e., there is no significant negative influence on the quantitative measure of signals in signaling pathways), we can control the activation state of the major enzyme in order to enhance the performance in molecular computing. However, if coupling phenomena occur, we have to decouple these coupling pathways in order to carry out the biochemical reaction of the computing process according to the concentration we design and expect in advance. For example, when the phosphorylation pathway A is an exerted dephosphorylation function on the phosphorylation pathway B, we have to control these coupled structures of cellular signaling in order to obtain the phosphorylation result from pathway B if the function in the inverse direction exists. Briefly, the relationship between the input and output of the pathways can be represented by a matrix, in which each element corresponds to the coefficient of the biochemical reaction for each input-to-output relation. The ideal situation is a diagonal matrix in which the nonnecessary elements can be regarded as zero.

In mathematics, the decoupling task is reflected by a kind of processing done to realize the above target, which requires an additional control mechanism to be embedded in the system.

When we set parameters by the Monte Carlo method, the simulated result shows that the related computing process can be controlled with biochemical feasibility. The feasibility of the computation model could be verified by simulation of the parameters. This is the most direct way to realize the KC model in simulation. Nanobiotechnology is required for the molecular manipulation, the one-to-one mapping from R-molecules to S-molecules realized by the physical nano-operations. Heterogeneity is an obvious feature of kinase computing. The coefficients are different so their speeds are inherently different. How to control them physically for computing is an important topic to be studied. The simulation using the Monte Carlo algorithm is expected to estimate the coefficients of the biochemical reactions in the designed control schemes.

Control of the molecular switches can be designed based on primitives in the form of reconfigurable pathway operators. An asynchronous system for generic I/O information requires that programmability be satisfied by the biomolecular computing systems.

Compared with the reconfigurable semiconductor hardware in which programming model and execution model are two important factors, the program is loaded and stored by the medium of moleware, and execution is carried out by biochemical reactions. The program that updates the data in moleware is realized by pathways and can be reprogrammed by the control of pathways. Consequently, pathway series/pathways need to be programmed, and we use a meta-program for the reconfigurable pathway structure.

Concentration versus time is the main measurement for the computing unit. The synchronous and asynchronous modes are considered according to the situation of the biochemical reactions. Clustering is helpful to quantitatively measure the variation range of the signals in the pathway units. From physical calculation, we select the quantities of the protein enzymes corresponding to the parameters and find the pattern features of the pathways that are clustered into different clusters. The biochemical reactions are carried out in the clustered domain. A controller based on feedback is designated to regulate the signaling into the expected values. When the signaling can be kept within the threshold, the synchronous signaling process by cell communication is set as the computing mode. When the signaling cannot be kept within the threshold, the asynchronous mode is set for the computing by cell communication.

The major processes are pathway clustering by pattern features, the controlling process, and the communication needed for information processing. The performance (fan-in, fan-out) is simulated by the Monte Carlo method for studies.

The quantitative analysis of the molecular computing process is briefly described as follows:

```
(parallel) for i = 0 to pathway-number
pathway-efficient (i) = set by Monte Carlo method
pathway (i, standard-input, pathway-efficient)
clustering (output(pathway(.)), thresholding)
```

design-controller (parameter(clustering))
set communication mode (synchronous/asynchronous mode)

The basic constraint we study is the numerical algorithm for calculation. The quantity of concentration reflects the accuracy of the computing process, and it is used for designing a prototype MC.

6.4.2 Materials for Information Processing

The kernel units of biomolecular computers based on the engineered signal transduction of molecular switches and their control mechanism are signaling pathways that carry out the required computing tasks and are implemented in biochemical reactions with interactions. Here, the engineered pathways of cells refer to those signaling pathways of cells that are controlled for computation in an engineering way and are employed for constructing computing units.

For information processing, the designed words of programs and the data for biomolecular computers are appropriate for the two kinds of signaling molecules in cellular molecular switches. One is used for instruction, the other is used for data.

As an instance of a design scheme in informatics, GTPase and SPKs are used to represent instruction and data.

The set of Rho family GTPases {Rho, Rac, Cdc42} and the set of GEFs are activated initially by thrombin through cell membranes.

The symbols/variables used for designing the words of the programs of the biomolecular computers described above are encoded by the signaling molecules in phosphorylation and dephosphorylation. Here the phosphorylation state represents 1 and the dephosphorylation state represents 0. The regulation functions are used to activate the Rho family GTPases by GEFs. For example, the regulation direction from GEFs to Rho family GTPases can be made to construct 23-bit words as the set of:

{Db1 for Cdc42 and Rho, Lbc for Rho, Lfc for Rho, Lsc for Rho, Tiam for Rac, p115 Rho GEF for Rho, Vav for Rac, Cdc42 and Rho, Fgdl for Cdc42, Trio for Rac and Rho, Ost for Rho and Cdc42, Bcr for Rac, Cdc42 and Rho, Pix for Rac, Smg GDS for Rac, Cdc42, and Rho}

In the case of in vitro biochemical experiments, the GEF/GAP can be regulated separately.

Although the above evidence concerns GEFs/GAPs and GTPases, the situation of GAPs is similar. The molecules of kinases/phosphatases offer many choices. For example, a 23-bit word can be designed by the set of kinases and phosphatases and other related target molecules of

{PIP5-kinase, Rhophilin, Rhotekin, PKN, PRK2, citron, citron-kinase, Rho-kinase, MBS, MLC, p140mDia, p140 Sra-1, Por1, PI3-kinase, S6-kinase, IQGAP, PAKs, MLK3, MEKK4, MRCKs, WASP, N-WASP, and Ack}

whose biological faithfulness and experimental feasibility are supported by the evidence reported in [11, 12].

It is necessary to ensure the feasibility of kinase computing in terms of the stability of the molecular switches and related operators. A feedback mechanism is needed to preserve stability in molecular computers by keeping the pathways controlled within the threshold. Accordingly, the biological faithfulness of the schemes in molecular computers can be inferred. For the formalization of signaling pathways in living cells, differential equations are commonly used to quantitatively study the biochemical reactions. Theoretical studies on the stability of biologically inspired computing have proven to be useful in designing the related systems. Concerning the issue of programmability, achieving stability comes to involve structurally controlling the related molecular computers.

6.4.3 Controllability Under Protocols in Bioinformation

Controllability of pathway units for molecular computing is an important theoretical aspect of the computing process. In the case of biological cells, the analog value of concentration is a factor that must be considered for molecular computing. Information is represented by the molecules, that is, the names of the molecules are interpreted as symbols. The symbolic form is the main information form for molecular computing based on cellular pathways. To realize the symbolic form, concentration is the measure used for detection and control. Theoretically, the concentration of a certain kind of molecule that is larger than a certain threshold will be detected and regarded as nonzero. If it is smaller than the threshold, it will be regarded as zero. The two types of molecular switches are binary switches that are used as the basis of information processing. As opposed to electrical semiconductor technology, in which the standards for reliability and testing are well established, the quantity of the concentration needs to be studied for control and computing, since this quantity in the pathways must reach the threshold for signal detection. When the biochemical reactions of the molecular switches can provide a stable and detectable signal, the functional level of this process can be used for computing. The functional level refers to the activated state of the signaling proteins. In logic, these signals are denoted by yes or no in binary logic.

The analog form of the signaling needs to be regulated to comply with the symbolic process. The concentration is expressed in a real value [0,1] for pathways. For a single pathway, it can be controlled at certain degrees of quantity. When multiple pathways are involved in parallel computing, we need to provide a generic scheme for the molecular computing process. Here, we discuss controllability.

From empirical studies by calculation (for example, the Monte Carlo method and simulation by computational biology), a computer-aided tool can be used to design (that is, to find the parameters of biochemical reactions that can carry out the pathways for computing). A nonautomated process can be devised through experiments in current biochemical technology and the use of empirical parameters.

In order to design, test and build a molecular computer, a systematic study needs to find the control structure that can regulate the signals of the pathways. The states are the state molecules in molecular switches. The feasibility of the states for the computing process needs to be studied from the viewpoint of achieving systematic controllability.

6.4 Blueprint of a Kinase Computer

Now let us look at the computing mechanism itself. The structural features of kinase computing determine the derived computing behavior. The two parts in the computing process—molecular operators and the corresponding biochemical reactions—are linked by higher-level computing processes in abstract form. An automaton is a kind of commonly used form for abstract description of computing. We discussed the automaton form of kinase computing in previous sections. Therefore, we can focus on the state of the biomolecular computing process. The state molecules $\{Q(l)\}$ and $Ggef \cup Ggap$ are the state sets that are relevant to computing. The quantitative relation between $Q(l)$ and $Ggef \cup Ggap$ is represented in the matrix $B_{L \times L}$. Because the total number of activated GTPases is L, the activated GEFs in Ggef and the activated GAPs in Ggap satisfy the condition that L1+L2 = L, where L1 is the number of activated GEFs in Ggef and L2 is the number of activated GAPs in Ggap. The relation represented in the matrix corresponds to the coefficient set of the biochemical reactions in kinase computing. The parameters of these coefficients are selected according to certain criteria, which can be empirical from bench-work experiments, theoretical by biophysics calculation, or estimated from software simulation. The feasibility of the integrated pathways can be inferred from studying the pathway of signaling molecule $\{Q(l)\}$, whose input is the $\{Q(l)\}^{(t)}$ and output is GEF/GAP in $\{Ggef \cup Ggap\}^{(t+1)}$, and the pathway of GEF/GAP, whose input is GEF/GAP in $\{Ggef \cup Ggap\}^{(t+1)}$ and output is $\{G(l)\}^{(t+1)}$.

The biological pathways of a GTPase switch are regarded as a molecular system when we view the control structure. Now, we consider the equation for a quantitative description of the GEF/GAP pathways that control the GTP-bound/GDP-bound states of GTPases:

$$d/dt\ [X] = A(t)\ X + B(t)\ U(t) \qquad (6.54)$$

where

> X refers to the vector represented by activated GEFs/GAPs whose total number is L.
>
> A(t)—a matrix—refers to the cross-talk in the above integrated pathways among the GEFs/GAPs, whose size is L × L.
>
> U(t) is a vector constructed from the set of $\{Q(l)\}$.
>
> B(t) is a matrix constructed from the cross-talk of the integrated pathway between $\{Q(l)\}$ and GEFs/GAPs, where the interaction is physically possible among the engineered-pathways. Its rank is L.

Here, controllability of the computing process means the ability to make the states of the m Rho family GTPases fall into ranges corresponding to all of the related combinatorial forms. This is achieved by utilizing the GEFs/GAPs through the pathways designed for the required clauses of the problems.

Let A(t) = A and B(t) = B. The existence of controllability of related pathways is dependent on the rank of

$$B:AB:\ldots:A_{L-1}B \qquad (6.55)$$

The rank of $B:AB:\ldots:A_{L-1}B$ is L because the GTPase pathway units encoded for GTPases whose number is L always generate the maximum number of L × L cross-talked pathways. Therefore, the signaling processes between GEF/GAP and $\{Q(l)\}$ pathways are quantitatively faithful according to the necessary and sufficient condition of controllability requiring that the related system [i.e., for (6.54)] be controllable if and only if the rank of $B:AB:\ldots:A_{L-1}B$ is L. (For more details on related controllability theory for control systems, see [10]. Some of the notations used for controllability here are different from those used in [10].) It is obvious that coupling only affects the GTP-bound/GDP-bound cross-talk among GTPase pathways. Consequently, we find that the control-space complexity of GEFs/GAPs is O(L).

One of the major conclusions from the discussion in this chapter is that the theoretical result of the kinase computing paradigm based on the biochemical faithfulness of cells is beneficial for building a biomolecular computer from signaling pathways. The biochemical information processing mechanism is an important bridge between the molecular implementation of the molecular computing process and abstract models of molecular computing in mathematics. Furthermore, the studies made on the tools of an automaton, computability, and formal language are helpful in exploring the massive parallelism from signaling pathways in cells. The biochemical features of the pathway units in a parallel MIMD architecture require us to design novel reprogramming and to develop a moleware-compiling technology.

This chapter presented the fundamental but crucial steps toward developing a complete kinase computing system. The most important experience we obtained from studies on kinase computation is that the reconfigurable pathway unit can be used for computing under the condition that the engineered pathways are somehow kept under control. The idea behind the reconfigurable pathway units is that the natural existing mechanism of biological cells [13, 14] can be employed systematically for massively parallel computing [15, 16]. The bottom-up method of nanobiotechnology [17] is mainly used in assembling molecules for computing systems. For the manufacture of a large-scale molecular computing system, the development of molecular manufacturing technology based on the self-assembly principle is imperative. The top-down method of biological technology cannot control the molecules as directly as can the bottom-up method of nanobiotechnology. The motivation of kinase computing is to learn about a naturally existing nanobiomachine—the cell—that is used as the basic framework of biomolecular computing. The philosophy we adopt here is to make the best use of the existing complex system of cells in nature and to control the underlying biochemical information processing system for the task of computing. The scalability and robustness of living cells are expected to be employed to give the designed architecture of kinase computing the scalability of computing in terms of signaling pathway units and the corresponding adaptive regulation mechanism of cell signaling. This mesoscopic methodology provides our guideline for the systematic design and testing of future prototypes of kinase computers.

References

[1] Etienne-Manneville, S., and A. Hall, "Rho GTPases in Cell Biology," *Nature*, Vol. 420, December 2002, pp. 629–635.

[2] Hafen, E., "Kinases and Phosphatases—A Marriage Is Consummated," *Science*, Vol. 280, May 1998, pp. 1212–1213.

[3] Isaacs, F. J., W. J. Bake, and J. J. Collins, "Signal Processing in Single Cells," *Science*, Vol. 307, March 2005, pp. 1886–1888.

[4] Hundley, H. A., et al., "Human Mpp11 J Protein: Ribosome-Tethered Molecular Chaperones Are Ubiquitous," March 2005, http://www.sciencemag.org/cgi/content/abstract/1109247v1?maxtoshow=&HITS=10&hits=10&RESULTFORMAT=&titleabstract=ubiquitous&searchid=1114132306774_11553&stored_search=&FIRSTINDEX=0&fdate=1/1/2005&tdate=4/30/2005.

[5] Helmreich, E. J. M., *The Biochemictry of Cell Signaling*, New York: Oxford University Press, 2001.

[6] Kumar, P. R., and L.-L. Xie, "Ad Hoc Wireless Networks: From Theory to Protocols," in *Ad Hoc Wireless Networking*, X. Cheng, X. Huang, and D.-Z. Du, (eds.), Boston, MA: Kluwer Academic Publishers, 2003, pp. 175–196.

[7] Lewis, H. R., and C. H. Papadimitriou, *Elements of the Theory of Computation*, Englewood Cliffs, NJ: Prentice-Hall, 1981.

[8] Berstel, J., and D. Perrin, *Theory of Codes*, Orlando, FL: Academic Press, 1985.

[9] Beaudry, M., et al., "McNaughton Families of Languages," *Theoretical Computer Science*, Vol. 290, 2003, pp. 1581–1628.

[10] Ogata, K., *Discrete-Time Control Systems*, 2nd ed., Upper Saddle River, NJ: Prentice-Hall, 1995.

[11] Kaibuchi, K., S. Kuroda, and M. Amano, "Regulation of the Cytoskeleton and Cell Adhesion by the Rho Family GTPases in Mammalian Cells," *Annu. Rev. Biochem.*, Vol. 68, 1999, pp. 459–486.

[12] Kawano, Y., et al., "Phosphorylation of Myosin-Binding Subunit (MBS) of Myosin Phosphatase by Rho-kinase *In Vivo*," *The Journal of Cell Biology*, Vol. 147, 1999, pp. 1023–1037.

[13] Ausubel, F. M., et al., (eds.), *Current Protocols in Molecular Biology*, New York: John Wiley & Sons, 1998.

[14] http://www.med.nagoya-u.ac.jp/Yakuri/.

[15] Tahoori, M. B., and S. Mitra, "Defect and Fault Tolerance of Reconfigurable Molecular Computing," *12th Annual IEEE Symposium on Field-Programmable Custom Computing Machines*, 2004, FCCM 2004, April 2004, pp. 176–185.

[16] Shukla, S. K., et al., "Nano, Quantum, and Molecular Computing: Are We Ready for the Validation and Test Challenges?" *Eighth IEEE International Conference on High-Level Design Validation and Test Workshop*, 2003, pp. 3–7.

[17] Dwyer, C., A. R. Lebeck, and D.J. Sorin, "Self-Assembled Architectures and the Temporal Aspects of Computing," *Computer*, Vol. 38, No. 1, January 2005, pp. 56–64.

[18] Courcelle, B., "The Expression of Graph Properties and Graph Transformations in Monadic Second-Order Logic," in *Handbook of Graph Grammars and Computing by Graph Transformations, Vol. 1, Foundations*, G. Rozenberg, (ed.), Singapore: World Scientific, 1997, pp. 313–400.

[19] Milner, R., "Bigraphical Reactive Systems: Basic Theory," 2001, http://www.cl.cam.ac.uk/users/rm135/bigraphs.ps.gz.

[20] Landweber, L. F., and L. Kari, "The Evolution of Cellular Computing: Nature's Solution to a Computational Problem," *BioSystems*, Vol. 52, October 1999, pp. 3–13.

[21] Ehrenfeucht, A., et al., *Computation in Living Cells: Gene Assembly in Ciliates*, New York: Springer-Verlag, 2004.

[22] Weiss, R., and T. F. Knight, Jr., "Engineered Communications for Microbial Robotics," *DNA6, Lecture Notes in Computer Science*, Vol. 2054, A. Condon and G. Rozenberg, (eds.), Berlin, Heidelberg: Springer-Verlag, 2001, pp. 1–16.

[23] Basu, S., et al., "A Synthetic Multicellular System for Programmed Pattern Formation," *Nature*, Vol. 434, 2005 pp. 1130–1134.

[24] Magnasco, M. O., "Chemical Kinetics Is Turing Universal," *Physical Review Letters*, Vol. 78, No. 6, 1997, pp. 1190–1193.

[25] Liu, J. Q., and K. Shimohara, "A Biomolecular Computing Method Based on Rho Family GTPases," *IEEE Trans. on Nanobioscience*, Vol. 2, No. 2, June 2003, pp. 58–62.

[26] The P Systems Web Page http://psystems.disco.unimib.it/.

[27] Paun, G., *Membrane Computing: An Introduction*, Berlin: Springer-Verlag, 2002.

CHAPTER 7
Comparison of Algorithms for Biomolecular Computing and Molecular Bioinformatics

There might be important links between biomolecular computing and molecular bioinformatics, where the collective mechanism of molecules in cells provides a good example for molecular computer designers. Using the cell as a reference, we may consider a broader scope of biomolecular computing, in which various types of molecular information processing mechanisms are employed for the best possible computing. This landscape is growing and no longer confined to the domain of DNA computing. Accordingly, we need to study the bioinformatics in molecular biology because the fields of proteomics, systems biology, computational biology and molecular bioinformatics encompass a rich body of knowledge on how the natural biological systems work through the biochemical activities of molecules. The field of molecular bioinformatics is multifarious, and within it the objects and methods in research vary greatly. Bioinformatics, with mathematics and computer science at the core of its informatics theory, also involves stochastic theory, searching technology and machine learning. The relationship between biomolecular computing and molecular bioinformatics is bidirectional, sometimes interacted or coupled [1–36] (Figure 7.1).

1. *From biomolecular computing to molecular bioinformatics:* Currently, theoretical biomolecular computing provides some formal operators and algorithms for molecular bioinformatics. In the future, we can expect biomolecular computers to be implemented in wetware forms of bioinformatics-oriented molecular-data-processing systems.
2. *From molecular bioinformatics to biomolecular computing:* Biological knowledge of molecular bioinformatics reveals how biomolecules behaves in their natural states. Moreover, it provides the biological concepts and biochemical references of natural molecular systems, which form the basis of biomolecular computing in theory. Enlightened by biomolecular knowledge, artificial design and engineering control methods are introduced in biomolecular computing systems through experiments.

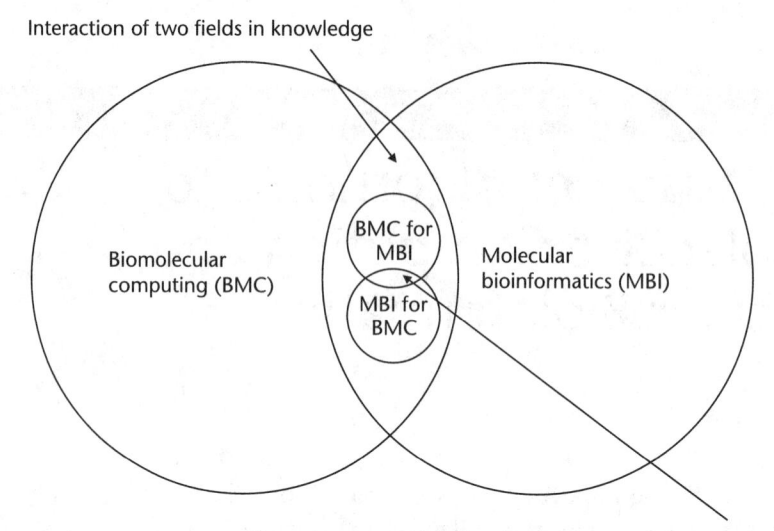

Figure 7.1 Relationship between the domains of biomolecular computing and molecular bioinformatic

3. *Intersection of biomolecular computing and molecular bioinformatics:* The intersection of the two fields resides in the algorithms that can be used for the common purpose of exploring the informatics in biomolecular signaling, which is the basis of developing experimental systems that work as moleware for computational cell biology.
4. *Different parts of the two fields:* The purposes of the two fields are different, the major differences can be found in following aspects:
   - Some biomolecular computing processes are realized by artificial molecular systems and behave differently from the natural biological systems in their biological function. Accordingly, molecular computing cannot be considered strictly a natural biological process.
   - Some molecular bioinformatics processes are implemented by computer programs based on a kind of mathematical inference, and thus they behave differently from the biochemical transition processes of natural biological systems.

Major segments of the algorithms of biomolecular computing are aimed toward computing rather than biological function discovery, and most of the bioinformatics algorithms employ mathematics skills to solve bioinformatics problems.

7.1 Formal Characteristics of Algorithms for Biomolecular Computing

By extracting the abstract form of the algorithms for biomolecular computing, we can clarify the informatics description of biomolecular computing. This is done by formalizing the informatics structure as conceptual objects based on the biochemical reactions. From the molecular operation process of biomolecular computing, the informatics form of the abstract operators can be extracted. At the mathematics level, the abstract operators are independent, that is, they are defined at a high level above the quantitative measurement of biochemical reactions. We can investigate the methodology of computing and the philosophy of problem solving in an unconventional way by using representative biomolecular computing algorithms.

7.1.1 DNA Computing

In Adleman's DNA computing method for solving HPP, the constraint of HPP is represented by the information of nodes and edges. The local operators connect the nodes in parallel according to the constraints on edges. The operation process produces a solution in the form of a hyperedge connecting the required nodes. The basic components of the problem's target are provided in advance. The operator is made based on the idea that it should merge the building blocks to form the solution, which is a kind of self-assembly in concept.

DNA computing can be formalized as a process in which a solution is obtained from a pool of candidates. The constraint on the expected computing process is embedded in the molecular operation mechanism explicitly or implicitly. The molecular operators are coded to satisfy the specific requirements of the computing task. The molecular level of the DNA computing process is dependent on the specific design and empirical experiences of tailor-made biochemistry experiments.

In Lipton's method of Adleman's DNA computing for solving 3-SAT, the constraints are first divided into clauses and then into literals. The operators check sequentially the state of candidates by using the literal. At the beginning, all of the combinatorial forms of the variables are put into a set (test tube). The serial operators select the candidates that satisfy the literal constraint in a stepwise way. Only the candidates that satisfy the constraint on the literal are kept for checking in the next step, and the candidates that do not satisfy this constraint are removed from the record. The sieving process for candidates described above is iterated to maintain or reduce the number of remaining candidates. The total number of steps for the entire process is $3 \times$ clause number.

7.1.2 Surface-Based DNA Computing

All of the combinatorial forms of the variables are prepared in the pool at the beginning. The operators are designed and used to delete the candidates that do not satisfy the clauses. The operator is serial and iterated on the set of remaining candidates. The final state of the set is the solution. The operation uses the clause as the unit. The number of operation steps is the number of clauses. It is obvious that the step number here is smaller than the step number in Lipton's SAT algorithm based on Adleman's DNA molecular operators.

7.1.3 H-Systems

H-systems use the rewriting operator on the data structure of strings, trees, circles and their hybrid forms. The operator is temporal. The data structure is spatial.

7.1.4 P-Systems

P-systems use multisets as the data structure to represent the spatially located elements, which can be strings or other forms.

7.1.5 DNA Computing Method by Ciliates

By mapping the gene operations in ciliates, string rewriting can be achieved by Kari and Landweber's method [1]; graph rewriting—spatial rewriting on strings—can be achieved by the method of Erhenfeucht et al. [2].

The rewriting-style operators are procedural. The objects in the set of candidates allow the rewriting operator to handle them in serial or in parallel. The process formulated by formal models in theoretical computer science is directly related to the abstract description of problem-solving algorithms.

In the problem space consisting of all possible samples, the operation process produces routes that lead to a final solution. The constraint on the problem is described in the rewriting rules and is divided into distributed representation.

The information of the constraint is explicitly represented by the operators in a parallel and distributed mode, which degenerates into serial form when the size of the problem to be solved is very small.

Derived from the above parallelism of biomolecular computing, a unified description of biomolecular computing (BMC) can be summarized as

$$X_0, X_1, ..., X_n \in X\text{-set}$$

where n is a natural integer.

The parallel operator of BMC is defined as

$$\phi: BMC(X_0, X_1, ..., X_n)(t) \to BMC(X'_0, X'_1, ..., X'_n)(t+1)$$

which is a kind of mapping from the set of molecules before the computing operation to the set of molecules after the computing operation. The biochemical features of the molecular signaling process correspond to the above computing operator.

Consequently, we can obtain this rule:

IF $(X_0, X_1, ..., X_n)(t)$ satisfies the constraint on the computing process,
THEN operator ϕ produces the output of BMC $(X'_0, X'_1, ..., X'_n)(t+1)$

For the logic-level representation of the generalized operator in biomolecular computing, the unified form of biomolecular computing can take various concrete forms at the molecular signal level described in Chapter 4.

The informatics-level description of biomolecular computing is based on a molecule-constrained signaling mechanism. The biomolecular signaling processes where the molecular messengers are performed by DNA, RNA and proteins are heterogeneous. The interaction of molecules not only reflects the matter flow but also determines the information flow. However, there are gaps between formal modeling and benchwork experiments. Programming the moleware is a way to bridge the gap. Successful examples of biomolecular computing systems have been developed for specific problems, but it is necessary to design a kind of programming language for the generic purpose of biomolecular computing systems. The data structure obtained from the formalization and abstract operators mentioned above is compatible with the requirements of developing a moleware programming language. The moleware programming language should provide both the primitive-level operations for describing the moleware signals, which we have already introduced in previous chapters, and the informatics-level function for information processing. An operating system is needed for monitoring the biomolecular computing processes, where molecular programs are adapted for different applications. With such an operating system, the automated molecular operators at the biochemical reaction level are expected to be synchronized with the computing processes at the informatics level. This is because the molecular dynamics of biomolecules and the three-dimensional structure of molecular complexes obey the principle of complex systems, which are essentially different from computational systems. If automated technology for controlling the biochemical reactions of moleware were developed using informatics programs designed in advance, a molecular computer could become a reality as a completely defined concept.

7.2 Interactions in Molecular Bioinformatics Algorithms

The objects of molecular bioinformatics cover a broader range, including DNA, RNA, proteins, and cells. Knowledge of genomics and proteomics can contribute to the development of applications such as drug design. The methods of molecular bioinformatics originated from the mathematical study of biological systems, where information-processing technology is applied to simulate biological signals. Basically, molecular information is extracted and used to analyze, estimate, and infer the features, mechanisms, and functions of the biological systems.

The abstract operator form for a bioinformatics system functions in the following way:

known data -> calculation/inference -> new information

The amount of data in bioinformatics is often huge, and the multiple relations among the data often give hints for understanding the biological meaning. The data structures for spatial description of relationships among the data items—trees and graphs—are commonly used for the algorithms in molecular bioinformatics. The space complexity and time complexity of bioinformatics algorithms refer to the computer source (hardware) and the time consumed by bioinformatics computing tasks. The advanced technology of high performance computing in hardware makes space complexity less of a problem, but time complexity is still a critical issue.

Reconstructing phylogenetic trees is a typical algorithm in molecular bioinformatics. From the set of partial trees, networks/graphs can be estimated. A tree is a data structure used for studying the relationships among genes resulting from evolution. The split graph method extends data structure from tree to graph.

The "conflict graph" is a representative graph in bioinformatics. In a conflict graph, the route/path is used to find knowledge on the relationships of the genes in evolutionary history. Splitting the graph is a way of studying the generalized concept of phylogenetic trees.

To handle the graph structure, the operators are exerted on the nodes and edges. The operation on a graph is aimed at finding the path that directly influences the complexity and leads to the solution in problem space. The feature and pattern identified by the bioinformatics knowledge make the algorithms more efficient over time. NP completeness often occurs in algorithms, and bioinformatics is not an exception.

In genomics and bioinformatics, the relationships among genes are studied through available data, and a commonly used data structure for the description of this relationship is a kind of partial tree. Reconstruction of a phylogenetic network by partial trees is a significant task. In the method of Huson et al. [3], the concepts of "splits" and "splits graph" are introduced. Intuitively, the data of the partial trees are represented by different sets. From the information on the relations of these sets, networking is achieved by stepwise computing.

The taxa set is denoted as

$$X = \{X_0, X_1, ..., X_{n-1}\},$$

where $n \in N$.

The corresponding tree is denoted as

$$T = (V, E, z, w).$$

where

> V—the set of nodes, the number of nodes is n
> E—the set of edges that connect the nodes by links
> z—the label for the element in X
> w—the weight for edges

Split refers to the configuration of the two separated parts of set X. The separation operator creates two subsets of X. The separation operation is binary. Mapping from the subsets, the derived tree is also a binary tree.

$\Sigma(T)$—the set of all splits from X, the number of splits is m

The method of networking the genes by Huson et al. [3] constructs Σ^* as the set of all full splits corresponding to the partial tree T in which the rule called Z-rule is used:

Let split $S_i = A_i/B_i$, $S_j = A_j/B_j$, where $S_i, S_j \in \Sigma$, $i, j \in N$.

IF $(A_i \cap A_j \neq \phi)$ AND $(A_j \cap B_i \neq \phi)$ AND $(B_i \cap B_j \neq \phi)$ AND $((A_i \cap B_j \neq \phi))$

THEN
> Update:
> $S_i \rightarrow S'_i = A_i / (B_i \cup B_j)$
> $S_j \rightarrow S'_j = (A_i \cup A_j) / B_j$

From the method [3] based on the above kernel operator, the following efficiency of operating processes can be obtained:

Time complexity $\leq O(nm^3)$

Space complexity $= O(nm)$

Number of splits $\leq \leq n + m$

To handle the graph structure, the operators are exerted on the nodes and edges. In order to update the graph, the nodes and edges are added or deleted. Accordingly, the process can be represented as

old graph -> new graph

From data-driven methodology, it is possible to strengthen the classification/clustering of data in bioinformatics in order to strengthen our understanding of the features of biological objects. However, to enhance the efficiency of bioinformatics algorithms, a great deal of background knowledge is needed. Machine learning, signal processing, data mining, and pattern recognition are thus useful.

The form of the binary tree can be obtained when the object is limited to the set of pathways of binary molecular switches. The length of the tree is the logarithmic order of the leaf number. When all of the choices are listed in the leaf layer of the tree, the leaf number is 2^n, where n is a natural number that denotes the branch in the logical judgment. The solution explicitly represented in the tree is the shortest path. Since the knowledge is incomplete, we have to estimate the possible path that approaches the solution path. The cost of searching varies depending on the paths passed. The cost ranges from polynomial (if possible) to exponential (in the case of labor-exhaustive search).

7.2.1 Example 1: Interaction of GTPases

The biochemical-reaction relationship among the molecules—via a cellular pathway—is represented by a graph. We use pathways as an example to discuss the algorithmic features, which are mapped from the set of rules to a network that takes the form of a graph.

In bioinformatics, two kinds of pathways—metabolic and signaling pathways—are important. The former directly determines the ability to sustain the cellular system, and the latter acts as the main mechanism for signal transduction and cell communication. From biochemical experiments, simple pathways can be recognized. However, complex networks formed by the interactions of the known simple pathways cannot be recognized easily owing to the existence of interactions among these known pathways. Networking based on proteins is a kind of problem-solving process for discovering protein interactions.

The problem-solving process is carried out by operating on the nodes in the graph. The complexity is calculated by the data structure in the process. The space complexity is the number of nodes, and the time complexity is the steps used in the solving process.

From the experiments, one can easily obtain the single pathways for specific input and output. The pathway set as a whole contains interactions. At the biophysical level, there are various types of pathways from the viewpoint of their molecular structure, biological function, biomedical information and experimental operation. We study the topic of specific protein interactions on the pathways that are constructed by molecular switches. The molecular switches were introduced in Chapter 6. Here, we use the term "molecular switch pathway."

The basic action of the molecular switch pathway consists of the following parts:

The regulator molecules set the state of the signal molecule to the positive state;
The regulator molecules set the state of the signal molecule to the negative state.

Under the regulation of the molecular switch pathway, the transition process between two states of molecules are reversible. The molecular switch of GTP-bound GTPase/GDP-bound GTPase is an example in cells. In order to systematically understand the behavior of GTPase pathways in cells, we have to study the structure of a GTPase pathway network by reconstructing it from the atomic (indivisible) pathways of GTPase switches. These atomic pathways refer to those GTPase pathways that determine the binary values of GTPases and that cannot be divided further.

There are matrixes connecting the different nodes/molecules in the networking process of pathways. The matrixes used to describe the connection from GEFs to GTPases (one direction), from GAPs to GTPases (one direction), and from GTPases to GTPases, are denoted as

Matrix-GEF-GTPase(n×n);
Matrix-GAP-GTPase(n×n);
Matrix-GTPase-GTPase(n×n).

Two directions can be considered in the matrixes of GEF/GAP-GTPase as well.

When no prior knowledge is taken into consideration, the total number of connections is $3n^2$. The connections in the same type of pathways involving the above three types of signaling molecules are denoted by

Matrix-GEF-GTPase i (n×n);
Matrix-GAP-GTPase i (n×n);
Matrix-GTPase-GTPase i (n×n).

Connections among the different clusters are

$(3n^2) \times (3n^2) \times ... \times (3n^2)$.

The interaction number we assume is n because the basic number of molecular switches is n. If other molecules in cells are considered, the number of connections will increase. The number given above can be regarded as the lower bound of the interactions. The information unit is assumed to be n at the initial moment of the algorithm.

The algorithm for constructing the interactions can be given in a pseudo-programming-language form:

```
for (i=0 to n; i++)
{
    for (j=0 to n; j++)
    {
        for (k=0 to n; k++)
        {
            Matrix-GEF-GTPase (j,k);
            Connection (GEF/GTPase);
            Matrix-GAP-GTPase (i,k);
            Connection (GAP/GTPase);
            Matrix-GTPase-GTPase (k,k);
            Connection (GTPase/GTPase);
            IF (constraint (connection of three parts)=TRUE)
            THEN connection (three matrixes)
            concentration-calculation (three matrixes);
            networking (GEF, GAP, GTPase);
        }
    }
}
```

Let PN-set = $\{PN_0, PN_1, ..., PN_{L-1}\}$, where PN_i (i = 0, 1, ..., L-1, L $\in$ N) refers to the atomic pathway of a GTPase switch. Assuming that the interactions among these pathways exist, we make the following conjecture.

CONJECTURE

If we can find an algorithm ϑ for reconstructing GTPase pathway networks, the time complexity of the reconstruction process by ϑ is NP under the condition that no knowledge on the interactions of pathways is introduced.

In addition to GTPase pathways, the interaction of kinases and proteins is also a topic that deserves study. As [4] reported, several representative combinatorial forms of the interactions of kinases, proteins and transcription factors (TF) have been discovered. If all of the nodes are connected without any constraint introduced for interaction, a complete network will take an exponential number if the number of kinase and phosphatase is 2000 and 1000, respectively, in the reconstruction pathway network of kinases/phosphatases, the complete number of connections of the K/P pathways is exponential: 2^{2000} for kinase and 2^{1000} for phosphates, 2^{3000} for interactions among kinases/phosphatases. It is obvious that this is an NP-complete problem for searching all of the nodes in this network without any constraints on interactions.

7.2.2 Example 2: Interaction of Kinases/Phosphatases

With respect to the interactions of kinases (denoted as "k"), photoproteins ("p"), and TF in yeast cells [4], the protocols of the related signaling processes are the keys to understanding the biological function. Photoprotein refers to the protein that is influenced by kinase for phosphorylation. As [4] reported, TF is also interacted with kinases and proteins. The interactions of intercellular signal pathways, proteins in membrane, extracellular signals, and cellular nuclear receptors are types of constraints that can reduce the size of the problem space. The effect of a scheme to reduce problem space is significant, especially in search algorithms. In addition to the concept-oriented approach, the data-oriented approach is also effective for reducing the size of the pathway network with interactions. As one kind of quantitative description method, plotting patterns of concentration versus time curves provides rich information on the quantitative relationship of the multiple influences of the signal molecules. From an analysis of the patterns, we can delete some assumed interactions in a candidate pathway network. A topic deserving of attention is to innovate a time-efficient algorithm for reconstructing kinase/phosphatase pathway networking by introducing knowledge on the signaling protocols and using high-speed parallel computers.

The notations of molecules involved in our discussion are given as follows.

Kinase (i) ∈ kinase-set, Nk—the number of kinases
Phosphatase (j) ∈ phosphatase-set, Np—the number of phosphatases
Photoprotein (k) ∈ photoprotein-set, Nr—the number of photoproteins, denoted as protein (k) for short
TF (*l*) ∈ transcription-factor-set, N*l*—the number of transcription factors in cells

SM (w) is the synchronization measure, where w refers to the logical description of pathway signaling in symbolic form. The logic description for the operator can take the predicate form that corresponds to the primitive in programming.

The constraint on synchronization comes from the nature of the biochemical reactions owing to the nonlinear characteristics of concentration versus time, where the phase and peak delay vary and depend on the enzymatic activation and environment. The synchronization of multiple pathways can be achieved by controlling the coefficients of the differential equations of biochemical reactions. The biological signaling processing method is used for quantitative description of the interaction effect of pathways in cells.

7.2 Interactions in Molecular Bioinformatics Algorithms

According to the principle of phosphorylation/dephosphorylation and experimental evidence [4], the predicates for describing the interactions are formulated here:

Interaction of kinase-to-protein: Phosphorylation (kinase(i), protein(k))
Interaction of phosphatase-to-protein Dephosphorylation
　　(phosphatase(j),protein(k))
Interaction of TF and protein: TF-regulation (TF(l), protein(k))

The interaction of phosphorylation/dephosphorylation from networking becomes a combinatorial form of the above predicates.

The interaction of kinases/phosphatases is presented by the following logical formula in which "(X)" refers to AND or OR operator:

(
Phosphorylation (kinase(i_1), protein(k_1))
(X)
Phosphorylation (kinase(i_2), protein(k_1)
(X)
...
(X)
Phosphorylation (kinase(i_n), protein(k_1))
)
(X)
(
Phosphorylation (kinase(i_1), protein(k_2))
(X)
Phosphorylation (kinase(i_2), protein(k_2)
(X)
...
(X)
Phosphorylation (kinase(i_n), protein(k_2))
)
(X)
...
(X)
(
Phosphorylation (kinase(i_1), protein(k_m))
(X)
Phosphorylation (kinase(i_2), protein(k_m)
(X)
...
(X)
Phosphorylation (kinase(i_n), protein(k_m))
)

(X)
(
Dephosphorylation (phosphatase(j_1),protein(k_1))
(X)
Dephosphorylation (phosphatase(j_2),protein(k_1))
(X)
...
(X)
Dephosphorylation (phosphatase(j_p),protein(k_1))
)
(X)
(
Dephosphorylation (phosphatase(j_1),protein(k_2))
(X)
Dephosphorylation (phosphatase(j_2),protein(k_2))
(X)
...
(X)
Dephosphorylation (phosphatase(j_p),protein(k_2))
)
(X)
...
(
Dephosphorylation (phosphatase(j_1),protein(k_m))
(X)
Dephosphorylation (phosphatase(j_2),protein(k_m))
(X)
...
(X)
Dephosphorylation (phosphatase(j_p),protein(k_m))
)

The matrix is updated by introducing the logical and quantitative constraints on protein signals.

Kinase-to-protein [n × m]
Phosphatase-to-protein [p × m]

Many-to-one mapping is a generalization of dual kinase regulation [5].
The synchronization primitive is defined as:

Syn (pathway)—phase and peak delay are below the allowed threshold
Asyn (pathway)—phase and peak delay are above the allowed threshold

The primitive function () is defined as logical values T or F.

Let n1 and n2 be the numbers of the synchronized pathways and asynchronized pathways, respectively. The signal constraint will reduce the problem space of the interactions by subtraction of n2. The number of resultant pathways is n1-n2.

The basic configuration in temporal orders and spatial distributions (topological locations of proteins) represents the logical relations among the interacted pathways of phosphorylation/dephosphorylation with synchronized signals. The exponential number of the interactions could be cut into a relatively small order by introducing the earlier constraints on signal synchronization.

The algorithm for generating a hypothesis route is the theoretical preparation for experimental verification in benchwork. Provided that the speed of the pathway response is preferred for evaluation of a phosphorylation event (in experimental molecular biology, the term "phosphorylation event" refers to the biochemical reaction process of phosphorylation), the following algorithm is given for finding a solution to interaction (in pseudo-programming language):

Algorithm

Input: activation matrix of kinases, phosphatases, array of existing photoproteins

Output: a route that describes a hypothesis pathway determined by the biochemical reaction speed

```
Time = t;
Index =0 ;
route [] ; /* string */
K-count [n] = P-count [p] = Q-count [m] = 0 ;
/* Q - photoprotein */
For (i = 0 to n)
{
for (j= 0 to p)
{
for (k= 0 to m)
{
  IF response (kinase(i), protein(k)) < threshold(t)
  THEN
    { /* if */
    IF kinase (i) is activated
    THEN route [index] = K|the-number-of-i;
IF protein (k) is activated
THEN route [index] = P|the-number-of-k;
    syn (pathway(kinase(i), protein(k)) = T;
    } /* end if */
  ELSE
  {
```

```
            asyn (pathway(kinase(i), protein(k)) = T;
        }
    IF response (phosphatase(j), protein(k)) < threshold(t)
    THEN
        { /* if */
        { /* if */
        IF phosphatase (j) is activated
        THEN route [index] = P|the-number-of-j;
        IF protein (k) is activated
        THEN route [index] = Q|the-number-of-k;
        syn (phosphase(i), protein(k)) = T;
        } /* end if */
    ELSE
        {
        asyn (pathway(phosphatase(j), protein(k)) = T;
        }
        /* end if */
    }
}
}
```

In living cells, multiple relationships among GEFs, GAPs, and GTPases have been discovered by molecular biologists from experiments [23]. Some of the relationships are shown in Figure 7.2.

The basic rules for description of interactions among kinases, proteins and transcription factors (TF) are well established [4]. Kinase, phosphatase, and photo protein are three elements for constructing a tree as a building-block for interacting phosphorylation/dephosphorylation pathways. The Rho-MBS-MLC pathway is an example of such a building block. Figure 7.3 gives a description of the Rho-MBS-MLC pathway. As shown in Figure 7.4, a unit of two phosphorylation/dephosphorylation pathways is transformed into a binary tree in which K refers to kinase, Pt refers to phosphatase, and photo-proteins refer to the set of {Xi} (i=1,2,..,n, n ∈ N).

As shown in Figure 7.5, the binary tree T0 that covers two photo-proteins X1 and X2 is selected as the starting point of the interaction processes. The remaining photo-proteins X2, X3, ..., Xn are covered by the set of binary trees {Ti}. The number of possible states of T0 is n(n-1)/2. The number of possible states of Ti (i=1, ..., n-1) is (n-2)(n-3)/2. Within the single binary tree, the phosphorylation/dephosphorylation can be verified by experiment. Under the condition of an in vitro cell, it is feasible to activate two proteins in the cell culture and to observe their effect by biochemical technology. The phosphorylation/dephosphorylation effect can be verified by tools such as immunofluorescence analysis. The possible effects of the phosphorylation/dephosphorylation relation within a binary tree are explicitly described in Figure 7.6.

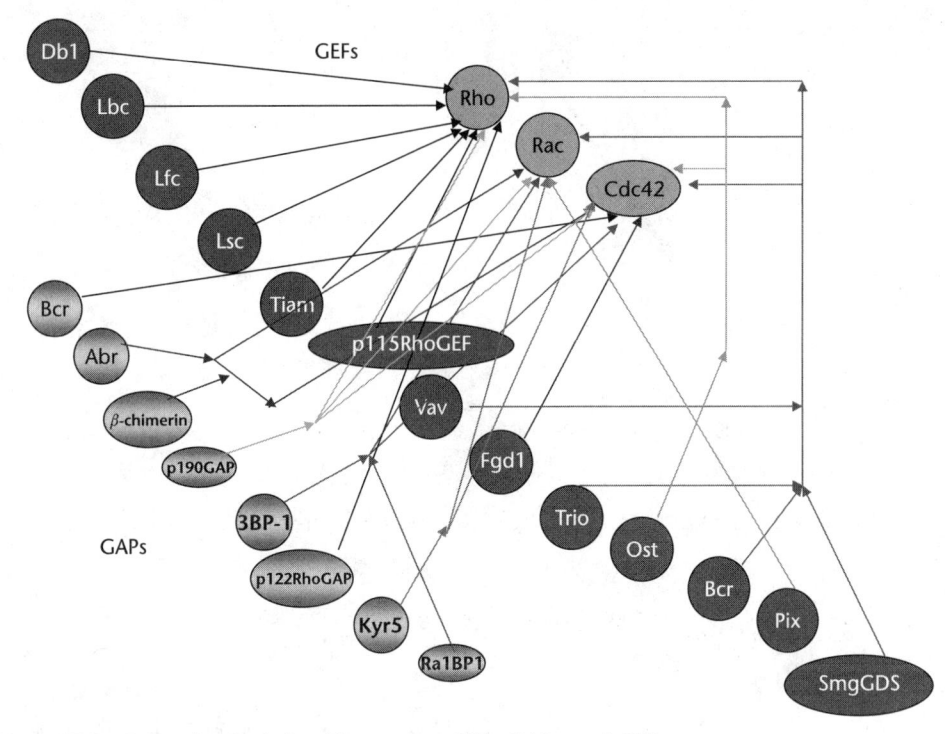

7.2 Interactions in Molecular Bioinformatics Algorithms

Figure 7.2 Example of relationships among GEFs, GAPs, and GTPases.

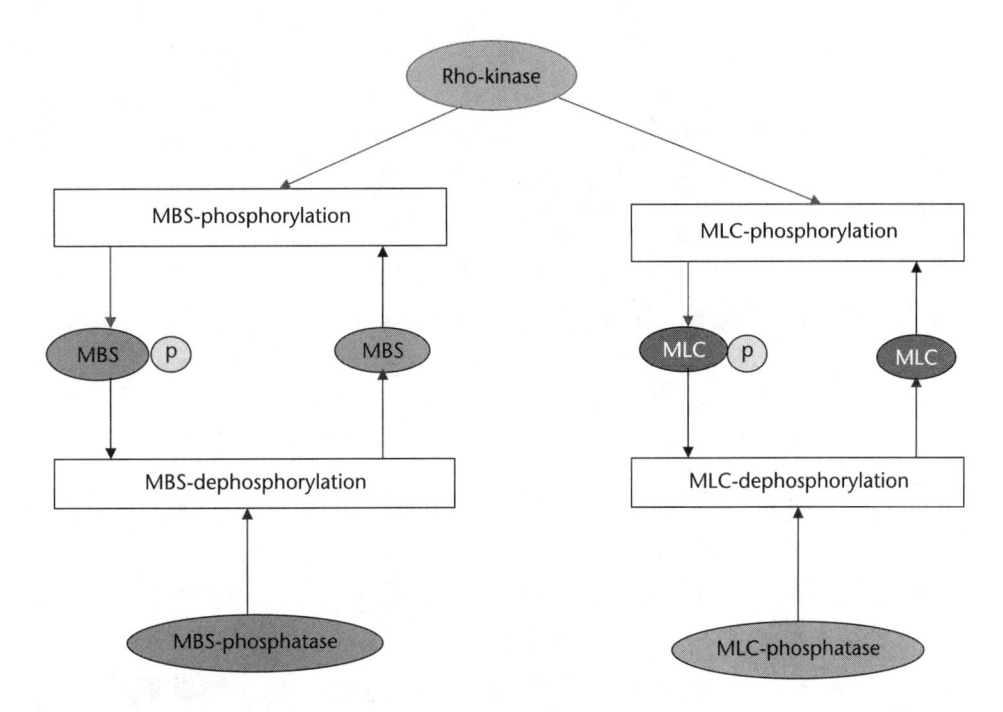

Here only the signaling molecules that directly represent the information of tree structure are given.

Figure 7.3 Biochemical relations in a Rho-MBS-MLC pathway.

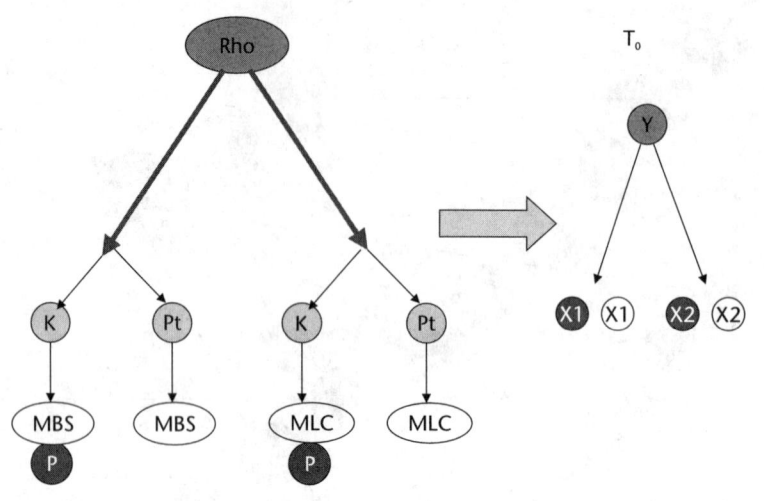

Figure 7.4 Transformation from a pathway to a binary tree.

Inferring interaction based on synchronization effect

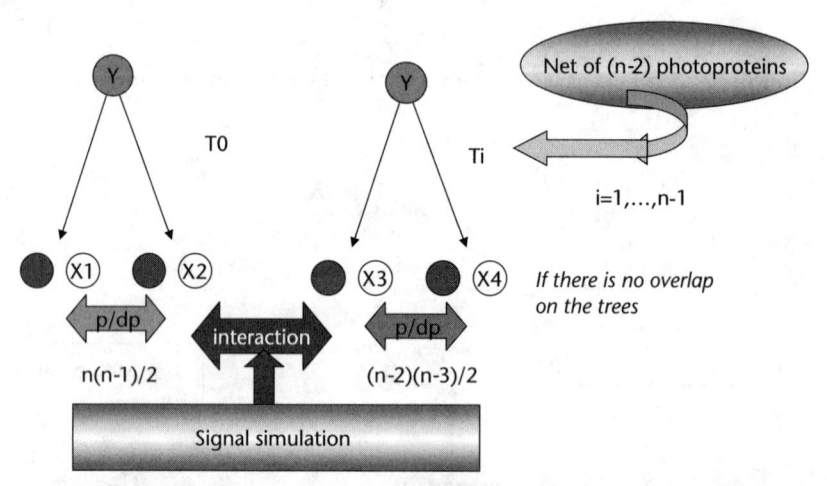

Figure 7.5 Schematic explanation of inference operation on binary trees.

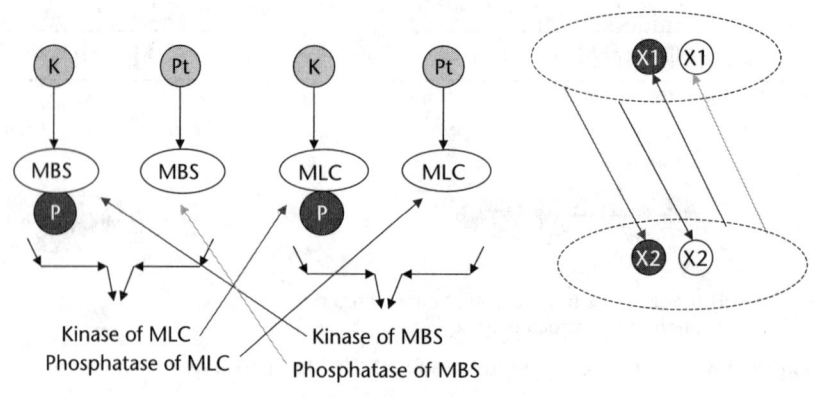

Figure 7.6 Possible mutual effect within the same binary tree.

The time complexity of the above inference process is

$$O(P(n))$$

where $P(n) = a_0 n^4 + a_1 n^3 + a_2 n^2 + a_3 n + a_4 n$ and a_0, a_1, a_2, a_3, and a_4 are empirical constants.

In the case of applying bioinformatics to biomolecular computing, knowledge on the interaction of kinases/phosphatases can also help us to select and design the motif of pathways of phosphorylation/dephosphorylation according to the protocols discovered by experiments where the phosphorylation/dephosphorylation signaling process is expected to carry out computing tasks. The signaling mechanism of natural molecular biological systems [6–9] provides us a window for observing the nonlinear behavior of cells and can contribute to the controlling of cells for expected effects, including computing results of biomolecular computing.

7.3 Common Points of Biomolecular Computing and Molecular Bioinformatics for Algorithms

Building molecular computers depend on several factors. One is the knowledge of the material itself, in which biological information is important. Complementary to experimental biology, bioinformatics is useful for discovering knowledge in life science by using IT technology. We expect to use computational molecular biology for exploring new phenomena in life science through engineered systems of molecular computing. The key technologies for connecting biomolecular computing and molecular bioinformatics are shown in Figure 7.7.

In the field of biomolecular computing, the studies on algorithms of biomolecular computing are very active. The informatics of the algorithms is located at the information processing level based on the mechanism of molecular biology. The abstract form extracted from the methods of biomolecular computing can be studied in informatics, in which the kernel task is to study the biomolecular information representation and operators. In brief, this study centers on the concept of molecular information flow, which consists of a natural biomolecular process and an engineered moleware process.

To solve real-world problems in molecular biology, molecular bioinformatics focuses on understanding the information processing mechanism of biological systems at the molecular level. The models in molecular bioinformatics describe the natural phenomena of biological systems, and the algorithms in molecular bioinformatics describe the relations among the molecules in biological systems. Some of the phenomena in molecular bioinformatics can be controlled in an engineering way. The molecular information flow either exists naturally or is produced by molecular engineering.

The rapid progress in experimental molecular biology is providing us with rich knowledge on the theory and tools for more precise biological cell experiments, where biochemical reactions are carried out with an enormous number of bio-molecules. Abstraction is not only a way to represent data for theoretical study but also a necessary method for simulation, especially in large-scale bioinformatics

data-driven/bottom-up: nanotechnology for moleware

Biomolecular computing →
Enumerating all the samples in the problem solving
This is supported by the molecular manufacturing

Operator works for taking the constrained objects
and removing the unuseful objects

space complexity NP
time complexity P

operators that give traces of the path
in the partial problem space
leading to the solution,
i.e.,
information is updated
on set, string, tree, graph
inference by calculating from data partial problem
space in logic

enumeration is impossible
(because the database of genomics
and proteins at bioinformatics is
huge, and parameters of
biochemical reactions are not
sufficient in many cases)

concept-driven/top-down:
knowledge is used to infer,
operator/operation is not necessary to be consistent with the biological mechanism
inference (logical processing, thinking/thought)

← molecular bioinformatics

Figure 7.7 Conceptual relationship between biomolecular computing and molecular bioinformatics.

systems. From experiments on molecular computers, we may gain some insight. In DNA computers, the operators of biomolecular computing are carried out in test tubes to accomplish specific computing tasks so that the relationship between input and output can be precisely determined. The process acts as a kind of bottom-up wetware-level inference machine for synthesis of the biochemical reactions in vitro. Based on the formalization of the physically feasible computing process, the formal models, operators, and algorithms extracted will be beneficial for extending our knowledge of the mechanism behind the underlying biological objects verified from experimental data to a broader domain of biological systems that include the molecules in reactions. This knowledge can act as a bridge toward top-down knowledge discovery.

Many examples indicate a relationship between biochemical reactions and the molecular information mechanism. A kind of biochemical reaction among the DNA molecules, called hybridization of DNA strands, can be controlled to carry out cycles of reactions in a test tube. When this process produces a specific output according to a specific input, it will become a kind of computing if the input and output are encoded as specific information. The input and output are DNA sequences in Adleman-paradigm of DNA computing, splicing systems (H-systems), and aqueous computing. Through multiset structure, the model of P-systems (membrane computing) can design string-rewriting processes for DNA computing. Ciliate-based computing can also provide string rewriting and graph reduction systems, based on real biological ciliates. What we want to emphasize here is that formal models can be extracted from the biological processes of DNA computers, and this is useful for bioinformatics because it is a bottom-up method where each operator can be physically realized in biological experiments. This approach is complementary to the popular formal models in the field of bioinformatics, which are top-down and designed based on knowledge-to-knowledge inference.

The common parts of biomolecular computing and molecular bioinformatics involve the algorithmic aspect of molecular information processing, that is, the information representation and related operators whose abstract form is formalized as a kind of problem solving in which some of the problems are NP problems. The states and the transition among the states are quantitatively described in order to understand the underlying complex mechanism of molecular biological systems.

When the models and algorithms of molecular computing are modified for bioinformatics, the systematic level of biology provides the methodology that unifies the two aspects. The common parts of the two fields can be studied at different levels, from individual molecules to cells. The analysis of cell systems in terms of nanobiotechnology is the best approach taken from systematic bioinformatics for understanding the global signaling of complex molecular interactions that take the form of networks.

The relationship between the two aspects mentioned above tells us that the informatics methods need to be developed based on molecular biological features. Examples are the related data structure, information representation, informatics operators, formalized algorithmic mechanism, logic operation and processes for bioinformatics knowledge, and the working principle of a cell that acts as a machine built by nanobiotechnology.

Bioinformatics basically handles the informatics issues of biological systems. Here, we study it in the molecular domain. The boom of molecular bioinformatics was triggered by the Human Genome Project, with the Human Proteome Project following it. Its contents include studies on the informatics issues involving DNAs, RNAs, proteins, cells and other biological functions at the molecular level. The methodology is systems biology. The computational tools are grounded in the radical changes in biology: "Transformation from a data-poor to a data-rich field began with DNA sequence data" [10]. The huge amount of data needs to be efficiently processed by powerful computers and IT technology. Networking is an important tool for bioinformatics, examples being gene regulatory networks, modeling of metabolic pathways, signal transduction pathway networks, and interactome networks. Devices such as microarrays and automated machines are also advancing the operations of biochemical experiments. Of course, studies on evolution and the work in the field of mathematical biology have also contributed to molecular bioinformatics. Findings on the organic moleware structure of DNA, RNA, and proteins are mainly obtained by the structural biology technology such as molecular imaging.

The phylogenetic network is a concept that refers to the gene relationship in living beings. We can use a generalized structure for describing the different molecules—DNAs, RNAs, proteins, and many other molecules—in complex cellular systems. The algorithms in molecular bioinformatics are diverse owing to the essence of the different problems that need to be solved. This section gives a formal description of pathways in cells, which bridge the abstract operator in biomolecular computing and the data structure in molecular bioinformatics. The tree and graph data structure is frequently used in molecular bioinformatics. By unifying the data structure of molecular bioinformatics as a graph, the extracted common part is formulated as the abstract form of formalized operators in the graph representation. One explicitly presented example is the cellular pathway, which is the basis of logical inference and informatics modeling in the next section.

In a word, the intersection of biomolecular computing and molecular bioinformatics is the abstract operators in the graph data structure under the condition of formalized constraints extracted from the biochemical signaling process in moleware systems.

7.3.1 Example: Describing Cellular Pathways by Graph Rewriting

In biomolecular computing, a feasible experimental process is carried out by biochemical reactions. The process is a kind of information updating on the nodes and edges, where the nodes denote the molecules and edges denote the reactions. Normally, the graph is a directed graph. Extended from the biochemical mechanism of biomolecular computing, the formal model of biomolecular computing inspired by molecular features is regarded as a generalized parallel information processing system in the graphs that describe the information flow. The graph structure is adopted for generalization. In Chapter 6, graph rewriting was presented as a formal model of biomolecular computing. Here, we discuss how graph rewriting is used to describe pathways in cells.

As a highlighted theme in the field of genomics and proteomic bioinformatics, the modeling of pathways with complex behaviors due to nonlinear interactions remains one of the most important tasks. With the goal of obtaining a systematic understanding of pathway-centered "systems biology," "Pathway logic," and the MAUDE system. Among these methods, only "pathway logic" directly handles pathway modeling. Its application in EGFR pathways is given in [17]. The Maude system [18] can realize pathway logic by adopting the function of "rewriting." In the Maude system, symbolic forms are used to describe the pathways, and the corresponding graphs are displayed based on the logic functions.

Because abstract modeling strategy can be adapted for various detailed descriptions in bioinformatics, an abstract form is beneficial for systematically modeling pathways and for studying the logic relationship among pathways, such as ontology. Formal methods for abstract machines such as automata can provide a continuous computing process in theory and are easy to program based on the powerful technologies of AI, which have been continuously augmented for many years. When equipped with biological verification and high-speed computer hardware, automatic logic systems derived from formal models and formalized operators may efficiently work for a fast bioinformatics analysis, and in a sense, this logical system may act as a software-robot for an automatic assembly line.

The biochemical reactions of signaling pathways can be conditionally described in the form of graph rewriting. The cellular pathways can be formalized as follows:

Object ::= molecules
Operation ::= interaction
(Data) Structure ::= graph or hypergraph

The interactions of molecules include the pathways used to describe the input-output relationship among molecules. The operation of rewriting is mainly used to describe, construct, and update the related molecular networks, and logic forms can be applied to pathway modeling. One-dimensional representation is a string, and two-dimensional representation is a graph or hypergraph. The relation among different molecules can be modeled as a kind of rewriting (e.g., string rewriting or graph rewriting).

Toward the goal of pathway analysis in bioinformatics by computer simulation, we need to develop conceptual knowledge at the symbolic level and quantitative calculation at the numeric level. Considering the common points of algorithms in biomolecular computing and molecular bioinformatics, data structure and operators on a graph are the major forms for pathway description.

A formal system defined as an instance of graph rewriting is helpful for unifying pathway models, since graphs can provide efficient representation and related semantic explanation. Also, graph rewriting permits dynamic operations on graphs. In theory, graph-rewriting systems have strong advantages in computability, operability, and explicitness for representing pathway knowledge, as explained below.

1. *Computability:* From graph-rewriting systems, certain kinds of equivalent string rewriting systems with Turing computability can be conditionally generated to provide universal computation. This is important and implies that computation can be carried out in this kind of model. Pathway simulation can then be conducted on a sound basis of theoretical computer science.
2. *Operability:* The basic operators of graph rewriting systems are edge replacement and vertex replacement. They directly operate on the pathways and can be easily used to interact pathways together in the synthesis of pathway networks.
3. *Explicitness:* Graph and graph rewriting are explicit representations of pathways and their relationship via networks. Equivalent automata can be designed to realize autonomous computing forms when logical operators are defined for graphs and networking.

With these advantages, formal descriptions in theoretical analysis and object-oriented programming designed for the development of a practical system can support bioinformatics methodologies in terms of theoretical computer science.

In the study of signaling pathways in cells, it is worthwhile to work out a unified formal method based on a molecular mechanism, where the signaling mechanism of pathway models and the bioinformatics structure of simulated pathways are consistent. To achieve such consistency, a graph-rewriting method for pathway simulation is discussed in terms of corresponding formal characteristics, representation schemes, and simulation issues. This is one of the efforts being made toward automatic analysis for bioinformatics by integrating elements of multi-disciplinary fields into computer-aided systems.

Here, we discuss the basic concepts of graph rewriting from the molecular level to the abstract level. At the molecular level, the biochemical reactions of pathways in cells can be described in the following graph series. In Figure 7.8(a), two atomic (indivisible) pathways are illustrated. In Figure 7.8(b); the corresponding representation of the integrated system is given, but the internal structure of this system remains in a black-box state. After the internal structure is defined as shown in Figure 7.8(c), the interaction can be designed as that in Figure 7.8(d).

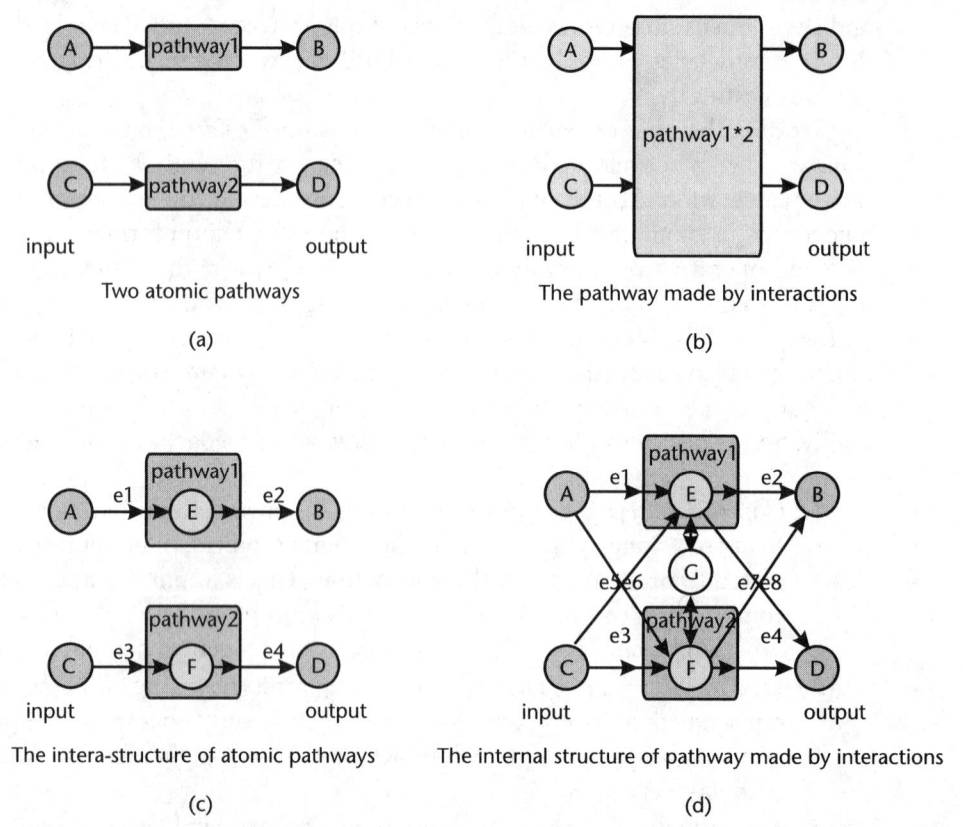

Figure 7.8 Graph-based representation and operation: (a) two atomic pathways, (b) the pathway made by interactions, (c) the internal structure of atomic pathways, and (d) the internal structure of the pathway made by interactions.

On the abstract level, the corresponding formal model can be constructed as a graph rewriting system.

We define the graph rewriting in a formal way as

Let $G_r = \langle V_r, E_r \rangle$

where

A—the alphabet set;
V_r—the set of vertexes;
E_r—the set of edges.

Consequently, from the above model, we can derive the following writing processes in graphs:

$G_r \rightarrow G'_r$ s.t. Q

where G_r and G'_r are graphs.

This formula is limited to the graph rewritten process by Q. Then, we can obtain the following rewriting rules for graph rewriting:

The rewriting operators:
    ER for e5–e8
    VR for G_r,

where
        ER = edge replacement
        VR = vertex replacement

Based on the data structure of graphs and the operators of graph rewriting, spatial and temporal hypotheses on signal transduction can be inferred from known experimental results, and a corresponding simulation result is clearly needed for experimental verification. The theoretical analysis and simulation of pathway networks is a bridge between bioinformatics and experimental biology. In a more generalized meaning, data structure is still one of the central topics in molecular bioinformatics [10–12].

7.4 Exploring Logical Description for Molecular Bioinformatics Based on Formalization and Abstract Operations

The common parts of biomolecular computing and molecular bioinformatics are mainly involved in the abstract level of molecular information processing. By using the abstract form from formalization, the operators and data structure are conceptually integrated into a unified logical description. With logical means, it is easier for us to deeply think about the interaction among molecules, which is helpful for giving us a complete image of systematic knowledge on the common mechanism in informatics.

Introducing logical methods and reasoning technology to molecular bioinformatics is a new idea. As [13, 14] suggested, the "abduction" logical method is helpful to proteomics study in terms of complex systems. If we briefly review the history of logic, the origin of thought on logic can be traced to Aristotle's time. The representation form, logical value, and inference model are concrete logical components to choose for specific applications. Normally, commonly used logic forms include induction, reduction, and deduction from the viewpoint of methodology. Three forms are substantially different in the inference mode. Induction is a process to summarize the set of truth from the set of all collected data. Reduction is a process to infer a conclusion from the general verified truth. Deduction is a process to produce hypotheses from partial truth and partial data. In bioinformatics studies on cellular signaling, a difficulty we often face is the lack of experimental data to describe the entire pathway network in the cell. Abduction-logic brings us hopes on solving this problem. The mathematics definition of abduction logic or reasoning method can be found in [15]. Before the abduction-logic method is used to discover new knowledge from proteomics data, we have to (1) obtain theoretical knowledge on the rules that describe the object, and (2) quantitatively extract the features from the data that fit the description of the objects.

In molecular bioinformatics, logical processes are used as an abstract way to generate new hypotheses based on the informatics description of the biological objects. The logic description is constructed for modeling complex cellular systems. The cellular signaling mechanism is discussed for the purpose of applying a nanobiomachine to molecular computing. The pathway network is the data structure, and a rule-form is used for the description of signaling protocols. The derived informatics issue here becomes how to infer unknown rules by interacting known pathways. The mathematics problem is the probability measurement for the reasoning operations, which can presumably be studied by applying methods from different disciplines: data mining or knowledge discovery in database (KDD) and machine learning. Data mining is a technology that extracts the desired knowledge from a huge amount of data. Machine learning is a kind of artificial intelligence method that imitates the human learning function. Data mining or KDD and machine learning have broad-ranging applications.

From the mathematics definition in Wikipedia [15] and the sense of biological applications by Hiroyuki Matsumoto, Sadamu Kurono, and Naoka Komori, several conceptual different forms of "abduction logic" have been given [13, 14]. The example presented by Umberto Eco [16] makes it easy for us to understand how the abduction-logic process works for inference. He said, "In the case of hypothesis or abduction there is the inference of a case from a rule and a result: All the beans from this bag are white—These beans are white—These beans are from this bag (probably)."

Matsumoto et al. [14] asked, "What is abduction? Rule: gene X is translated, protein Y is produced, protein Y produces peptidmassfingerprinting from (y1, y2, y3, ...) Result: The product of protein A produces the peptidmassfingerprinting from (y1, y2, y3, ...) Case: protein A is coded by gene X."

Here the case refers to a hypothesis. The abduction inference process explained in the second example is aimed at inferring new knowledge from experimental data. The data provided by genomics and proteomics is rapidly increasing. Facing the huge amount of experimental data, it is natural to develop computer-aided logic tools to accelerate the data processing. In the situation of pathway study, the signaling pathways in cells discovered by experiments are still part of the whole pathway network of cells. Normally, knowledge of atomic pathways observed under the in vitro condition is completely known to us. The logical relation of signaling proteins involved in these pathways is clear. The word "atomic" means that each component of the pathway with known structure in the biochemical reactions is a signaling protein that can be detected and regulated. According to the current state of pathway networking, an entire pathway network is not available. The networking process depends on hypothesis and verification. The hypothesis is made by empirical speculation and/or theoretical inference, and the verification is made by experiments through benchwork. Computer simulation provides a reference for making hypotheses.

In computational bioinformatics, we can represent the programming form of the abduction-logic that corresponds to the above example as follows:

Rule A: X->Y, Y->(y1,y2,y3,...)

where X is gene, Y is protein, (y1,y2,y3,...) is peptidmassfingerprinting, -> denoting the causal relation.

Result A: A-> (y1,y2,y3,...)

where A is protein.

Case A: X->A

From these items, the inference process is formulated as:

Result A → Case A

where → denotes the "infer" operation.

The generalized form of abduction-logic is formulated as follows:

Rule set = {rule1, rule2}

rule1 = X -> Y
rule2 = Y -> Z

A new rule A->Z is added to the rule set. Then, the new hypothesis is represented in a rule form:

X -> A

Obviously, this is only an inferred result.
To extend this rule to a generalized form, the rule set is defined as

{rule-1, rule-2, ..., rule-n}

rule-1: X1->Y1
rule-2: Y1-> Z1
...
rule-l: A1->Z1
...

rule-n: ...
It is inferred that

X1 -> A1

In a general description, the formal system of abduction reasoning in terms of rules is a link graph. This logic method is applied not only to the above inference process but also to other data structures in molecular bioinformatics.

By considering the reasoning represented in a rule form IF condition THEN action, X -> Y in the example of the abduction-logical process, the node can be described.

The operation X -> Y is described as a link. The logical inference routes consist of a direct graph for the network-like information flow.

The logic process is related to the structural information representation at the conceptual level. Accordingly, the data structure of a graph needs to be embedded in the logical description. The integration of logical inference and graph operation is adopted for the abduction logic process. Considering that the tree is degenerated from a graph, a general form of the graph inference process is formulated as follows:

graph (t+1,n+1) = graph(t,1), graph(t,2), ..., graph(t,n)

s.t. constraint

where graph (t+1,n+1), graph(t,1), graph(t,2), ..., graph(t,n) refer to graphs.

The constraint limits the operations of graph inference. The probability of the rule created by hypothesis is estimated by the empirical data. The logical process based on the graph structure is helpful for programming.

In applications, the rule x -> y can be adopted as different functions of the inference. If we define the branch of inference as binary, the binary tree can be generated from the graph structure of inference corresponding to all of the samples in the search space. This assumption on the application connects the rule to the data structure.

In the generalization form of logic systems, the rules correspond to signaling pathways in which the nodes are defined as items.

For example, assume that there exist the rules:

X1 -> Y1

X1 -> Y2

Y2 -> Z1

Y2 -> Z2

where the nodes of X1, Y1, ..., Z2 refer to the items in the rules of generalized abduction logic.

The path

X1 -> Y2 -> Z1

thus refers to the solution we expect.

Hypotheses will be implied according to the corresponding abduction-logical operator. The graph is assigned estimated probability, which is regarded as weight.

With the weighted graph, the inference becomes a nondeterministic decision-making process based on probability measurement.

As an instance of the logical operation based on the graph, we briefly explain the inference route in informatics as shown in Figure 7.9. Let G be a graph and G1 and G2 be subgraphs. Assume that G1 in G is known and describes the known pathway denoted as net (X, W, Y), whose input set is X, output set is Y, and the set of internal nodes is W, where x, y, z belong to X, Y and Z. If we have y and z in hand, it is inferred that x in G2 is a new hypothesis according to the rule X -> Y. G2 is reconstructed through the sets of X, Y and Z. The link X -> Y implies the relationship between x and y. The hypothesis on G2 is assigned by a probability obtained through estimation or prediction.

Extended from the previous example of Matsumoto et al. on the logical inference at the molecular level, the rules are formulated as:

IF
(two pathways A and B show similar features in a certain pattern)
AND
(pathway B is regulated by pathway C)

THEN
A hypothesis on the regulation from pathway C to pathway A is inferred.

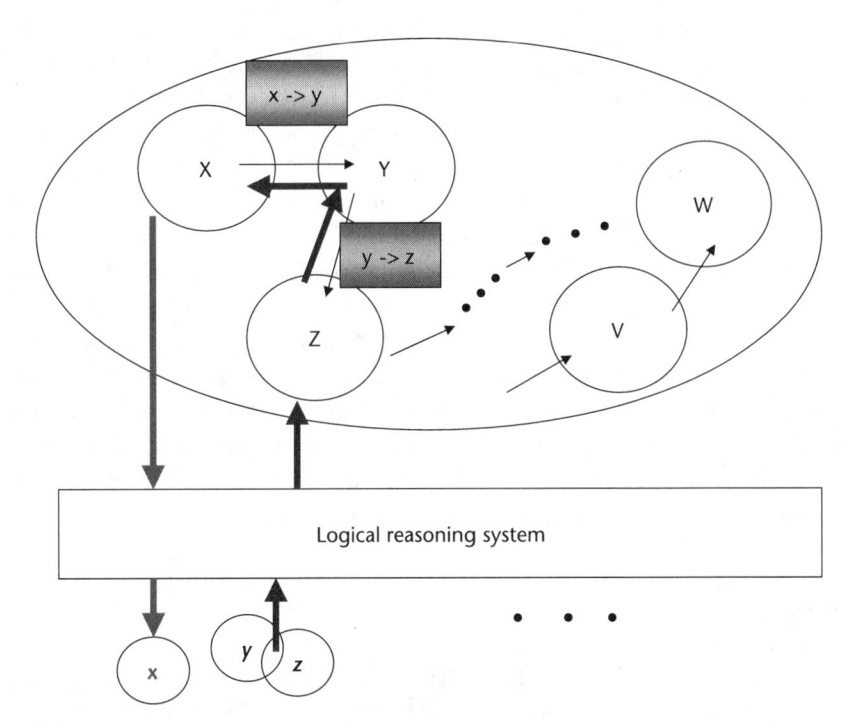

Figure 7.9 Set description of rule-based abduction inference.

For interaction of three pathways with molecular switch function, i.e. pathway i, pathway j, pathway k, we have:

IF
(pathway i is in the downstream of pathway j)
AND
(there is an intersection of the input pathway i and the input of pathway k)
AND
(there is an intersection of the output pathway i and the output of pathway k)
where pathway i and pathway k share same or similar function

THEN

a hypothesis that pathway j and pathway k share the same or similar function is inferred.

Actually, nanobioscience covers various types of molecular information processing. Molecular bioinformatics and biomolecular computing are only two aspects of the systematic study of signaling mechanisms for molecular information processing. The nonlinearity in moleware systems makes it very difficult to thoroughly understand the cellular functions. For example, the robustness [19] mechanism of cell signaling [20] is a key to the development of anti-cancer drugs. Rich knowledge on signaling pathways is rapidly increasing, providing bioinformatics resources such as STKE [21] and KEGG [22]. Our vision of signaling pathways is not limited to the G-proteins [23, 24]. It has become much more global, extending to a protein interaction network [25–27]. The quantitative analysis [28] of biochemical reactions for pathways is supported by experimental evidence on the regulation of cell signal transduction. Confirmed through measurement, computational methods such as probabilistic graphical models for inferring cellular networks [31] have joined the mainstream of bioinformatics. The issues of modeling, controlling and signal processing of moleware [30–36] need to be explored more deeply. In terms of nanobioICT, nanobioinformatics can be classified into four areas: systematic methodology based on systems science that regards a cell as a nanobiosystem, control and communication of nanobiosystems based on cybernetics, information processing based on computer science, and moleware communication based on information theory. As we have seen in this section, molecular signaling in cells is modeled as a nonmonotonic logic process. Through biomimetic engineering, synthetic biology will become feasible for building a molecular logic inference in wetware.

References

[1] Landweber, L. F., and L. Kari, "The Evolution of Cellular Computing: Nature's Solution to a Computational Problem," *BioSystems*, Vol. 52, October 1999, pp. 3–13.

[2] Ehrenfeucht, A., et al., *Computation in Living Cells: Gene Assembly in Ciliates*, New York: Springer-Verlag, 2004.

[3]   Huson, D. H., et al., "Phylogenetic Super-Networks from Partial Trees," *IEEE Trans. on Computational Biology and Bioinformatics*, Vol. 1, No. 4, October–December 2004, pp. 151–158.

[4]   Ptacek, J., et al., "Global Analysis of Protein Phosphorylation in Yeast," *Nature*, Vol. 438, December 2005, pp. 679–684.

[5]   Zeng, X., et al., "A Dual-Kinase Mechanism for Wnt Co-Receptor Phosphorylation and Activation," *Nature*, Vol. 438, December 2005, pp. 873–877.

[6]   Papin, J. A., et al., "Reconstruction of Cellular Signaling Networks and Analysis of their Properties," *Nature Reviews Molecular Cell Biology*, Vol. 6, February 2005, pp. 99–111.

[7]   Mueller, B. K., H. Mack, and N. Teusch, "Rho Kinase, a Promising Drug Target for Neurological Disorder," *Nature Reviews Drug Discovery*, Vol. 4, May 2005, pp. 387–398.

[8]   Kollmann, M., et al., "Design Principles of a Bacterial Signaling Network," *Nature*, Vol. 438, November 2005, pp. 504–507.

[9]   Woodson, S. A., "Assembly Line Inspection," *Nature*, Vol. 438, December 2005, pp. 566–567.

[10] Gusfield, D., Introduction to the *IEEE/ACM Trans. on Computational Biology and Bioinformatics*, Vol. 1, No. 1, January–March 2004, pp. 2–3.

[11] Gusfield, D., *Algorithms on Strings, Trees and Sequences: Computer Science and Computational Biology*, Cambridge, U. K.: Cambridge University Press, 1997.

[12] Dress, A. W. M., and D. H. Huson, "Constructing Splits Graphs," *IEEE/ACM Transactions on Computational Biology and Bioinformatics*, Vol. 1, No. 3, July–September 2004, pp. 109–115.

[13] Matsumoto, H., S. Kurono, and N. Komori, "Proteomics Approach to Complex Signaling Systems," *Sebutsubutsuri*, Vol. 43, No. 6, 2003, pp. 270–274, in Japanese.

[14] Matsumoto, H., S. Kurono, and N. Komori, "Proteomics and Abductive Inference by C.S. Peirce," *Sebutsubutsuri*, Vol. 43, 6, 2003, pp. 291–294, in Japanese.

[15] http://en.wikipedia.org/wiki/Abductive_reasoning.

[16] Eco, U., *A Theory of Semiotics*, Bloomington, IN: Indiana University Press, 1979.

[17] Talcott, C., et al., "Pathway Logic Modeling of Protein Functional Domains in Signal Transduction," *PSB 2004*.

[18] http://www.csl.sri.com/projects/maude/.

[19] Kitano, H., "Systems Biology: A Brief Overview," *Science*, Vol. 295, March 2002, pp. 1662–1664.

[20] Special issue, Mapping Cellular Signaling, *Science*, Vol. 296, May 2002. Various types of signaling pathways have been discovered.

[21] *Science's* Signal Transduction Knowledge Environment (STKE), http://www.stke.org.

[22] KEGG PATHWAY Database, http://www.genome.ad.jp/kegg/pathway.html.

[23] Kaibuchi, K., S. Kuroda, and M. Amano, "Regulation of the Cytoskeleton and Cell Adhesion by the Rho Family GTPases in Mammalian Cells," *Annu. Rev. Biochem.*, Vol. 68, 1999, pp. 459–486.

[24] Etienne-Manneville, S., and A. Hall, "Rho GTPases in Cell Biology," *Nature*, Vol. 420, 2002, pp. 629–635.

[25] Han, J. D. J., et al., "Evidence for Dynamically Organized Modularity in the Yeast Protein-Protein Interaction Network," *Nature*, Vol. 430, 2004, pp. 88–93.

[26] Li, S., et al., "A Map of the Interactome Network of the Metazoan C. Elegans," *Science*, Vol. 303, 2004, pp. 540–543.

[27] Giot, L., et al., "A Protein Interaction Map of Drosophila Melanogaster," *Science*, Vol. 302, 2003, pp. 1727–1736.

[28] Koshland, D. E., Jr., "The Era of Pathway Quantification," *Science*, Vol. 280, 1998, pp. 852–853.

[29] Friedman, N., "Inferring Cellular Networks Using Probabilistic Graphical Models," *Science*, Vol. 303, February 2004, pp. 799–805.

[30] Wolkenhauer, O., et al., "Control and Coordination in Biochemical Networks," *IEEE Control Systems Magazine*, Vol. 24, No. 4, August 2004, pp. 30–34.

[31] Rodriguez, S., et al., "Modular Analysis of Signal Transduction Networks," *IEEE Control Systems Magazine*, Vol. 24, No. 4, August 2004, pp. 35–52.

[32] Shvartsman, S. Y., et al., "Epidermal Growth Factor Receptor Signaling in Tissues," *IEEE Control Systems Magazine*, Vol. 24, No. 4, August 2004, pp. 53–61.

[33] Khammash, M., and H. El-Samad, "Systems Biology: From Physiology to Gene Regulation," *IEEE Control Systems Magazine*, Vol. 24, No. 4, August 2004, pp. 62–76.

[34] Paliwal, S., et al., "Responding to Directional Cues: A Tale of Two Cells [biochemical signaling pathways]," *IEEE Control Systems Magazine*, Vol. 24, No. 4, August 2004, pp. 77–90.

[35] Schmidt, H., and E. W. Jacobsen, "Identifying Feedback Mechanisms Behind Complex Cell Behavior," *IEEE Control Systems Magazine*, Vol. 24, No. 4, August 2004, pp. 91–102.

[36] Ezzell, C., "Proteins Rule," *Scientific American*, Vol. 286, No. 4, April 2002, pp. 26–33.

CHAPTER 8
Emerging Nanobiotechnology in Multiple Disciplines

The direct motivation of research on new molecular information processing systems by nanotechnology is to make things smaller and smaller. The physical mechanisms at the nanolevel are the direct target for us to shoot at, and the material science and technology are direct tools toward this eternal target. The biological systems in nature provide us some outstanding examples in nanobioscience and nanobiotechnology. The applications of nanotechnology have been extended from the most commonly mentioned field of molecular electronics by nanotechnology to the medical applications of nanobiotechnology. Along this line of thinking, several examples of new progresses in the multiple disciplines involved in nanobiotechnology are singled out in this chapter [1–31], which is the main supporting technology for molecular computing. It is arranged into two major parts. One part consists of Sections 8.1, 8.2, and 8.3 and it is on interfaces of biomolecular computing with molecular electronics along the tendency of nano-electronics and miniature. The other part consists of Sections 8.4 and 8.5 and it is on the perspectives of biomolecular computing inspired by the emerging nanobiotechnology, which provides functional configuration of spatial structure for us to explore nanobioICT systems that may go beyond the limitation of traditional technologies.

8.1 The Tale of Two Media: Molecular Electricity and Biomolecular Signaling

Nanotechnology originated from the material science that is oriented to miniature materials. Developing nano-size electronic circuits is consistent with the tendency of semiconductors, in which the size goes down to less than 100 nm. Nanotechnology for molecular electronics is to innovate new circuits as small as possible. Their application is so broad that it is impossible to enumerate all of them.

In concept, molecular electronics refers to the electronics where the molecule is used to design, test, and implement circuits. At the nanolevel, inorganic/metal molecules and organic molecules have shown the electricity effect. Biomolecules also conditionally connect to the electricity mechanism. This contributes to the medical engineering of nanobiotechnology owing to the fact that chemical signals of organic molecules are related to electricity in moleware and can be controlled for biomedical engineering at the moleware level. Basically speaking, a transistor built by moleware is the first step for molecular electronics. At the circuit level, the electrical signaling of logical gates is generated by molecular mechanisms. AND, OR, and NOT are the basic units for combinatorial circuits. The most fundamental device for molecular electronics is the molecular transistor because the transistor is one of the most basic building blocks that are normally used for designing electronic circuits (see Figure 8.1).

8.2 How Small Can an Information Processing System Be Made?

Our thinking should not stop at molecular electronics. We should continuously go closer towards the bottom of the material world than R. P. Feynman ever predicated. As the size of the material is decreased into nanoware, atom is a component that could be the nanomedium where the fundamental particles within atoms are the smallest elements that consist of matter known to humans. The fundamental particles—quarks—have interactions among them in physics. String or super-string theory is one of the most promising theories that quantitatively describes the field whose measurement determines the physical behavior of the material structure at the very bottom of matter structure.

The quarks are the elements in an atom, which are connected by a string. The force exerted on these elements is produced by the quantum field within atom. The fundamental particles within the atom are the most elementary materials for us to consider the extreme small information processing system in nature that is possibly capable of being observed technically.

How to use the subatomic physical structure of quarks to represent information and to define the operators on the information are the prerequisite for subatomic computing (see Figure 8.2). The structure of the string is shaped in a topological form. The computing form considered here is the information updating

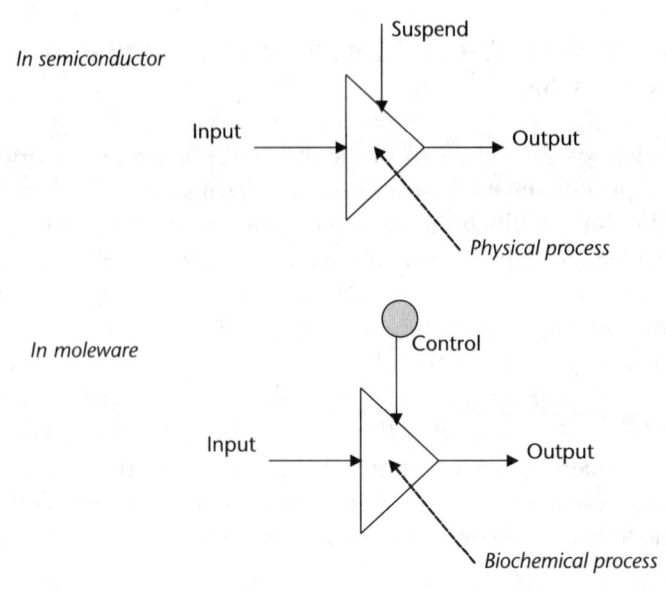

Figure 8.1 Schematic comparison of a transistor in semiconductor and moleware.

8.3 Informatics of Porphyrin Systems

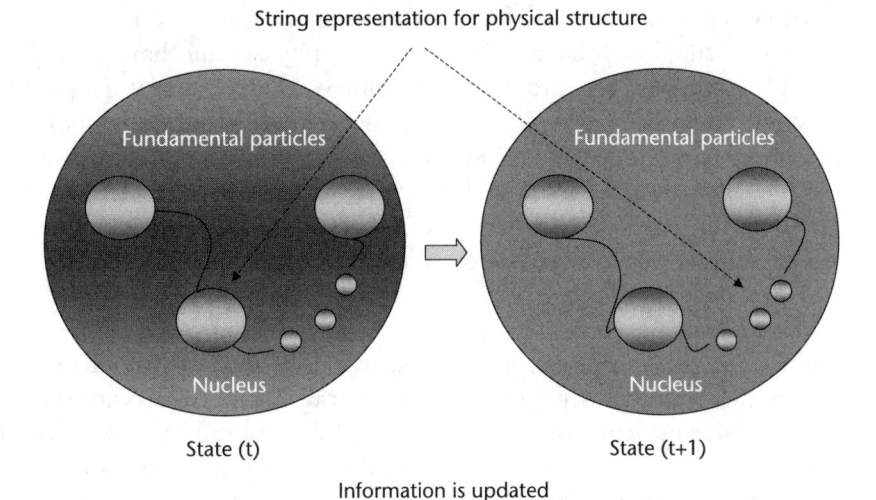

Figure 8.2 Information representation by fundamental particles and its updating.

in terms of the measurement of the string state. The formula of physics comes from quantum filed theory. The issues on detection, measurement, and control are still challenging.

We should mention here that quantum computing is promising and is involved in the level of physics where quantum mechanics is the main principle of the information processing. Experimental processes on quantum computing devices, mathematics-based computing models, and algorithms are the supporting basis for the discipline of quantum computing as a mature science and a realistic technology. Electron is the major medium in quantum computing where the rotation directions of an electron—clockwise and counterclockwise—are defined as the binary values 1 and 0. The physics theory for quantum computing and the subatomic computing we presented above is different. Quantum mechanics is the physics principle of quantum computing. The quantum field theory is the physics principle of subatomic computing where the string theory is helpful for scientific hypothesis and speculated models.

If quarks-based subatomic computing becomes available in the future, it is natural to integrate and/or interface the computing within an atom and computing outside an atom. In this sense, the unit size of an information processing system including computing will be the most crucial factor for the development of supernanosystems, where information description is the basis of the whole idea.

8.3 Informatics of Porphyrin Systems

Porphyrin is a wonderful material that has drawn lots of attention in a broad scope of material science. The reason of employing porphyrin in nanotechnology is mainly due to the fact that porphyrins and metalloporphyrins can be used to fabricate nanowires and build nanocircuits for molecular electronics. One of the most obvious physical phenomena brought forth by porphyrin technology is photosynthesis, which can contribute to the studies on natural systems for the purpose of realizing optical communication processes by molecular information processing. The computing potential of the naturally originated physical mechanism of porphyrin system [1] is the key for us to design the moleware structure of the porphyrin system by connecting the porphyrin-made circuits for molecular electronics with nanowire for molecular computers and molecular communication systems.

The physics structure and chemistry features of porphyrin have been intensively studied in chemistry experiments [1]. Porphyrin is an organic molecular compound and its shape is shown in Figure 8.3. The elements of porphyrin are carbon (C), hydrogen (H), and oxygen (O). Metalloporphyrin is porphyrin complex where a metal element is embedded into the center of the porphyrin strcture. As illustrated in Figure 8.3, the metal element is embedded into the center of porphyrin that makes the organic porphyrin capable of transferring electricity. This is the basic biochemistry principle of nanowires made by porphyrin. In the viewpoint of chemistry applications of porphyrins and metalloporphyrins, the combinatorial form of the metalloporphyrins has been used to build nanowire.

The potentials of porphyrin structure in applications include (1) molecular electronic; (2) photonics; and (3) possible medical applications.

1. *Molecular electronics:* In the application of porphyrin to building nanowire, the direct merit of the nanowire made by porphyrin is the capability of electricity transferring. One of the important reasons for applying porphyrin in electronics is that the porphyrin system is good at flexibility of changing the structure of the molecular compound and controllability of the derived specific structure configuration. In unconventional information processing paradigms, information is represented in moleware form in which the biochemistry characteristics determine the tools of conceptual design and

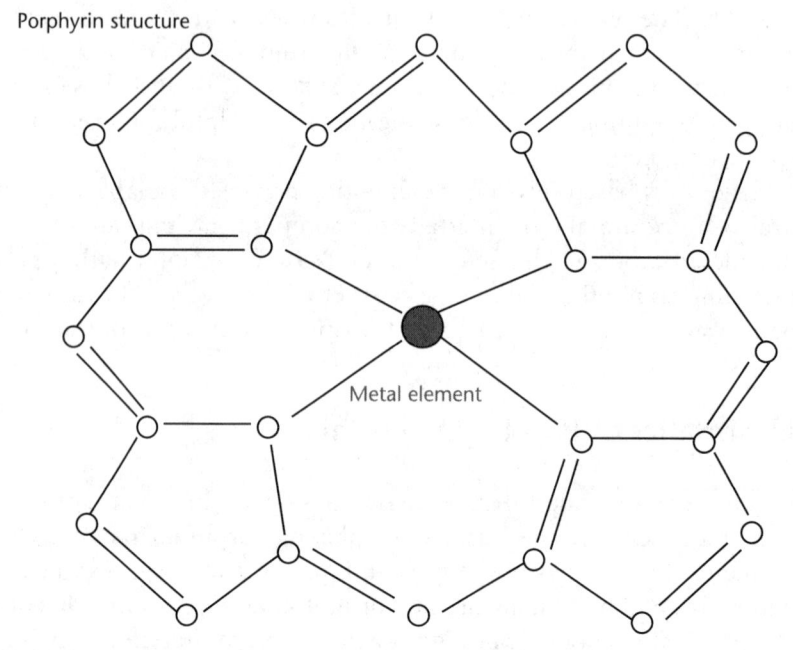

Figure 8.3 Basic porphyrin structure.

experimental test. Thus, the rich flexibility of porphyrin structure allows various kinds of configurations of information processing systems by releasing the great power of the organic moleware architecture to technological applications needed by electronics industry. The porphyrin-based electronics could become a kind of green electronics to protect the natural environment. This kind of artificial photosynthesis is a kind of nature-inspired nanosystem, which is similar to the photosynthesis in plants.

2. *Photonics:* Artificial photosynthesis systems can generate light under control, development of self-assembled porphyrin systems may lead to connecting moleware systems with light communication. The molecular electronics and molecular signal processing systems built by porphyrin can directly work for the communication engineering, which could enhance the performance of moleware systems especially in the speed of molecular information processing and communication. The motivation of the application of porphyrin systems to optical communication is the photosynthesis function of artificially synthesized porphyrin systems is beneficial for the porphyrin circuits connected to electronic circuit and light transportation. Since light communication has the merits in speed of signal transmission, it is amazing that the high-speed molecular devices for light communication can synthetically work with the commonly adopted IT infrastructure.

3. *Possible medical applications:* Porphyrin is also a kind of molecular compound and is involved in the medical studies and is expected to be applied in drug discovery and medical therapy. The result reported by Hudson et al. shows the porphyrin's role in tumor-developing mechanism of cells [2]. The basic principle of molecular manufacturing of the porphyrin structure is self-aggregation. The term self-aggregation refers to the assembly process of connecting the basic functional building block of molecular complexes into desired specific macro-structures by nanotechnology. The water soluble porphyrin has good performance in flexibility of supramolecular manufacturing and moleware system synthesis. As reported in [3–5], synthesis of artificial prophyrin systems under the water-soluble experimental condition is important for applying porphyrins to wetware systems. Cyclodendrix (CD)—a kind of molecular structure—can be used as a template for molecular design and can also be integrated with the template of porphyrin structure we already discussed previously with certain modification. The topological configuration of molecular complexes is the technological basis for us to work out structural solutions to the supramolecular manufacturing. A schematic description of the molecular manufacturing process for synthesizing porphyrin complexes is illustrated in Figure 8.4.

The patterns that can be generated by the current available synthesis technology for porphyrin systems [6, 7] decide the information processing capability of porphyrin systems constructed in engineering [8, 9]. The shape and size of porphyrin is controllable. The spatial pattern of porphyrin provides rich amounts of structural information for molecular bioinformatics. By using the

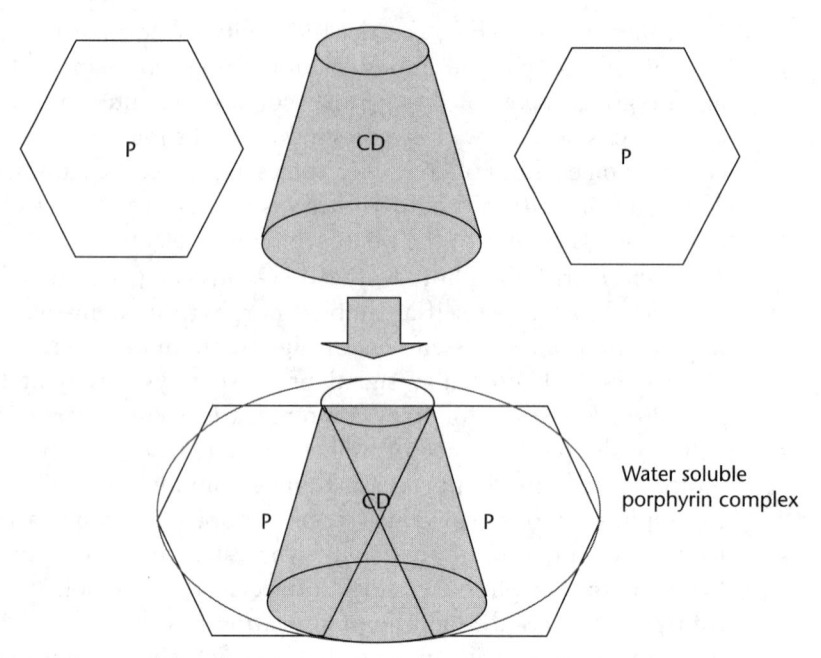

Figure 8.4 Water soluble porphyrin complex from structural templates.

information of the mapping from the physical mechanism mentioned above to the abstract representation of information processing, the informatics model of porphyrin systems will reflect the spatial patterns of complexed porphyrin structures where the surface is involved with metal elements (e.g., Au) and the aligned molecular structure herringbone. The chemistry technology in benchwork is the main method for porphyrin synthesis. Metalloporphyrin is regarded as a pattern template by which the embedded metals derive various types of information forms by molecular structures. No doubt that the material of porphyrin can be used to fabricate nanowires wherein metal elements such as Mn, Fe, and others act as components for transferring electricity through the backbone-like component within the corresponding porphyrin system. The structure of porphyrin with embedded metal elements also shows a kind of physical structure for information storage if we notice that the symbols represented in the notation of the metal elements are located in three-dimensional structures of porphyrins. This porphyrin structure is constructed by the self-assembly in material science. The reason for the difference of this kind of porphyrin-based information processing system from electronic computers is that the topological shape of the molecular complexes affects the physical attributes in which the semiconductor-made circuits are homogeneous. In contrast to existing electronic computers, the porphyrin structure embedded with different metals is heterogeneous. In the aspect of molecular operations, the informatics unit of the metalloporphyrin complex is defined as the configuration of the porphyrin structure in which a metal element is embedded in the center of the porphyrin complex. The porphyrin metal complex is formed by the integrated mechanism of the organic structure and metal element. There are two factors that directly influence the feasibility of building porphyrin synthesis systems, which are: (1) informatics patterns extracted from the integrated structure of porphyrin metal and (2) the availability status of experimental nanotechnology for building specific structure equipped with desired functions. Considering the current status of porphyrin-oriented technology, informatics analysis of the porphyrin patterns is a theme for informatics study on synthesized porphyrin systems. After we work out the information representation scheme by the complex system of porphyrins and corresponding operators required by the derived molecular structure, computer simulation may act as a design tool for molecular manufacturing in laboratories. Although a gap exists between the idealized model of the porphyrin-based model and the reality of feasible technologies of porphyrin synthesis, the theoretical work on the issue of designing theoretical models and possible algorithms for porphyrin-based information processing is still challenging.

The most direct result obtained from the combinatorial porphyrin structure is formal language. The set of metals used in the porphyrin structure includes Zn, Fe, Cu, and other metal elements [1]. If two metals are selected from the above set and denoted as A and B, two porphyrin structures where metal A and B are embedded can be denoted as P(A) and P(B). The structural components P(A) and P(B) are connected in turn as

$$P(A)—P(B)—P(A)—P(B)— \ldots —$$

The sequence extracted from this self-assembly process is

ABAB...

which is also denotaed as $(AB)^n$.

Through the three dimensional structure of the porphyrin-metal complex, the spatial location of metals in porphyrin structure generate certain patterns corresponding to formal languages such as the one formulated above—$(AB)^n$. Although a gap exists between possible information patterns derived by the porphyrin structure and practical technology for manufacturing the patterns speculated, the molecular nano-informatics analysis of the porphyrin structure is a necessary step for building porphyrin-based nanobioICT systems in the near future. The self-assembly of porphyrin for nano-architecture and detection of the molecular signals from the metal elements embedded in the related porphyrin structures can be used to represent information and derived information flow within porphyrin systems.

A porphyrin computer in which the molecular operators are designed for connecting structural porphyrin units will have the nanometer size and high controllability of the reconfigurable organic/inorganic structure. The experimental evidence for supporting the feasible self-assembly and self-coordination operation in porphyrin systems can be found in [7]. Anyway, it is one of the technological choices for us to develop the porphyrin structure embedded with the metal elements [1], which is regarded as the building blocks to design complex patterns of structural metalloporphyrin systems when we are exploring future nanobioICT technologies.

8.4 Transition from the Supporting Points to Integrations of Different Aspects of Molecular Information Processing

From the eyes of structuralism, the atom-to-atom relation for molecular structure and observing single molecules are one of the typical examples of bottom-up studies on physical matter structure. Atom-to-atom bonds [e.g., carbon-carbon (C-C) bonds] could be designed by physics theory where the molecular structure is desired to obtain the shortest bond length [11]. The atom level structure is also useful information for understanding protein-to-protein interactions. In signaling molecules in cells, the activation status of GTPases is reflected on the location of molecular complexes where the GTP/GDP is attached. The physical principle in a more generalized sense is on the level of topological shape of the molecular complexes for biochemical reactions according to the knowledge of material science [12].

A molecular motor built by dynesin is a moleware device functionally similar to the one made by myosin we once discussed in Chapter 4. The working principle of nanomachines made by molecular motors is that the molecular movement of molecules is enforced by biochemistry operations and is sustained by energy feeding mechanisms. When we broaden our vision to the naturally existing molecular movement mechanism in cells, molecular movement controlled by the Golgi apparatus can be regarded as a post office for monitoring the molecular "postman" we once discussed in Chapter 4. To build an interface between the above nano- and bioaspects of nanobiomachines is a difficult task for molecular system integration. The self-assembly in nanotechnology and self-organizing in biotechnology are two key technologies for connecting the natural molecular mechanism for nanoengineering usage and the artificial manufactured molecular mechanism for biofunction application. Single molecule technology is one of the bridges between nanotechnology and biotechnology. In biomolecular computing, a molecule is the unit for information representation and operation. In the case of DNA computing, the DNA molecules directly describe the information by A, T, C, G, which can be mapped into codes. The situation in RNA computing is similar to the DNA ones. As we know, protein has a more complicated spatial structure than DNAs, from which protein functions are often predicated. There are two routes to investigate functional proteins on their bioinformatics features. One route is to go to the direction to explore a much smaller size of moleware as we have discussed above in nanotechnology. The other route is to go to the direction of systematic biology in biotechnology. The nanobiotechnology is the strong-bond for integrating the two aspects. The current technology on a single molecule measurement and controlling is rapidly advancing [13]. The single Ras molecule is normally observed by imaging technologies. When activated in cells, the fluorescence color of the Ras molecules changes from green to red. The interactions of Ras and other molecules happen in less than one minute [13]. The control of timing is useful in nanobiotechnology. The technology of single molecules has been developed for membrane receptors and signaling molecules in living cells [13].

The signaling processes are explained by the fence model and the picket model for dynamics of the corresponding cellular activities. Since the relation of Ras to cancers, the medical application of single molecule technology on Ras is significant and also contributes to the neuroscience study [14]. Backtracking to the molecular

8.4 Transition from the Supporting Points

motor built by proteins, a nanobiomachine is also a possible tool for developing a kind of nanobiorobots within living cells in the future.

As mentioned, the principle of fluorescence imaging on molecules, the models and working hypothesises, as well as the technology that helps us understand the basic idea of the single molecule technology should be merged into a more complete framework of knowledge on quantitative measurements of signaling proteins in cells. Ras is one of the signaling proteins in cells that can be detectable by the fluorescence method whose principle is to make the protein emit light and then to detect the light in moleware mode. The fluorescence method is to use fluorescence molecules to combine the molecules and detect the fluorescence effect of these molecules for measurement. The immunofluorescence method is the fluorescence method where the immunomechanism is applied by using the antigen to combine with the molecules to be detected. In the fluorescence method for detecting molecules, some specific molecules also may be used (e.g., the gold particle—goldcoroid). The probe is the kernel in the instrument technology whose principle is shown in Figure 8.5.

The term "domain" in structural biology, more exactly "structural domain," is the part of the molecular complex that shows the features for the biochemical reactions between the molecular complex directly used for molecular signaling and other concerned molecules in cells. The protein domain is the key for protein interaction in the structural biology level. By connecting the domain part of protein kinases with photoproteins, a phosphorylation effect is generated. The parts of known domains are: SH2 (Src homology region 2), PTB (phosphotyrosine binding domain)/PID (phosphotyrosine interaction domain), SH3, PH (pleckstrin homology), and so forth.

Here, the components such as module, motif, and other patterns are helpful in understanding the structure and function of proteins. Alpha, beta, and gamma helix are three types of the protein structure where major and minor structures exist. The patterns look like building blocks from the viewpoint of bioinformatics.

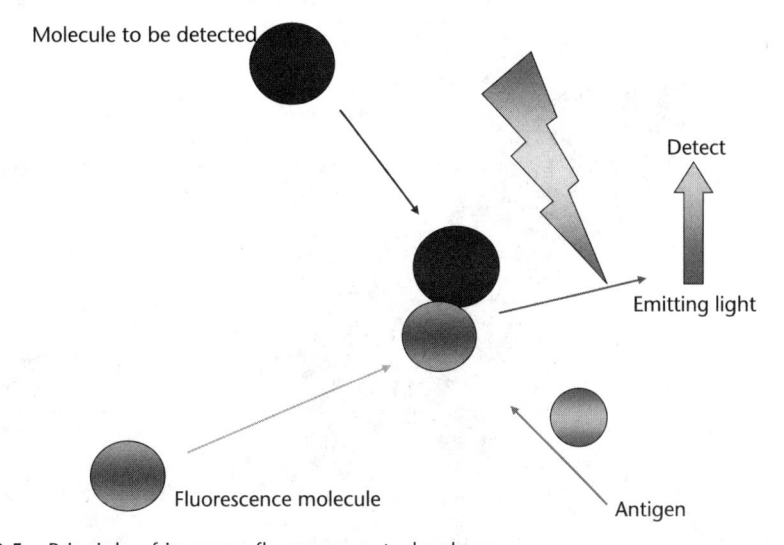

Figure 8.5 Principle of immuno-fluorescence technology.

In the biophysics level of structural stability in proteins, there exist only two types of protein structures according to the capability of the protein reaction with the water—easy or difficult. The imaging method for obtaining the 3D structure of proteins has changed the landscape of computer-based informatics methods for searchability and prediction of moleware.

The high levels of abstract models are also helpful in the exploration of molecular bioinformatics. Graph rewriting on topological structures could be used as an abstract modeling strategy for generating spatial structures to represent the amino acid (sequence) in three dimensions. Here, "3D" refers to spatial locations of the molecules/atoms of the molecular complexes; temporal/time refers to the biochemical reaction process for biofunction/bioinformatics/computation; and function refers to symbols of molecules and the activation states of the corresponding molecules in signal transduction. Designing the rewriting operation on the topologically constrained graph relies on the bioinformatics knowledge.

The three dimensional structure of proteins (see Figure 8.6) that includes kinases and G-proteins are often imaged and measured in order to understand the biochemical features of the proteins in structural biology. The amino acid sequence with domains/motif is one dimensional in symbols and with biological functions for the cells. The two-dimensional description of the bioinformatics mechanism is an operable tool for understanding the coding mechanism of signal transduction. Here, coding refers to the information that is reflected by the protein structure mapped to the corresponding biological function in cells. The bioifnormatics modeling is useful to control processes needed for the Golgi apparatus and vesicle transport in cellular trafficking [23–25]. To summarize this reported evidence, protein molecules are controlled by the Golgi apparatus and move from ER near cell membranes towards nucleus which is a transportation process in cells. In order to transport the molecules, vesicles carry the cargo molecules that are uploaded on

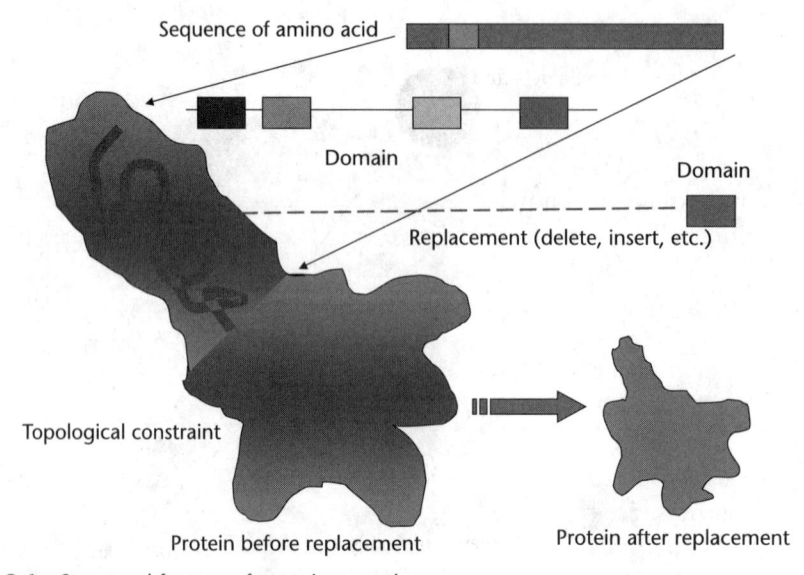

Figure 8.6 Structural feature of protein complex.

vesicles and deliver them to the destination within the cell. The phosphorylation pathways are involved in the molecular transportation processes. In some cells, G-protein is also involved. Two different structural parts of vesicles—*cis* and *trans*—give the different function for begin/end of working of cargo carriers and loading/unloading of the cargos [25]. The cargo transportation process is accompanied by diffusion mechanism from higher intensity to lower intensity in physics. Parallelism in informatics is generated by the simultaneous activation of multiple molecules of molecular complexes.

8.5 Cell Communication for Engineering Purpose

Telecommunication technology is practical and sufficiently mature for real-world applications. Messages are often transmitted by certain units according to the quantity of information contained in the messages. Packet is a kind of unit for this usage. The term "packet" defined by IEEE 802.11 [26] refers to the set that consists of the messages transmitted in the communication processes and is used in wireless ad hoc networks where the standards for networking are well understood when we observe the network performance and behavior. The situation in cell communication is different.

Telecommunication:
   The communication processes are well understood.
   The communication protocols are designable and operable.
   The signaling mechanism for information processing and communication is controllable.

Cell communication:
   The communication processes are less understood.
   The entire communication protocols can not be designed completely.
   Part of communication protocols mainly at the local level is possible.
   The signaling mechanism for information processing and communication is less operable and less controllable.

To some extent, the knowledge of telecommunication processes is the reference for modeling cell communication processes. The biological features of cell communication mechanisms are also beneficial to new ideas for designing telecommunication systems. For the purpose of communication by nanobioinformatics, it is meaningful to compare the telecommunication protocols with the cell communication behavior where information representation and transportation are made by moleware. The kernel signaling mechanism of a wireless ad hoc network described by the protocol of IEEE 802.11, may give us some hints on the logical description of the signaling processes for cell communication that starts from the beginning of sending messages to the end of receiving messages within cells.

The signal transduction simulation for cellular functions can be modeled regardless of the locations (average concentration rather than local concentration except the single molecule). In simulation, the Michealis and Menten equation for reaction, the diffusion equation, and the reaction-diffusion equation are needed for quantitative measurement. The simulated result reported by Kuroda et al. shows that interaction of known pathways can be modeled in order to explain new phenomena appearing in wetware experiments [27]. It is necessary to use computer simulation as a tool for theoretical analysis and empirical synthesis. It is also necessary to introduce the mechanism of the signaling process ranging from sender to receiver into the modeling schemes of biochemical reactions around the membrane. The hypothesis for the communication protocols and "information-theoretic" measures requires the quantitatively description of the performance of communication processes represented in curves. The temporal features of the signaling processes are observed in the signaling process of phosphorylation/dephosphorylation. The applications of coding schemes are oriented to objects (e.g., the technologies are different for images and speech signals). In the case of molecular communication by cellular signaling pathways, the signal transduction of GEFs/GAPs and kinases/phosphatases are crucial for cellular activities. The historic role of signal transduction is witnessed in the field of biology [28]. In engineering, transducer refers to changing the form of the molecular signals (e.g., light to electricity). In mechatronics, signaling processing is also related to the bioinformatics system whose features include molecular motors and other biomechanics processes. Here, the feature could be a kind of generalized pattern that is more generalized than the term "pattern" in pattern recognition and classification.

In the level of protocol, the MAPK (mitogen activated protein kinase) is typical for a signal cascade in systematic vision for cellular communication. The single molecule detection technology makes it possible to observe the single molecule. The information about the locations of the biochemical reactions has to be achieved from physical measurement. The well-defined behavior of the IEEE 802.11 protocol tells us how to make communication processes efficient [29] and those protocols are hoped to possibly be used to study the moleware communication systems. In signaling processes of cell communication, especially the intra-cell communication, normally the curves concentration versus time obtained by experiments or calculated by software simulation directly give the information of the protein interaction. This means that in concept a protein can have biochemical reaction with a protein under the condition of specific enzymes (GEFs/GAPs) in cells. Under the condition in vitro and in vivo, the performance of individual proteins can be studied relatively easily. The location of molecules can be observed by advanced imaging technology. Normally, local concentration of molecules in cells is seldomly taken into consideration. In the future, the detection problem of local concentration will be solved by developing accurate measuring tools. By a single molecule technology, we can theoretically discuss communication processes between molecules in the biophysics level.

In Chapter 7, Figure 7.1 describes the information flow among molecular signaling systems. If we go further on this, we may notice that the message transferring process is carried out through the molecules. It is interesting to describe the molecular information processing behavior as a kind of protocol-like way. The communication behavior of molecular communication processes depends much on the protocols that are represented as the rule form to define the message-transmission behavior. The molecules moving from the membrane boundary of cell to the internal space of cells are well studied in the field of biology. The cargo molecules in concept refer to the molecules in the cell that carry materials that look like cargo. During the molecular signaling process from membrane to the nucleus, the cargo molecules are involved in the cell communication process through the biochemical reactions. The Golgi apparatus is a kind of controller or monitor for the molecular movement processes. The primitive description for molecular operation and logical description for molecular communication are the formal informatics terms to understanding the related functions of molecular signaling in cells. The basic idea of molecular communication protocol is to describe the causal relation of the signal cascade. Normally, the pathways are regarded as a kind of hierarchical structure. The terms "upstream and downstream pathways" refer to the different position of the pathways in the signaling cascades. The meaning of "up" and "down" is relative. These two kinds of directional information are normally assigned from a viewpoint of signaling flow. In the signal cascade, the relative locations of molecules in the upstream and downstream pathways are the prerequisite condition for us to understand the functions of the molecules involved in the pathways. If we define the start point or information flow source at certain location of membrane, the flow will be formulated from membrane to the nucleus where the membrane is up and nucleus is down. Here, the other parts of the cell are omitted for explicit description. A description at the level of single molecule could be presented as follows:

membrane-molecule (): activated;
molecule-move (from:=membrane, to:=location-X);
reaction (reactants:= molecules, location:=X).

In the case of single molecule, the molecule that represents the message moves towards the destination. The signal is denoted by the corresponding molecule. The corresponding signaling process could be studied in terms of the communication protocol model. This means that the signal represented by a single molecule expresses the information in the information sequence under the control of moleware communication systems built by cells. From information in moleware, we could analyze the behavior of cellular communication in terms of protocol.

8.5.1 From Bit Level of Information Representation to Observe Cellular Communication

At the level of single molecule for cellular communication, the unit of message is represented by one bit. By this mode, the communication protocols in telecommunication can be lent for us to analyze the behavior of the message-sending process. By borrowing the commonly used IEEE 802.11, we can model the cellular communication in a quantitative way. One of the direct cellular communication routes is defined as the process where a molecule is sent from the place around the membrane to the place outside the nucleus. Checking the messages received in the destination place is a necessary task. It is noticeable that feedback signals like response signals need to be produced to confirm whether the message is correctly received or not at the sender side. This is expected to be generated by molecular recognition and control in the in vitro cell culture whose simplest form is using purified protein and individual control in test tubes. The message for communication per molecule is accumulated in a certain period to form a package in information. The bitwise communication mechanism is used as the fundamental unit of parallel molecular information processing in cells.

8.5.2 The Biophysical Effectors of the Molecular Information Flow

In the sense of biophysics, the reaction phenomenon can be directly simulated by the Michaelis-Menten equation where the location information within the cell is often neglected. The diffusion is a persuasible factor to explain the location-dependent signaling process. The constraint of the diffusion speed is exerted on the molecular movement. In the aspect of informatics, we regard the diffusion process as a passive process. The gradient factors of the molecule concentration and the fluid dynamics of the cargo molecules affect the speed of the molecule movement. Basically, quantity of molecular flow in the cell communication is crucial for the molecular movement in moleware communication. The medium is regarded as channel in the sense of information theory and communication engineering. The delay of the molecular delivery and jam of the directed molecular movement should be prevented. The saturation of the reaction in concentration constrains the speed of the molecular communication as well as the result of the message-sending in moleware. In the case of in vivo cells if possible (as it could be), the concentration based moleware communication requires the external control exactly manipulated in engineering. Originated from the essence of the informatics on the molecular level, the symbolic and analog descriptions are expressed by the names of molecules and the quantity of molecular concentration, respectively. The symbolic mode is good for efficient coding and the analog mode is good for introducing digital signal processing into molecular communication. The discrete form given by the molecules fits the scope of the coding theory that is well known for us. The filtering function for concentration signals will contribute for the quantization, synchronization, and "hand-shaking" signal in protocol. The reaction effect is mainly observed in the location of destination and diffusion effect is mainly observed in the channel. The derived biochemical characteristics we have to consider for efficiently monitoring the communication process is the capacity of the moleware channel. In the moleware channel, the stochastic process theory and calculation of probability theory are useful for estimation of molecular channel measurement and control of the communication function. For example, the queue theory and Poisson process are beneficial for approximation of the proper moment of controlling/monitoring in the sense of martingale.

8.5.3 Effects of Molecular Protocols by the Internal Components of Cells

We can observe the behavior of molecular communication by two extreme examples of the cell communication inspired framework including single molecule and molecular aggregation. The natural information mechanism in the cell is open to us at many instances. In the sense of cargo carrier, the Golgi apparatus and mitochondria are two outstanding instances. When we observe the physical process of molecular signaling in the Golgi apparatus, the information flow can be described by signaling pathways (e.g., Rab and ARF pathways). The Golgi apparatus is interacted with rough-surfaced endoplasmic reticulum. The vesicle—a molecular complex with 0.1-µm diameter is the medium that is regulated by the Golgi apparatus in order to monitor the intercell communication as a postoffice. The major function of postoffice is to collect, classify, and deliver the message. The Golgi apparatus is a kind of information processing device in the sense of nanomachine where the information is coded by moleware. The related primitive can be given in logic form:

Vesicle (): message

This program-like representation is mapped from the biochemical reaction of the molecular signaling processes. The direct informatics rules used to design the protocols are the proper arrangement of the information flows in moleware. The constraints to the corresponding processes are biochemical. Any efforts on improving performance of controlled molecular signaling processes need both good solutions to optimize the information channel in moleware and experimental techniques of noisy reduction for molecular signaling processes.

8.5.4 Control Nodes in Moleware Communication Networks

In moleware communication, the codes are interpreted as the signaling cascades in cells with diverse biological functions. Based on the internal mechanism of cell communications in nature, molecular traffic has to be controlled in order to regulate moleware communication among nanobiomachines. The behavior of signaling mechanisms of cellular communication whose medium is molecular movement looks like a person who's walking and talking to others and some of the words could be lost in the communication channel. Here the solution will be a kind of compromise of the expected performance and realistic technology.

Owing to the high degree of unknown dynamics in cell communication, some parts of the information may be lost caused by the mobile feature and ad hoc characteristics of the sender or the receiver in the molecular signaling network. The self-organizing mechanism of cell growth makes the cellular signaling robust and adaptive for the biological functions. The concentration of signaling molecules can be controlled to the "required" quantity within the domain that corresponds to normal physiological functions of organisms. The engineered moleware communication and related computing is a little different from this in the instinct mechanism in which the specific message have to be delivered in a specific moment. In order to realize the nanobiocommunciation process in the sense of communication engineering, the control issue is the first compulsory work for us to tackle. The complete information is necessary to be defined and transported in moleware. To check the status of the molecules in communication, feedback, or something like a sensor is expected to employ the signaling pathways. The monitoring function is a prerequisite for the engineering operation in nanobiotechnology. This idea is part of our theoretical consideration on computational moleware cybernetics The protocol for telecommunication covers multiple layers of the communication processes. However, moleware communication and molecular information processing differs at the lower layer of communication protocols. The solution should come from the natural regulation of signal cascade in cells.

Tracing the origin of the signaling mechanism of cells, different forms of cellular activities are related with different functions such as the differentiation, development, regulation, and sustaining of cells. The signal transduction of phosphorylation/dephosphorylation and GTP/GDP-bound switches directly set the states of the signaling proteins at the biochemical states with structural characteristics of molecular complexes. In these two processes, the energy transformation between ATP/ADP and GTP/GDP (in some sense) is involved, respectively. In the signal transduction of the phosphorylation/dephosphorylation process, the electron transfer phenomenon also occurs. This electron transfer does not form any regular electricity as observed in electrical circuits. However, this phenomenon offers a kind of possibility of electricity generation in moleware. The oxidization is coupled with phosphorylation. The oxidative phosphorylation exists in mitochondria, which is one of the major organelles in cells for energy production by external input. In cells, the biochemical effects of cellular signaling processes such as phosphorylation/dephosphorylation are influenced by the biological function of organelles. The Golgi apparatus is capable of relaying the molecular signals between the membrane and the nucleus. Of course, this function happens in signal cascade also. Any codes and meaningful information for computing and communication should be selected in a specific order and controlled according to the states of moleware. The different molecules that consist of information flow are the basis of the MIMD architecture for parallel information processes.

Naturally, we can conclude that the central problem is how to curb the natural molecule flow and tailor it into a kind of engineered moleware flow to build a nanobioICT system. Decoupling the crosstalks among cellular pathways in the theory of information science and in the practice of nanobiotechnology is the key to the success of controlling the cellular signaling processes. The controlled effect will be determined by engineering technologies and experiment skills in molecular biology and material science. It seems that the microreactor is powerful in the control of molecular signals under the in vitro condition. Under the in vivo condition, we probably need to develop a kind of biomolecular nanotube to extend the application of nanobiomachines to the field of molecular devices. Control of the molecular signals under the in vivo condition depends on the system-level nanobiotechnology for living cells. The nanotechnology in the synthesizing of molecular materials is mature, but the devices are still in the infancy, which leaves many themes open.

8.5.5 Collision-Avoid: An Issue on Efficiency of Moleware Communication in Cells

The kernel of the engineered cell communication process is how to design the communication protocols required by the moleware communication in nanobioICT. The communication engineering technologies are easy for us to understand the communication principle in concept. The biological world nowadays is new for communication sciences. Owing to the identity of communication in channel behavior, the signaling process of IEEE 802.11 could be used as a reference for us to quantitatively understand the intracell communication process in engineering. The channel performance of cellular communication is our concern here. The cell-to-cell communication process is mainly regulated by the signaling pathway within the cells and obeys the signal-molecules-to-membrane-receptor mode basically. When we focus on intracell communication processes, ad hoc networks could be references for modeling moleware communication systems. The signaling mechanism of the cell communication process is efficient for local signaling within cells as well as short distance communication of cellular nanobiomachine. This is consistent with the application of ad hoc network for local and short distance of telecommunications. In communication systems, the minimum requirement for the performance of communication processes is the correct message delivered from the sender to the receiver. The collision is a phenomenon that two nodes require the same resource within the communication network and the message can not be delivered to the destination node. This phenomenon is caused by the physical limitation of the same network resource required by the multiple services in the network (from users). The collision-avoid in telecommunication is crucial for releasing the stronger power of networking. In moleware communication, the collision problem is described as two molecular information flows entering the same node of a molecular information processing unit. In concept, two input signals occur in the same node. This phenomena is called collision in moleware communication, which means that the conflicted signals are caused by corresponding channels interacted in the node where the signals pass through.

Different from the physical conditions in telecommunication networks, the structure of the molecular signaling network in cell communication systems is changeable. The moleware mechanism of cellular signaling affects the biochemical features of cell communication. The molecular level of cell communication involves two aspects of information processing functions, which are single molecule and molecular complex. The single molecule message is used to describe the bit-information. One molecule corresponds to one bit of information. This mode is similar to the bit in the frame of information transported by packages in telecommunication. The molecular complexes and/or the aggregate set of the molecular complexes correspond to the information package in telecommunication. Thinking about the message in the moleware channel, the central idea of information collision-avoid schemes in moleware communication is efficient activation of the enzyme proteins and regulation of signaling pathways to maximally transmit the messages through moleware. In molecular bioinformatics, this problem becomes how to dynamically activate and regulate the pathway network for delivering the message coded by moleware.

The molecular message is transmitted through the molecular channel. The network structure of the pathway gives good collision-avoid schemes owing to the concurrency control of parallel molecular information processing in cells. It is a kind of channel intersection and information processing within the nodes of the communication system. In concept, the central idea of collision-avoid in telecommunication is how to allocate the information processing source to handle the transactions from users. In mobile wireless ad hoc networks, the dynamical features of information flow generate nonlinearity in the communication process.

As an example to explain the collision-avoid schemes, we briefly discuss the hidden node problem in ad hoc networking. The problem means that two terminals search for a node that is desired to provide their service of communication. The terminal's behavior is limited and looks like a kind of nearsighted vision. The node handling their request needs other nodes of the telecommunication network for service, which causes the collision when the same node is requested within the domain. The node here is a hidden node that is not aware of the network directly from the standpoint of terminal side. The reason for this phenomenon is mainly because of the dynamics of the telecommunication networking. We are required to study protocol design based on the features of source and channel codes to solve this problem because information representation is dependent on the dynamic characteristics of molecular signals.

A complete nanobioICT system that includes the information source and the channel needs well-developed nanotechnology. It is currently still in the stage of principle. In addition to the source codes in moleware in Chapter four, channel codes are crucial for improvement of communication processes in a nanobioICT system. The collision-avoid is one of the most fundamental protocol-level schemes for molecular networking. The conceptual description is given in the Figure 8.7. As it is shown, the information encoded in moleware is independent on the distribution of the channel signals in molecular denotation (symbols). This will be very helpful for efficient design of molecular channel.

The physical mechanism in cells depends on the internal cellular structure. The cellular organelle such as the Golgi apparatus has the function of base station BS (access point AP) in ad hoc network. The molecules around the cellular membrane are relatively easily handled by current detection and measurement technology. The knowledge of the cargo molecules as postmen and the Golgi apparatus as post offices in cells provide feasible solutions. The exact place and local concentration are direct

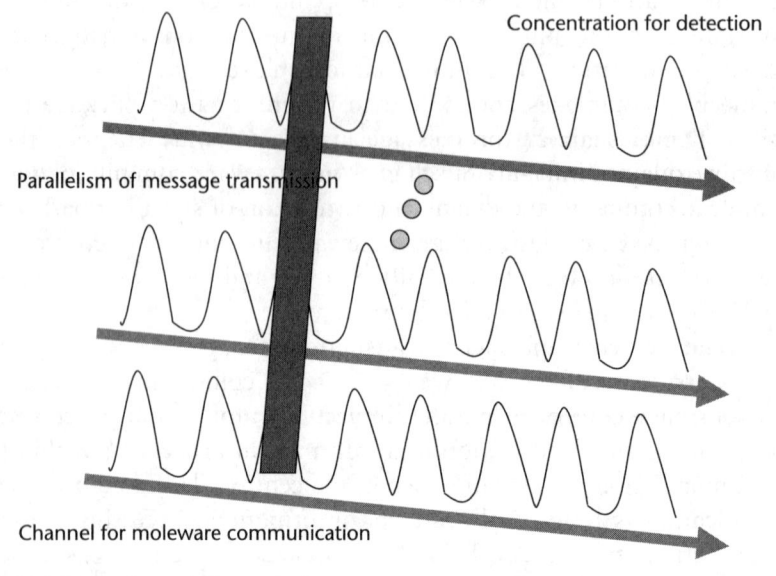

Figure 8.7 Multiple channels for moleware communication.

measurement for a single molecule. In order to describe pathways at the level of nanomachines, we need to make a mapping from a single molecule to the molecular set in which the molecular concentration can be controlled according to the specific mode of signal transduction obtained. The symbolic representation is made by the denotation of molecules. The different bit with order is assigned with molecules. The states of the molecules are used to represent information. The message transmission for sender-to-receiver process is a kind of key-hole mode. The information medium is heterogeneous, which is different from the homogeneous structure of telecommunication networking. We can detect a single molecule by current technology. Actually, the same type of molecules are recognized as the same entity to provide the same information. Basing the information representation schemes on the concentration level is sufficient for molecular operation and measurement. The physical location (within the cell) where molecular signals are controlled is defined as the node for the molecular networking process. Hinted by the collision concept in telecommunication, the collision in the molecular node refers to the phenomenon that the same molecular information processing unit has to handle two transaction of communication in moleware. This phenomenon can be described by the form of interaction of molecular channels. As we discussed before, the molecular message from the sender to the receiver needs a medium to support the molecular communication process. If the medium is the cell, the collision problem in moleware communication can be studied in terms of the interactions of moleware channels. One of the solutions we proposed here, concerning the merits of adaptation of signaling pathways in cells, is to reconfigure pathway structure by transforming an individual unit of a single node into an integrated unit constructed by multiple nodes of moleware communication systems, where each molecular signaling process for communication is modeled as a channel. The pathways in cells have a network structure and are capable of being combined into an integrated form of pathways. Thus, the individual node X in the pathways is defined as "macronode" and can be divided into two other nodes corresponding to the channels. This structure is a dualism of two subpathways for two channels. The node X is modeled as a macronode that consists of micronodes Y1 and Y2. This kind of integrated pathway structure allows two molecular signaling mechanisms under the activation of enzyme proteins. Each node consisting of pathways is extensible (extendable) to multiple channels under the condition that the channel is modeled as the temporal signaling process. The central idea of this node-division method is briefly explained in Figure 8.7. It is obvious that the structure of the node is reconfigurable according to the dynamic environment. The complexity (i.e., the number of the enzymes required for controlling the pathways) increases in a linear order because multiple pathways can be controlled by the common enzyme proteins when each unit is constructed by the same set of the molecules representing the informatics elements for moleware communication. The multiple channels for molecule flows are parallel where biochemical reactions are simultaneously carried out and can be regarded as a concurrent information processing and communication system in the form of MIMD/MIMO architecture. When each unit consists of multiple molecular processing modules, the enzymes for the activation of the pathways of the signaling molecules can activate multiple pathways so that the complexity of the control does not increase in bioinformatics. This signal activation mechanism in a cell is unique in the sense of the multiple biomolecular information flow for cellular pathways.

The biological features greatly influence the behavior obtained from cell communication. Although the bioinformatics knowledge of cellular pathways is much more rich than a decade ago and the complete pathway description can be found in the database such as KEGG and publications, the existence of various types of biological functions makes it difficult to formalize a certain abstract description for them. The informatics criterion for the bioinformatics description is normally suggested for functional analysis of cellular pathways. In the case of cellular pathways (e.g., Ras pathway), the signaling mechanism of the cancer generation should be studied and possible anticancer approaches could be innovated as well. The activation of Ras with mutation that causes the cancer is the direct hint for medical drug design and anticancer solutions. The inactivation operation of Ras is for anticancer that may contribute to the development for apoptosis—programmed cell death. The medical application of signal transduction is obviously because of the history of biology.

The high order description of nonlinear behavior of the cellular signaling fits the requirement of the signal transduction model for cells. This modeling is beneficial to control by analysis and synthesis. In principle, the engineered moleware control mechanism is the key to analysis-based synthesis. Technically, detailed work needs to be done for different parts of the above working model and derived prototype design.

How to connect the individual mechanism synthesized in material science and structural biological systems existing in nature efficiently for biomolecular information processing is a crucial task. Only the data structure and information description method is not enough. The artificial structure requires engineering control technology for nanolevel operations and biomolecular signaling. The methodology of computational nanobiomoleware-cybernetics is expected to make breakthroughs for integration of theory and implementation whose impact could be extended to medical applications. The complexity emerging from systemizing processes gives us a hint about the deeply hidden relationship among the nanobiosystems. Although different levels of biomolecular computation systems are physical, chemical, biological, and informational, systems of biomolecular computation can be unified in terms of informatics. Seeking a unified design method for biomolecular computation systems to go beyond the limitation of conventional engineering technology for designing moleware system becomes more methodological. "Tao Te Ching" says [30, 31]:

The Way [Tao] that can be told of is not an Unvarying Way [Tao];
The names that can be named are not unvarying names.

"Tao" refers to the "way," generally understood as "method/methodology," and "name" can be "description" or "explanation." As it implies, a new paradigm for information, communication, control and synthesis of nanobiosystems is needed when facing the unexpected phenomena of nanobioworld.

References

[1] Kadish, K. M., K. M. Smith, and R. Guilard, *The Porphyrin Handbook*, Vol. 8, Academic Press, New York: 2000.

[2] Hudson, R., et al., "The Development and Characterization of Porphyrin Isothiocyanate-Monoclonal Antibody Conjugates for Photoimmunotherapy," *British Journal of Cancer*, Vol. 92, 2005, pp. 1442–1449.

[3] Kano, K., "Molecular Complexes of Water-Soluble Porphyrins," *Journal of Porphyrins and Phthalocynines*, Vol. 8, 2004, pp. 148–155.

[4] Kano, K., et al., "Anion Binding to a Ferric Porphyrin Complexed with Per-O-Methylated ß-Cyclodextrin in Aqueous Solution," *J. Am. Chem. Soc.*, Vol. 126, 2004, pp. 15202–15210.

[5] Kano, K., et al., "Dioxygen Binding to a Simple Myoglobin Model in Aqueous Solution," *Angewandte* (A journal of the Gesellschaft Deutscher Chemiker, International Edition) (Angew. Chem. Int. Ed.), Vol. 44, 2005, pp. 435–438.

[6] *Proceedings of 2nd International Conference on Porphyrins and Phthalocyanines*, Kyoto TERRSA, Kyoto, Japan, June 30–July 5, 2002.

[7] Yokoyama, T., et al., "Selective Assembly on a Surface of Supramolecular Aggregates with Controlled Size and Shape," *Nature*, Vol. 413, October 2001, pp. 619–621.

[8] Weiss, P. S., "Nanotechnology: Molecules Join the Assembly Line," *Nature*, Vol. 413, October 2001, pp. 585–586.

[9] Yokoyama, T., et al., "Selective Assembly on a Surface of Supramolecular Aggregates with Controlled Size and Shape," *Nature*, Vol. 413, October 2001, pp. 619–621.

[10] Ogawa, K., and Y. Kobuke, "Formation of a Giant Supramolecular Porphyrin Array by Self-Coordination," *Angewandte, Chemie, International Edition*, Vol. 39, No. 22, 2000, pp. 4070–4073.

[11] Siegel, J. S., "Brevity for Bonds," *Nature*, Vol. 439, February 2006, pp. 801–802.

[12] Ostojic, S., E. Somfai, B. Nienhuis, Scale Invariance and Universality of Force Networks in Static Granular Matter, *Nature*, Vol. 439, February 2006, pp. 828–830.

[13] http://www.nanobio.frontier.kyoto-u.ac.jp/.
Related topics can be found in: http://www.nanobio.frontier.kyoto-u.ac.jp/lab/guideline.html, http://www.nanobio.frontier.kyoto-u.ac.jp/lab/gaiyo/j.html, http://www.nanobio.frontier.kyoto-u.ac.jp/#slide, and http://www.nanobio.frontier.kyoto-u.ac.jp/lab/slides/3/j.html.

[14] http://www.med.nagoya-u.ac.jp/Yakuri/.

[15] Blair, S. A., et al., "Epithelial Myosin Light Chain Kinase Expression and Activity Are Upregulated in Inflammatory Bowel Disease," *Laboratory Investigation*, Vol. 86, 2006, pp. 191–201, published online January 9, 2006.

[16] Lange, A., et al., "Toxin-Induced Conformational Changes in a Potassium Channel Revealed by Solid-State NMR," *Nature*, Vol. 440, April 2006, pp. 959–962.

[17] Papin, J. A., et al., "Reconstruction of Cellular Signaling Networks and Analysis of Their Properties," *Nature Reviews Molecular Cell Biology*, Vol. 6, February 2005, pp. 99–111.

[18] Mueller, B. K., H. Mack, and N. Teusch, "Rho Kinase, a Promising Drug Target for Neurological Disorder," *Nature Reviews Drug Discovery*, Vol. 4, May 2005, pp. 387–398.

[19] Kollmann, M., et al., "Design Principles of a Bacterial Signaling Network," *Nature*, Vol. 438, November 2005, pp. 504–507.

[20] Woodson, S. A., "Assembly Line Inspection," *Nature*, Vol. 438, December 2005, pp. 566–567.

[21] The Worldwide Protein Data Bank (wwPDB), http://www.wwpdb.org/, which includes RCSB PDB (USA) : http://www.rcsb.org/pdb, MSD-EBI (Europe): http://www.ebi.ac.uk/msd, and PDBj (Japan): http://www.pdbj.org/

[22] Albert, B., et al., *Molecular Biology of the Cell*, 4th ed., New York. Garland Science, 2002.

[23] Howe, C. L., "Modeling the Signaling Endosome Hypothesis: Why a Drive to the Nucleus Is Better Than a (Random) Walk," *Theor. Biol. Med. Model*, 2005.
http://www.pubmedcentral.nih.gov/articlerender.fcgi?artid=1276819.

[24] Kholodenko, B. N., "Four-Dimensional Organization of Protein Kinase Signaling Cascades: The Roles of Diffusion, Endocytosis and Molecular Motors," *Journal of Experimental Biology*, Vol. 206, 2003, pp. 2073–2082, http://jeb.biologists.org/cgi/content/full/206/12/2073.

[25] http://en.wikipedia.org/wiki/Golgi_apparatus.

[26] http://www.sss-mag.com/pdf/802_11tut.pdf, A Technical Tutorial on the IEEE 802.11 Protocol, http://www.ieee802.org/11/ IEEE 802.11, The Working Group Setting the Standards for Wireless LANs.

[27] Kuroda, S., N. Schweighofer, and M. Kawato, "Exploration of Signal Transduction Pathways in Cerebellar Long-Term Depression by Kinetic Simulation," *J. Neurosci.*, Vol. 21, No. 15, pp. 5693–5702.

[28] http://www.nobelprize.org.

[29] Reguera, G., et al., "Extracellular Electron Transfer Via Microbial Nanowires," *Nature*, Vol. 435, June 2005, pp. 1098–1101.

[30] http://en.wikipedia.org/wiki/Tao_Te_Ching.

[31] Waley, A., *The Way and Its Power: A Study of the Tao Te Ching and Its Place in Chinese Thought*, London, U.K.: Allen & Unwin, 1934.

About the Authors

Jian-Qin Liu is Expert Researcher at Kobe Advanced ICT Research Center (KARC), National Institute of Information and Communications Technology (NICT) in Kobe, Japan. He is the author of two books and numerous journal articles and conference papers. He earned his Ph.D. in Industrial Automation at Central South University of Technology, China and his Ph.D. in informatics at Kyoto University, Japan.

Katsunori Shimohara is Professor of Department of Information Design, Faculty of Engineering, Doshisha University in Kyoto, Japan. He received his Ph.D. in computer science and communication engineering from Kyushu University, Japan.

Index

μ-recursive function, 193
3-SAT computing, 158–59
    Chen and Ramachandran algorithm, 159
    Schöning algorithm, 158
    Suyama-Yoshida algorithm, 159
    See also Computing
3-SAT problem solving, 120–21
    by kinase computing, 162–65
    by kinase computing with interaction mechanism, 165–71

A

Abduction inference, 246, 247
    formal system, 247
    rule-based, set description, 249
Abduction-logic method, 245–46, 247
Abstract cellular pathways, 174–78
Adleman-Lipton DNA computing, 120–21
Adleman's computing system, 117–20
    bio-state description, 117–20
    logic operators, 120
    outline, 119
Amino acids, 24
Arabidopsis thaliana, 7, 35, 107
Artificial intelligence (AI), 3
Artificial life (AL), 3
Artificial neural networks (ANN), 3, 85
Atom force microscope (AFM), 40
Atomic pathways, 243–44
ATP
    defined, 84
    proteins, 85
ATPase, 94, 105
Automaton, 145–46
    computing formalized as, 188–90
    deterministic fine state, 189
    DNA, 154
    input, 189
    MSP, 191, 192, 205
    schematics description, 146
    state updating mechanism, 145
Autonomic computing (AC), 4
Axioms, 151

B

Being Digital (Negroponte), 92
Bigraphs, 203–5
Binary trees
    interference operation on, 238
    mutual effect within, 238
    transformation from pathway to, 238
Binary words, 104
Biochemical reactions
    capability, 98
    in clustered domain, 216
    dynamic procession, 83–84
    in MDCK cells, 110–12
    in molecular communication, 58
    physical characteristics, 59
    self-assembly mechanisms, 117
    of signaling from X to Y, 62
    temporal process, 201
    TTxm, 61
Biochemistry
    concepts, 11–12
    informatics and, 178
Bioinformatics. *See* Molecular bioinformatics
Biological engineering technology, 132
Biological signaling mechanism, 33
Biomolecular computing, 141–78
    biochemical features, 19
    cells for, 161
    cellular, 181–220
    compiler, 210–12
    informatics operators, 142
    informatics structure, 153–57
    molecular bioinformatics commonality, 239–45
    molecular bioinformatics conceptual relationship, 240
    molecular bioinformatics links, 223–24
    nano-level, 178

Biomolecular computing (continued)
    pathway-based, 202
    pathway structure, 200
    performances, 13–15, 18
Biomolecular computing algorithms
    characteristics, 224–27
    comparison, 223–50
    DNA computing, 225
    DNA computing by ciliates, 226–27
    H-systems, 225
    informatics, 239
    P-systems, 226
    surface-based DNA computing, 225
Biomolecular signaling, 253–54
Biomoleware, 91–136
Biophysics, 83
BlueGene/L System, 2
Bottom-up design, 163
Brute-force searching, 119–20

C

Cargo transportation process, 263
Cell communication, 129
    channel performance, 269
    for engineering purpose, 263–72
    molecular switch, 184–99
    observe, 265–66
    schematic description, 183
    single molecule level, 265
    ubiquitous, 182–84
    unknown dynamics, 267
Cells
    ATP, 84
    for biomolecular computing, 161
    defined, 29
    dynein dynamics, 31
    internal components, molecular protocol effects, 266–67
    MDCK, 102
    as nanobiomachine, 31–35
    pathways, studying, 43
    receiver, 128
    robustness, 220
    scalability, 220
    sender, 128
    signaling pathways, 35–38
    signaling protein memory in, 102–4
    signal transduction, 35–38, 84
    in vitro, 192
    in vivo, 192
Cellular biomolecular computing, 181–220
Cellular structure
    elements, 29–30
    illustrated, 30
    molecular biology viewpoint, 29
Chen and Ramachandran algorithm, 159
Chlamydomonas reinhardtii, 94
Chloroplast, 29
Chomsky hierarchy, 143
Chromosomes, 29, 130
Ciliates
    defined, 129
    DNA computing by, 226–27, 240
    gene operations, 129–31
Circadian rhythm, 107
Clustering, 216
Collision, 269
Collision-avoid schemes, 270
Communication
    DNA motor, 126
    information terms, 183
    information theory, 54
    microbial cell, 127–29
    molecular, 61
    molecular motor/molecular computer, 95
    moleware, 59, 67–76
    nanobio process, 54
    See also Cell communication
Compiler, 210–12
Complexes
    binary states, 192
    DNA, structure, 144
    FTPase, 103
    GTPase, 57, 58, 70
    kinase/phosphatase, 164
    relations among, 73
    SPK, 70, 103
Complexity
    control-space, 162
    in difficulty explanation, 157
    observing from benchmarks, 157–60
    problem, 157–58
    table, 160
Composition function
    designing, 196–97
    minimum configuration, 197
    realization, 196
Computational bioinformatics, 246
Computing
    3-SAT, 158
    automaton-based, 191–92
    autonomic (AC), 4
    by biochemical reactions in microbes, 127–36

Index

biomolecular, 13–15, 18, 19, 141–78
ciliate-based, 240
controlled, 97
DNA, 115–21
by gene operations in ciliates, 129–31
general parallel, 199–214
kinase (KC), 161
in mathematics scripts, 161
membrane, 150
nanobioICT, 51
by nucleic acids, 114–27
pathway structure for, 175
quark-based subatomic, 255
RNA, 121–23
SAT, 158
surface-based DNA, 123–25
unconventional, 2
Context-sensitive languages (CSLs), 175
Controlled computing diagram, 97
Control-space complexity, 162
Cross-talk, 68
Cybernetics
    feedback, 68
    moleware, 267
    nanobiocybernetics, 56

D

Discrete mode modeling, 83
DNA
    automaton, 154
    defined, 11
    encryption, 68
    helix structure, 101
    machine-based handling, 116
    memory operations, 101
    processing elements (PEs), 126
    recombinant technology, 133
    size, 11
    structures, 147
    templates, 125
    tile model, 144
DNA-based coding methods, 108
DNA computing, 115–21
    Adleman-Lipton, 120–21
    Adleman-paradigm, 240
    algorithm, 121
    artificially manufactured DNA sequences, 116
    cancer diagnosis, 15
    by ciliates, 226–27
    development, 1
    DNA molecules, 260
    DNA motor communication, 126
    emergence, 116
    finite state machine, 117
    formalized, 225
    HPP constraint, 225
    HPP problem solution, 118
    informational structure, 117
    logic operators for moleware, 117
    nanobiotechnology, 1–2, 125–27
    operation illustration, 127
    parallelism, 14
    self-contained devices, 117
    strings, 143
    surface-based, 123–25, 225
    test tube, 115
    See also Computing
DNA molecules, 5, 6, 14
    in DNA computing, 260
    information encoding, 68
    limited number of, 14
    spatial location, 91
    symbolic sequence, 38
    in test tubes, 99
DNA sequences, 14
    artificially manufactured, 116
    circular structures, 149
    double-strand, 98–99, 149
    Hamming distance, 108
    mapping process from, 23
    message storage, 115
    nodes/edges representation, 98
    quantitative differences, 108
    in test tubes, 99
    in Watson-Crick arrangement, 98
DNA-strand sticker model, 49
Docking, 43
Double-strand DNA sequences, 98–99, 149
Downstream pathways, 265

E

Efficiency
    defined, 160
    moleware communication, 268–72
    nanobiomachines, 31
    from pathway designs, 160–71
EGFR pathways, 242
Empirical schemes, 132–33
Epidermal growth factor receptor (EGFR), 27
Euprymma scolopes, 128
Evolutionary algorithm (EA), 147, 158
Evolutionary computation (EC), 3, 147

F

Feedback, 68
Fence model, 260
Fluorescence protease protection (FPP), 9
Fluorescence resonance energy transfer (FRET), 8, 39, 85, 93
Formalized molecular computing, 146–47
Formal language, 143
FTPase complex, 103
FtsK, 126
Functional NMR (fNMR), 40

G

Gaussian distributions, 64
Gedanken model, 10–11
GEF/GAP pathways, 103, 165, 187
   algorithm, 165–66
   algorithm steps, 166
   complexity analysis, 168
   output, 166
   quantitative description, 219
   See also Pathways
Genetic algorithm (GA), 147
Genetic programming (GP), 147
Gene unscrambling, 130
Genome informatics
   clustering, 25, 26
   outcome, 25
   success, 25
   See also Informatics
Genomics, 24, 228
Gödel number, 205
Golgi apparatus, 262
   base station function, 270
   defined, 30
   as monitor, 265
   as post offices, 270
   protein molecule control, 262
G protein-coupled receptors (oGPCRs), 42
G-proteins, 250, 263
Graph rewriting, 203
   computability, 243
   definition, 244
   explicitness, 243
   operability, 243
   operations, 204
   rewriting rules, 245
   on topological structures, 262
Graphs
   bigraphs, 203–5
   conflict, 227
   hypergraphs, 199, 203
   inference process, 248
   splitting, 228
   transformation, 203
   weighted, 248–49
GTPase
   activation state, 215
   complexes, 57, 58, 70, 190
   GDP-bound state, 186
   GEF and GAP relationship, 237
   GTP-bound state, 186
   memory, 104
   nanobiomachine energy, 105
   pathways, 32, 106, 231
   Rho family, 171, 217
   state, determination, 102
GTPase switch, 69
   biological pathways, 219
   defined, 185
   informational representation, 186–88
   See also Molecular switches

H

Hamiltonian path problem, 119, 120
Hamming distance, 108
Helicobacter pylori, 23
Heterogeneity, 55
H-system, 133, 147
   informatics view, 153–57
   rewriting operator, 225
Human genome project (HGP), 12–13, 241
Hypergraphs, 199
   rewriting, 213
   transition to bigraphs, 203–5
   See also Graphs

I

Immuno-fluorescence technology, 261
Informatics
   biochemistry and, 178
   biomolecular computing algorithms, 239
   communication level in nanobioICT systems, 67
   engineered, 154
   graph-based, 199
   of molecular Viterbi algorithm, 76–80
   moleware communication, 53, 67–76
   for nanobiomachine, 34–35
   network coding, 80–84
   pathway unit in, 196
   of porphyrin systems, 255–59

Index

Information and communication technology (ICT), 4–5
Information compression problem, 75
Information processing
    aspects, integration, 260–63
    kinase computer materials, 217–18
    membrane computing, 150
    microbes, 127–29
    nanobioICT, 51
    nanobiomachines, 94
    pathway function, 201
    pathway units, 201
    system, size, 254–55
    VA, 77
Information storage, 71
Input/output graph (IOG), 200

K

KaiC, 106, 107, 108
KEGG, 250, 272
Kinase computer
    biochemical features, 214–17
    blueprint, 214–20
    controllability, 218–20
    future prototypes, 220
    materials for information process, 217–18
Kinase computing (KC), 161, 181–220
    3-SAT problem solving by, 162–65
    algorithm, 164
    feasibility, 218
    with interaction mechanism, 165–71
    key element, 181
    pathway design cost, 165
    process, 164
    structural features, 219
Kinase pathways, 44
Kinase/phosphatase complexes, 164, 170
Kinases/phosphatases
    interaction formula, 233–34
    interaction of, 232–39
    pathways, 103–4
    synchronization measure, 232
Kinase switch, 190–91
    configuration series, 190–91
    defined, 190
    See also Molecular switches
Knowledge discovery in database (KDD), 246
K-place projection function, 194–95

L

Logical inference, 248, 249

Long-term depression (LTP), 38

M

Many-to-one mapping, 234
MAPK pathway, 44–45
MAUDE system, 242
Maximum likelihood (ML) estimation, 64
McNaughton language, 205–6, 207
MDCK cells, 102
Measurement
    H(X,Y), 64
    by information theory, 62
    probability-based, 62–63
Medical applications, 257
Membrane computing
    flow chart, 152
    information processing structure, 150
    multi-set structure, 150
    for SAT problem solving, 159
    See also Computing
Memory
    DNA, 100
    DNA, operations, 101
    GTPase, 104
    molecular complex as, 100–104
    SPK, 104
Metalloporphyrin, 258
MgcRac GAP, 72
Michaelis-Menten equation, 75, 214
Microbes
    biochemical reactions in, 127–36
    communication mechanism, 127–29
    information processing mechanisms, 127–29
    logical programming, 128–29
    state-level description, 128
Microtubules, 29
Migration inhibitory factor (MIF), 44
MIMD architecture, 173
Mitochondrion, 29
MNL
    control feedback, 210
    defined, 205
    definition, 207
    process, 206
Model check, 174
Molecular bioinformatics
    abstract models, 262
    abstract operator form, 227
    algorithm comparison, 223–50
    biomolecular computing commonality, 239–45

Molecular bioinformatics (continued)
   biomolecular computing conceptual relationship, 240
   biomolecular computing links, 223–24
   cellular signaling, 245
   conflict graph, 227
   data amount, 227
   gene relationships, 228
   logical description, 245–50
   pathways, 229–30
   reasoning technology, 245
Molecular biology
   automated devices, applying/modifying, 132
   cellular structure and, 29
   central dogma, 24
   state-of-the-art, 23–45
Molecular Biology of the Cell (Alberts), 29
Molecular biotechnology, 5
Molecular clock, 105–8
Molecular communication protocols, 61
Molecular computing
   algorithm design, 157–71
   basic concepts, 142–46
   cell communication bridge, 184–99
   complexity from benchmarks, 157–60
   cutting-edge technologies, 5–9
   formalized, 146–57
   molecular motor communication, 95
   operators, 172
   process, 141
   quantitative analysis, 216–17
   See also Computing
Molecular dynamics (MD), 27
Molecular electricity, 253–54
Molecular electronics, 256–57
Molecular imaging technology, 85
Molecular informatics. *See* Informatics
Molecular information flow, 266
Molecular messages, 53, 269
Molecular motor, 31
   communication, 95
   information flow, 96
Molecular signaling
   biochemical characteristics, 66
   mechanism, 58
   synchronizing, 75
Molecular switches, 69, 184–99
   GTPase, 185, 186
   information representation, 186
   kinase, 190–91
   pathways, 230
   regulation, 188–89
   reversible, 186, 215
   SPK, 185, 186
Molecules
   cargo, 33
   complexes, 55, 74
   detection, 39
   index, 70
   information flow, 32
   labeling, 70
   measurement, 38
   MSP-automaton, 208
   R, 194, 204
   S, 194
   signaling, 63, 199
   systematic relationship, 28
Moleware
   cybernetics, 267
   language translation, 210–12
   logic operators for, 117
   mechanics, 33–34
Moleware coding, 65
   in nanobiomachines, 108–14
   network structure, 81
Moleware communication
   architecture, 59
   control nodes, 267–68
   defined, 69
   efficiency in cells, 268–72
   image compression by, 76
   informatics, 67–76
   information theory, 68
   kernel, 268
   multiple channels, 270
   nonlinearity, 250
   signaling mechanisms, 59
Moleware logic, 171–78
   defined, 173
   formalized method, 173–78
   process verification, 171–73
Moleware microarray, 132–36
   control, 134
   control unit, 135
   illustrated, 135
Monocentris japonica, 128
Monte Carlo method, 27, 216, 218
Moore's law, 1, 91
Motifs, 79–80
Motor neuron disease (MND), 34
Moveable, 94–96
MSP-automaton, 191, 192
   defined, 191
   molecules, 208

Index

See also Automaton
Multiple agent systems (MAS), 57

N

Nanobiocybernetics, 56
NanobioICT
    basis, 49
    biological function, 57
    communication, 51
    communication processes improvement, 270
    computing, 51
    elements, 50–51
    embryonic approaches, 56–67
    fault tolerance, 87
    fundamental principles, 49–87
    global informatics specifications, 67
    information processing, 51
    information space, 72
    information theory, 50, 53–55
    integrated systems, 51
    methodology principles, 58–59
    mission, 50–53
    molecular structure, 67
    objective, 51
    processing element (PE) for, 86
    progress, 52
    research, 51, 52
    as unconventional, 66–67
Nanobioinformatics, pharmaceutical, 41–45
Nanobiomachines
    action, 94
    adaptation, 31
    cell as, 31–35
    control, 94
    efficiency, 31
    energy, 105
    functional requirements, 93
    information processing, 94
    kernel, 94
    molecular informatics, 34–35
    moleware coding, 108–14
    moleware mechanics, 33–34
    moveable, 94–96
    robustness, 31
    structural control inspiration, 133–36
Nanobioscience
    molecular information processing, 250
    preliminaries in, 9–13
Nanobiosystems
    engineered computational, building, 92–97
    information processing, 97–114

Nanomachines, cells as, 31–35
Nanomolecular-bioICT, 51
Nanoscale science, engineering, and technology (NSET), 10
Nanotechnology, 5
    emerging, 253–72
    of nonbiomolecules, 53
    state-of-the-art, 23–45
National Nanotechnology Initiative (NNI), 2, 10
Natural computation (NC), 3
NBIC, 84–87
    cognome, 86
    defined, 84
    systems, 84
    technologies, 84
Network coding, 80–84
    algorithm, 82–83
    defined, 80
Networks
    GTPase pathway, 231
    moleware communication, 267–68
    phylogenetic, 241
    rewriting on, 203
Neural networks (NN), 3
Neurons, signaling in, 85
Nonbottom-up design, 163
Nuclear magnetic resonance (NMR), 8, 40
    functional (fNMR), 40
    for protein measurement, 40
Nucleic acids, computing by, 114–27

O

Operators
    composition, 197
    defined, 143
    empirical experimental implementation, 172
    graph-rewriting, 203
    logical, 177
    porphyrin computer, 259
    rewriting, 202–3
    splicing, 148
Oxidative phosphorylation, 268

P

Parallel computing, 199–214
Pathway-centered systems biology, 242
Pathway computing
    generalized form, 212–14
    interaction structure, 212
Pathway logic, 242

Pathway networks, 28
Pathways
    abstract cellular, 174–78
    atomic, 243–44
    biochemical meaning, 175
    bioinformatics, 229–30
    cell, 43
    cell, formalization, 242
    clustering, 216
    coefficients, 64
    controlled, 163
    defined, 185
    design, efficiency and, 160–71
    downstream, 265
    EGFR, 242
    GEF/GAP, 103, 165, 219
    GTPase, 106
    GTPase switch, 219
    information capacity, 64
    information processing function, 201
    input, 201
    input/output relationship, 215
    interaction of, 203
    kinase/phosphatase, 103–4
    MAPK, 44–45
    molecular switch, 230
    output, 176, 177, 201
    output states, 195
    representation, 185
    rewriting on, 203
    Rho-MBS-MLC, 236, 237
    signaling, 35–38, 60
    signaling process, 199
    test, 210
    upstream, 265
Pathway units, 175
    computing model based on, 193–99
    control, 208, 218–20
    in informatics, 196
    information processing, constructing, 201
    input, 208
    logic rules, 208
    multiple, construction, 197
    output, 208
    with Turing computability, 193–99
Pharmaceutical nanobioinformatics, 41–45
Pharmaceutics
    "naive" thinking, 41–42
    oGPCRs, 42
Photonics, 257
Phylogenetic network, 241
Phylogenetic trees, 227

Polymerase chain reaction (PCR), 118
Porphyrin
    chemistry features, 256
    complex from structural templates, 258
    coordination of, 76
    defined, 255
    metal complex, 258
    molecules, 16, 17–18
    spatial pattern, 257
    structure, 256
Porphyrin systems
    combinational structure, 259
    informatics of, 255–59
    molecular operators, 259
    potentials, 256–57
    synthesis technology, 257
Primitive recursion, 197–98
Probability-based measurement, 62–63
Proteins
    amino acids, 24
    analysis technologies, 8–9
    ATP, 85
    controlling, 15
    informatics, 9
    location in cells, 9
    motor, 33
    see-and-touch, 8
    similarity of, 28
    three-dimensional structure, 262
Protein-to-protein networking, 28
Proteomics, 26–29
    bioinformatics integration, 29
    defined, 26
    philosophy, 27
P-system, 150–53
    computing model, 151
    cooperation mechanism, 152
    defined, 150
    hierarchy, 151
    multisets, 226

Q

Quadruple convergence, 84–87
Quark-based subatomic computing, 255
Quarks, 254

R

Ras, 43, 261
Reaction-diffusion effect, 83
Receiver cell, 128
Receptor tyrosine kinases (RTK), 36

Index

Recognition, 143
Rewriting operators, 202–3
Rewriting process design, 209–10
Rho family GTPases, 171, 217
Rho-MBS-MLC pathway, 236, 237
R-molecules, 194, 204
RNA
 defined, 11
 detection principle, 39
 tree structure, 149
RNA computing, 121–23
 algorithm, 122–23
 benefit, 121
 biochemical features, 123
 DNA computing integration, 122
 logical programming, 123
 problem description, 123

S

Saccharomyces cerevisiae, 27, 28, 35
SAT-based search mechanism, 178
SAT computing, 158, 162
SAT problem solving, 159
Schizosaccharomyces pombe, 33
Schöning algorithm, 158
Sender cell, 128
Shannon's information theory, 67
Signaling mechanisms
 biological, 33
 biomolecular, 253–54
 for information flow, 82
 moleware communication, 59
 in neurons, 85
Signaling pathways, 35–38
 cellular biomolecular computing based on, 181–220
 encoding/decoding process, 109
 information representation and, 103
 molecular movement link, 37
 neuron function link, 37–38
 processes, 60
 protein network, 154
 recursive mechanism, 212
Signaling processes
 for codes, 110–11
 motifs, 79
Signal transduction knowledge environment (STKE), 35, 250
Single molecule technology, 13
S-molecules, 194
SPK complexes, 70, 103
SPK memory, 104
SPK switch
 defined, 185
 informational representation, 186–88
 See also Molecular switches
Splicing operator, 148
Splicing system, 133
Stem cell research, 13
Strings
 defined, 143
 MNL, 207
 rules for computing system, 174–75
Supercomputers, 2
Suprachemistry, 13
Surface-based DNA computing, 123–25, 225
 bioconditions, 124
 defined, 123
 operations, 124–25, 225
 readout, 124
 for SAT problem solving, 159
 strategy, 124
 word design, 123–24
 See also Computing; DNA computing
Suyama-Yoshida algorithm, 159
Symbols, 143
Synchronization measure, 232
Synchronous moleware, 105–8
Synthetic biology, 6–8
System of systems, 155
Systems biology, 12

T

T cell antigen receptor (TCR), 36
Telecommunications
 physical conditions, 269
 processes, 263
 technology, 263
Theoretical biomolecular computing, 141–78
Top-down method, 220, 240
Transcription factors (TF), 231, 236
Transducers, 104
Transition matrix, 72
Trellis structure, 78
TTxm, 61
Turing machine, 193
 pathway units and, 193–99
 power, 193

U

Unbounded minimalization function, 198–99
Unconventional computing, 2
Universal communication (UC), 4

Upstream pathways, 265

V

Vibro Fischeri, 127–28
Viterbi algorithm (VA)
    decoders, 77
    in electronic computer, 77
    informatics form, 76–80
    information processing structure, 77
    trellis, 80
    trellis structure, 78

W

Word, biomolecular computers, 13

Recent in the Artech House Titles

Advanced Methods and Tools for ECG Data Analysis, Gari D. Clifford, Francisco Azuaje, and Patrick E. McSharry, editors

Electrotherapeutic Devices: Principles, Design, and Applications, George D. O'Clock

Intelligent Systems Modeling and Decision Support in Bioengineering, Mahdi Mahfouf

Microfluidics for Biotechnology, Jean Berthier and Pascal Silberzan

Nanotechnology Applications and Markets, Lawrence D. Gasman

Nanoelectronics Principles and Devices, Mircea Dragoman and Daniela Dragoman

Nanotechnology Regulation and Policy Worldwide, Jeffrey H. Matsuura

Recent Advances in Diagnostic and Therapeutic 3-D Ultrasound Imaging for Medical Applications, Jasjit S. Suri, Ruey-Feng Chang, Chirinjeev Kathuria, and Aaron Fenster

Systems Bioinformatics: An Engineering Case-Based Approach, Gil Alterovitz and Marco F. Ramoni, editors

Text Mining for Biology and Biomedicine, Sophia Ananiadou and John McNaught

Time-Frequency Analysis of Biomedical Signals, Piotr Durka

For further information on these and other Artech House titles, including previously considered out-of-print books now available through our In-Print-Forever® (IPF®) program, contact:

Artech House
685 Canton Street
Norwood, MA 02062
Phone: 781-769-9750
Fax: 781-769-6334
e-mail: artech@artechhouse.com

Artech House
46 Gillingham Street
London SW1V 1AH UK
Phone: +44 (0)20 7596-8750
Fax: +44 (0)20 7630 0166
e-mail: artech-uk@artechhouse.com

Find us on the World Wide Web at: www.artechhouse.com